AF556895

SP
LLC

BIOLOGICAL CONTROL OF INSECT PESTS

Editor
Ninfa M. Rosas-García
Laboratorio de Biotecnología Ambiental. Centro de Biotecnología Genómica, Instituto Politécnico Nacional. Boulevard del Maestro S/N esq. Elías Piña, Colonia Narciso Mendoza, Reynosa Tamaulipas
México

2011

STUDIUM PRESS LLC
P.O. Box 722200, Houston, Texas 77072, USA
Tel: (281) 776-8950, Fax: (281) 776-8951
e-mail: studiumpress@gmail.com
Website: http://www.studiumpress.in

BIOLOGICAL CONTROL OF INSECT PESTS

ISBN: 1-933699-27-2

Published by:

STUDIUM PRESS, LLC

P.O. Box-722200, Houston, Texas-77072, USA

Tel. 281-776-8950; Fax: 281-776-8951

E-mail: studiumpress@gmail.com

Printed at:

Thomson Press India Limited.

Foreword

Insects and humans cohabit the earth and have developed complex relationships. Insect pests (less than 1% of all insect species) are those insects that feed on, compete for food with, or transmit diseases to humans and livestock. Ecosystems by human activities have provided opportunities for insects, and species that successfully adapt often become pests. Intensive agriculture encourages the development of insect pests by concentrating food items (crop plants and stored food) on which insects can feed. Food-plant concentrations, often in monocultures, also may reduce the effectiveness of natural enemies attacking pest species in natural environments. Insects may attack any part of the plant, at any stage of development.

Biological control is defined as the deliberate use of natural enemies such as parasitoids, predators, pathogens, antagonists, and competitors to supress a pest population, making it less abundant and thus less damaging than it would otherwise be. It is important to point out that not all nonchemical control methods are biological control. Traditional plant breeding, transgenic plants, cultural control, and use of semiochemicals, although intended to supress pests, are not biological control. Successful biological control operates at the population level, not the individual level, so biological control agents must be able to regulate populations of a pest, rather than just eating them when they are common. Biological control is applied population ecology, and successful biological control requires detailed understanding of species interactions across three or four trophic levels: The plant the pest feeds on; the pest; the biological control agent; the natural enemies of the biological control agent.

Benefits by successful biological control programs are well documented around the world. However, assurances regarding the environmental safety of biological control have been questioned during the last years. Negative consequences have been attributed to the lack of adequate legislative requirements for quarantine and post-release studies for biological control agents for insect pests. In recent years, however, more funding has become available for research into non-target impacts of biological control, and researchers have the responsibility to conduct good ecological research in this area to the benefit of environmentally safer biological control.

This book offers a good opportunity to get insight into the different biological control agents and methods. Use of bacteria, viruses, fungi, predators, and parasitoids are discussed. Also, some case studies of biological control of agricultural insect pests are included. Finally, the book discusses current important topics such as the rol of biotechnology and molecular strategies in biological control, and the environmental impact and cost/benefit analysis of biological control.

Luis A. Rodríguez-del-Bosque

About the Editor

Ninfa M. Rosas-García

Dr. Rosas-García was born in Monterrey, N.L. México. She holds a Master of Science degree in microbiology and a Ph.D. in biotechnology granted by the Universidad Autónoma de Nuevo León, Monterrey, Mexico. She currently serves as director of the Centro de Biotecnología Genómica of the Instituto Politécnico Nacional. Up to date she is head professor of the Environmental Biotechnology Laboratory at the same Center where she develops the research line of biological control of insect pests of economically important crops. She also directs research on genomics and molecular aspects of pests and their pathogens. Dr. Rosas-García is member of the National Researchers System level 1 and has published 20 research papers in international journals, has written 7 book chapters and has participated in over 45 national and international conferences. She has directed 11 research projects funded by Mexican institutions and has served as thesis director of 4 undergraduate students, 6 Masters and 1 PhD students. In 2002 she received the research award for her doctoral thesis by the Universidad Autónoma de Nuevo León. In 2006 and in 2009 she received the award for quality thesis by the Universidad Autónoma de Tamaulipas among others.

Acknowledgements

I wish to express my sincere gratitude to contributing authors for their support, dedication and confidence for the edition of this book. I also thank Alfredo Vélez from the Informatics Unit, CBG-IPN for his invaluable support in figures edition. I am especially grateful to the Comisión de Operación y Fomento de Actividades Académicas (COFAA) and to the Secretaría de Investigación y Posgrado (SIP) from Instituto Politécnico Nacional.

List of Contributors

Alejandro Sánchez Varela: Laboratorio de Biotecnología Ambiental, Centro de Biotecnología Genómica, Instituto Politécnico Nacional. Boulevard del Maestro S/N esq. Elías Piña, Colonia Narciso Mendoza, Reynosa Tamaulipas, México.
E-mail: asanchezv@ipn.mx. Phone: +52 (899) 924-3627 ext. 87728, Fax: ext. 87746.

Ali Mohammed-Ali: Centro de Biotecnología Genómica, Instituto Politécnico Nacional. Boulevard del Maestro S/N esq. Elías Piña, Colonia Narciso Mendoza, Reynosa Tamaulipas, México.
E-mail: alialimh@yahoo.com.

Alvin M. Simmons: USDA-ARS U.S. Vegetable Laboratory 2700 Savannah Highway Charleston, South Carolina 29414, USA.
E-mail: alvin.simmons@ars.usda.gov. Phone: (843) 324-6568 ext. 5307, Fax: (843) 573-4715.

Benito Pereyra-Alférez: Instituto de Biotecnología, Facultad de Ciencias Biológicas, Universidad Autónoma de Nuevo León. San Nicolás de los Garza, N.L. Mexico.
E-mail: bpereyra@gmail.com. Phone: +52 (81) 8329-4000 ext. 6415, Fax: +52 (81) 8352-2422.

Benjamin C. Legaspi Jr.: Florida Public Service Commission 2540 Shumard Oak Blvd. Tallahassee, Florida 32399, USA.
E-mail: blegaspi@psc.state.fl.us. Phone: (850)413-6277, Fax: (850) 413-6278.

Erick de Jesús de Luna-Santillana: Centro de Biotecnología Genómica, Instituto Politécnico Nacional. Boulevard del Maestro S/N esq. Elías Piña, Colonia Narciso Mendoza, Reynosa Tamaulipas, México.
E-mail: edeluna@ipn.mx. Phone: +52 (899) 924-3627 ext. 87747, Fax: ext. 87746.

Filiberto Reyes-Villanueva: Centro de Biotecnología Genómica, Instituto Politécnico Nacional. Boulevard del Maestro S/N esq. Elías

Piña, Colonia Narciso Mendoza, Reynosa Tamaulipas, México. *E-mail*: frv65@hotmail.com. Phone: +52 (899) 924-3627 ext. 87747, Fax: ext. 87746.

Hugo A. Luna-Olvera: Instituto de Biotecnología, Facultad de Ciencias Biológicas, Universidad Autónoma de Nuevo León. San Nicolás de los Garza, N.L. Mexico.
E-mail: hluna@fcb.uanl.mx. Phone: + 52 (81) 8329-4000 ext. 6415, Fax: +52 (81) 8352-2422.

Jaime Molina-Ochoa: Universidad de Colima, Facultad de Ciencias Biológicas y Agropecuarias, Laboratorio de Nematología Entomopatógena y Resistencia Vegetal a Insectos. Km. 40 autopista Colima-Manzanillo, Tecomán, Colima 28930, México.
E-mail: jmolina18@hotmail.com, jmolina@ucol.mx. Phone: +52 (313) 322-9405 ext. 52451, Fax: +52 (313) 322-9405 ext. 52252. University of Nebraska-Lincoln, Department of Entomology, 3B Entomology Hall, Lincoln, NE 68583-0816, USA. Phone: (402) 472-8686, Fax: (402) 472-4687.

Jesús DiCarlo Quiroz-Velázquez: Laboratorio de Biotecnología Experimental. Centro de Biotecnología Genómica, Instituto Politécnico Nacional. Boulevard del Maestro S/N esq. Elías Piña, Colonia Narciso Mendoza, Reynosa Tamaulipas, México.
E-mail: jquiroz@ipn.mx. Phone: +52 (899) 924-3627 ext. 87749, Fax: ext. 87746.

Jesús M. Villegas Mendoza: Laboratorio de Biotecnología Ambiental. Centro de Biotecnología Genómica, Instituto Politécnico Nacional. Boulevard del Maestro S/N esq. Elías Piña, Colonia Narciso Mendoza, Reynosa Tamaulipas, México.
E-mail: jmvillegas@ipn.mx. Phone: +52 (899) 924-3627 ext. 87744, Fax: ext. 87746.

Jesusa Crisostomo Legaspi: USDA-ARS-CMAVE / FAMU-Center for Biological Control 6383 Mahan Drive Tallahassee, Florida 32308, USA.
E-mail: jesusa.legaspi@ars.usda.gov. Phone: (850) 656 9870 x 10. Fax: (850) 656 9808.

John E. Foster: University of Nebraska-Lincoln, Department of Entomology, 3B Entomology Hall, Lincoln, NE 68583-0816, USA.
E-mail: jfoster1@unl.edu. Phone: (402) 472-8686, Fax: (402) 472-4687.

Jorge E. Ibarra: Departamento de Biotecnología y Bioquímica, Cinvestav Unidad Irapuato, Mexico.

E-mail: jibarra@ira. cinvestav.mx. Phone: +52 (462) 623-9643, Fax: +52 (462) 624-5996.

Katiushka Arévalo-Niño: Instituto de Biotecnología, Facultad de Ciencias Biológicas, Universidad Autónoma de Nuevo León. San Nicolás de los Garza, N.L. Mexico.
E-mail: karevalo@fcb.uanl.mx. Phone: +52 (81) 8329-4000 ext. 6415, Fax: +52 (81) 8352-2422.

Lilia H. Morales-Ramos: Instituto de Biotecnología, Facultad de Ciencias Biológicas, Universidad Autónoma de Nuevo León. San Nicolás de los Garza, N.L. Mexico.
E-mail: lilia.moralesr@uanl.mx. Phone: +52 (81) 8329-4000 ext. 6415, Fax: +52 (81) 8352-2422.

Luis J. Galán-Wong: Instituto de Biotecnología, Facultad de Ciencias Biológicas, Universidad Autónoma de Nuevo León. San Nicolás de los Garza, N.L. Mexico.
E-mail: lgalanw@mail.uanl.mx. Phone: +52 (81) 8329-4000 ext. 6415, Fax: +52 (81) 8352-2422.

Ma. Cristina Del Rincón-Castro: División de Ciencias de la Vida, Campus Irapuato-Salamanca, Universidad de Guanajuato. Irapuato, Gto. Mexico. ExHacienda El Copal Km 9.0 Carr. Irapuato-León. CP 36500. Irapuato, Gto., Mexico.
E-mail: cdelrincon@dulcinea.ugto.mx, mdelrinc@yahoo.com. Phone and Fax: +52 (462) 624-2484.

María Elena Márquez Gutiérrez: Instituto de Investigaciones de Sanidad Vegetal. Calle 110. No 514 e/ 5ta B y 5ta F. Playa. Ciudad de La Habana. Cuba. C.P. 11600.
E-mail: mmarquez@inisav.cu. Phone: (537) 202-3720, Fax: (537) 202-9366.

Mario A. Rodríguez-Pérez: Centro de Biotecnología Genómica, Instituto Politécnico Nacional. Boulevard del Maestro S/N esq. Elías Piña, Colonia Narciso Mendoza, Reynosa Tamaulipas, México.
E-mail: mrodriguez@ipn.mx. Phone: +52 (899) 924-3627 ext. 87719, Fax: ext. 87746.

Miguel A. Pérez-Rodríguez: Centro de Biotecnología Genómica, Instituto Politécnico Nacional. Boulevard del Maestro S/N esq. Elías Piña, Colonia Narciso Mendoza, Reynosa Tamaulipas, México.
E-mail: mrodriguez@ipn.mx. Phone: +52 (899) 924-3627 ext. 87747, Fax: ext. 87746.

Ninfa M. Rosas-García: Laboratorio de Biotecnología Ambiental. Centro de Biotecnología Genómica, Instituto Politécnico Nacional. Boulevard del Maestro S/N esq. Elías Piña, Colonia Narciso Mendoza, Reynosa Tamaulipas, México.
E-mail: nrosas@ipn.mx, ninfarosasg@yahoo.com.mx. Phone: +52 (899) 924-3627 ext. 87721, Fax: ext. 87746.

Patricia Tamez-Guerra: Dep. Microbiología e Inmunología, Fac. Ciencias Biológicas, Universidad Autónoma de Nuevo León. San Nicolás de los Garza, N. L. México. 66450.
E-mail: patamez@hotmail. com. Phone: +52 (81) 8329-4000 ext. 6453, Fax: +52 (81) 8352-4212.

Robert W. Behle: National Center for Agriculture Utilization Research. United Sates Department of Agriculture, Agriculture Research Service. Peoria Illinois, USA.
E-mail: robert.behle@ars.usda.gov. Phone: (309) 681-6310. Fax: (309) 681-6693.

Sergio M. Salcedo-Martínez: Departamento de Botánica, Facultad de Ciencias Biológicas, Universidad Autónoma de Nuevo León. San Nicolás de los Garza, N.L. Mexico.
E-mail: checosalcedo@yahoo. com.mx. Phone: +52 (81) 8329-4000 ext. 6456, Fax: +52 (81) 8376-4537.

Xianwu Guo: Centro de Biotecnología Genómica, Instituto Politécnico Nacional. Boulevard del Maestro S/N esq. Elías Piña, Colonia Narciso Mendoza, Reynosa Tamaulipas, México.
E-mail: xguo@ipn.mx. Phone: +52 (899) 924-3627 ext. 87752, Fax: ext. 87746.

Yu Ziniu: National Engineering Research Center of Microbe Pesticides and State Key Laboratory of Agricultural Microbiology, Huazhong Agricultural University, Wuhan 430070, Hubei, China.
E-mail: yz41@mail.hzau.edu.cn. Phone: 86-27-8728-0802, Fax: 86-27-8728-3882.

Zhang Jibin: National Engineering Research Center of Microbe Pesticides and State Key Laboratory of Agricultural Microbiology, Huazhong Agricultural University, Wuhan 430070, Hubei, China.
E-mail: zhangjb@mail.hzau.edu.cn. Phone: 86-27-6387-7181, Fax: 86-27-8728-0670.

Preface

Biological control of insect pests has been a reliable resource used formally since the last century, by those who wish to control insects and preserve environment at the same time. Since then, the development of bioinsecticides has shown substantial progress to attack numerous insect pests, which has allowed them to be positioned in the global market of insecticides. Among these control agents microbial agents have been used through time successfully. However, new agents with new properties have arisen so it is important to visualize in context, the effectiveness of these control agents when they are directed to a particular pest, and how they have been improved to offer a wide activity spectrum.

Recent technique advances in science and technology such as molecular biology highly useful for gene identification, genetic recombination, site-directed mutagenesis, chimeric-scanning mutagenesis to form chimeric proteins to enhance toxic activity, analysis of protein receptors in insect midgut in order to select suitable control agents, and others, have been strongly considered to generate ideas intended to improve the efficacy of control agents. In this way the development of recombinant organisms or proteins with new capabilities and new potential insecticidal action resulted successful, although much of this scientific work has not been commercially exploited. The new technologies derived form the traditional and modern biotechnology, offer the development of biodegradable insecticidal products. This makes possible for agents with high entomopathogenic potential, to be seriously considered by important biotechnological industries that made heavy investments, to produce bioinsecticides for field application.

In this way this book is devoted to explain in depth the use of various control agents especially those important in agriculture, emphasizing in their mode of action, activity spectrum, killing abilities, and suitable application according host range. As well as much relevant information of agriculturally important pests that cause economic losses to producers.

As environmental implications of biological control application have frequently been questioned and often are a matter of discussion, a clear explanation of the risks and environmental impact that may arise due to the use of different control agents, the relation among them and towards non target insects, flora and fauna, including human beings is given. This book is intended to give a wide vision of the great potential of this dynamic and interesting alternative called biological control.

Ninfa M. Rosas-García

Contents

Foreword v
About the Editor vii
Acknowledgement viii
List of Contributors ix
Preface xiii

1. Entomopathogenic Bacteria 1
MARÍA ELENA MÁRQUEZ-GUTIÉRREZ (Cuba)

2. Entomopathogenic Viruses 29
MA. CRISTINA DEL RINCÓN-CASTRO AND JORGE E. IBARRA (Mexico)

3. Entomopathogenic Fungi 65
NINFA M. ROSAS-GARCÍA, ALEJANDRO SÁNCHEZ-VARELA, JESÚS M. VILLEGAS MENDOZA AND JESÚS D. QUIROZ VELÁSQU (Mexico)

4. Recent Research Trends in the Use of Predators for Biological Control 95
JESUSA CRISOSTOMO LEGASPI, BENJAMIN C. LEGASPI, JR. AND ALVIN M. SIMMONS (USA)

5. Biological Control Agents for Lepidoptera Pests 123
LILIA H. MORALES-RAMOS, HUGO A. LUNA-OLVERA, KATIUSHKA ARÉVALO-NIÑO, BENITO PEREYRA-ALFÉREZ, SERGIO M. SALCEDO-MARTÍNEZ AND LUIS J. GALÁN-WONG (Mexico)

6. Biological Control of Coleopteran Pests 145
ZHANG JIBIN AND YU ZINIU (China)

7. Biotechnology and Derived Products 165
PATRICIA TAMEZ-GUERRA AND ROBERT W. BEHLE (Mexico, USA)

8. Improvement of Biological Control Agents through Molecular Strategies 233
NINFA M. ROSAS-GARCÍA (Mexico)

9. Enhancing the Virulence of Baculovirus as Biopesticides with Wasp Parasitoid Polydnavirus Genes to Control Lepidopteran Insect Pests 259
MARIO A. RODRÍGUEZ-PÉREZ, MIGUEL A. PÉREZ-RODRÍGUEZ, ALI MOHAMMED-ALI, ERICK DE JESÚS DE LUNA-SANTILLANA, XIANWU GUO AND FILIBERTO REYES-VILLANUEVA (Mexico, USA)

10. Environmental Impact and Cost Benefit Analysis of Biological Control Application 271
JAIME MOLINA-OCHOA AND JOHN E. FOSTER (Mexico, USA)

Subject Index 283

{1}

Entomopathogenic Bacteria

María Elena Márquez-Gutiérrez

ABSTRACT

Most entomopathogenic bacteria belong to the families Pseudomonadaceae, Enterobacteriaceae, Streptococcaceae and Bacillaceae. The most widely used classification group them in spore forming and non-spore forming bacteria. In this chapter relevant characteristics comprising the most important genera, their mode of action, morphology and potential for use as biological control agents are presented among other issues. The application of Bacillus sphaericus for control of dipteran larvae, the Cuban experience in relation to Bacillus thuringiensis products development based on selected strains with biological activity against lepidoteran, mites, nematodes and the quality system implemented through the Governmental Plant Health Network are topics addressed in this chapter. Other bacteria such as Paenibacillus (Bacillus) popilliae, Paenibacillus (Bacillus) larvae y Paenibacillus (Bacillus) lentimorbus are very specific and have a more limited use because some of them are obligate pathogens so in vivo production is expensive and industrial production is limited due to requirements for mass production. Serratia entomophila, the causal agent of amber disease in Costelytra zealandica larvae, has been widely studied in New Zealand and since 1992 several formulations containing this pathogen have been applied. The complexity of bacterial symbiotic species Xenorhabdus and Photorhabdus and their associated nematodes give a significant contribution to the mass production method to be chosen. At present, the microbial insecticides account for less than 2% overall insecticide sales. Actions must be taken urgently to make entomopathogenic bacteria to be widely used, to prospect new promising isolates, to carry out strain improvements, to produce more efficient formulations, to encourage local production, to achieve higher quality standards and to assess registration requirements in the world.

INTRODUCTION

Today, more than 1500 microorganisms among bacteria, fungi and viruses pathogenic to insects are known. But they represent less than 2% overall insecticide world sales due in great part to difficulties for mass production, the influence of biotic and abiotic factors, the inability to exert a desirable control as well as the adverse effects on man and non target organisms.

Bacteria are prokaryotic microorganisms living in practically all terrestrial and aquatic habitats. Their reproduction regularly occurs by binary fission in aerobic and anaerobic environments, in hot and cold climates, in dark and in light, in dry or humid ambient and they are found in niches ranging from completely saprophytes to obligate parasites (Tanada and Kaya, 1993).

Despite the great bacterial diversity, those causing infectious diseases to insects (entompathogenic bacteria) bear a poor diversity if compared to other groups as protozoan and fungi. About 90 bacterial species cause infectious diseases in insects, but only few of them have a great potential as biological control agents (Cloutier and Cloutier, 1992). Other species behave as opportunistic pathogens. They constitute a natural alternative and the man consciously can use it to mitigate pest damage together with other biological control agents.

Infectious diseases caused by bacteria are present in all organisms including the very simple and the more complex ones. In the same way the vertebrates are subject to the infection by different bacteria (cholera, tuberculosis, anthrax, tetanus), and the insects do too. These pathological interactions can be simple or even accidental (as a wound infection caused by *Serratia marcescens*) until complex interactions where a total host dependency for bacterial reproduction occurs. The latter is the case for *Paenibacillus larvae* to bees (Matheson and Reid, 1992).

The primary invasion route of bacteria is the oral cavity and occurs when larva feeds. After this primary process, bacteria invade the insect haemocel causing septicemia altering homeostasis in the infected individual. These bacteria are extracellular pathogens (except rickettsias) and their first effects on their hosts are lack of appetite, diarrhea, vomiting, and a complete host invasion.

The symptoms often depend on the bacterial genus attacking. As a rule, the cadaver becomes dark due to hemolymph oxidation and a great amount of bacterial saprophytes can grow coming from the insect digestive tract (midgut flora and food accompanying flora) and from the surrounding environment. The cadaver gets a foul-smelling and the body decomposes in a flocculent way with the exception of the integument (exoskeleton) which sometime later becomes dry and hardened. The most successful bacteria as biological control agents are those which produce toxins and kill the insect (Ibarra, 1998).

Unlike entomopathogenic fungi, bacteria produce epizootics when there is a high host density (stored products, insect colonies, freshwater populations subjected to particular physical or chemical conditions); in other circumstances epizootics are rare or are not detected (Castillo *et al.*, 1995). The latter is not critical when we refer to both biopesticide commercial production and application.

BRIEF HISTORICAL REVIEW

The insect diseases are known since ancient times. In the days of Aristotle diseases affecting bees were studied since the beehives decreased remarkably. By 4700 B.C., Chinese people had implemented some techniques to control pests and after the first practical application, they faced problems with the silkworm which suffered of soothes in the massive rearing they had in communities. The report made by Reaumur in 1726, about a pathogen of the *Cordyceps* that was developed on a caterpillar of the family Noctuidae, perhaps constitutes the first report of a biological control agent. Two important facts were registered in the chapter of diseases described in 1826 by Kirby and Spence in a text of Entomology, and the treaty of 1836 written by Agostino Bassi on the disease affecting the larvae of *Bombyx mori* (Lepidoptera: Bombycidae), the muscardine, of which causal agent was the fungus *Beauveria bassiana.* This fact marked the beginning of the Insect Pathology (Carballo *et al.*, 2004).

Nevertheless, many authors agree that the knowledge of bacterial diseases in insects was described with greater consistency in 1870 by Louis Pasteur, who studied the silkworm disease. In 1853 Cheshire and Cheyne described *Bacillus alvei* as the causal agent of a disesase in bees.

Important contributions were done by Shigetane Ishiwata in 1901 when he discovered the bacterium *Bacillus thuringiensis* which he called *Bacillus sotto*, 10 years later in Thuringia, Germany, Berliner rediscovered *Bacillus thuringiensis* (Glare and O'Callaghan, 2000). The study of the effect caused by a bacterium in the insect pest *Schistocerca* sp. (Orthoptera: Acrididae)) was revealed by D'herelle, for the first time in 1911. This causal agent was known as *Coccobacillus acridiorum,* later was identified as *Cloaca cloacae* var *acridiorum* (d'Herelle). The called "milky" infections are known since 1921 in the Japanese beetle *Popillia japonica* (Coleoptera: Scarabeidae), and in 1940 it was determined that two species of *Bacillus* caused the two forms in which the disease appears. They were classified as *Bacillus popilliae* (now *Paenibacillus popilliae*) and *Bacillus lentimorbus*. These bacteria along with the *Serratia marcenses* species were the most important cause of death of several groups of insect pests (Ibarra, 2007).

It is important to emphasize that among the most well-known entomopathogenic bacteria, the species of *Bacillus sphaericus*, are highly toxic to *Anopheles* (Diptera: Culicidae) larvae.

Bacillus thuringiensis (Bt) is undoubtedly the most important pest control agent worldwide, due to its wide diversity, wide spectrum of action and successful massive reproduction by low technology and industrial methods.

CLASSIFICATION OF ENTOMOPATHOGENIC BACTERIA

The majority of the entomopathogenic bacteria belong to the families *Pseudomonadaceae*, *Enterobacteriaceae*, *Streptococcaceae* and *Bacillaceae,* according to Bucher criteria (1960). They can be aerobic or facultatively anaerobic, and opportunistic. The facultative anaerobes are the most known and studied entomopathogenic bacteria, their pathogenicity is frequently low, but the virulence is very high (*e.g. B. thuringiensis*). They can multiply in the intestine of the hosts; they can produce toxins and/or enzymes and rapidly invade the haemocoel. They are natural inhabitants of grounds and they grow very easily in artificial media.

The obligate entomopathogenic bacteria are only associated to the host in nature. They cause specific diseases in a very limited number of insects, they are very demanding in their requirements to multiply and generally they are not well developed in artificial media. Finally, the potential opportunistic pathogens, normally do not present either significant levels of pathogenicity nor virulence. Its pathogenic activity is conditional to the stress situations confronted by the potential host (climatic and/or nutritional conditions, and causes weakening and fragility to the host defense mechanisms. They do not present any specificity and generally grow well in artificial media.

Other authors as Vergara and Varelas (1978) reported two classifications that group insect-associated bacteria. The first classification contains six categories: 1) Non entomopathogenic bacteria present in the external environment of the insect, 2) Bacteria present regularly or irregularly in the digestive tract of healthy insects, 3) Non spore-forming pathogens (generally facultative), 4) Spore-forming facultative pathogens, 5) Spore-forming obligate pathogens, 6) Crystal inclusion-spore-forming pathogens. The second classification contains four categories: 1) Obligate pathogens, 2) Crystal inclusion-spore-forming bacteria, 3) Facultative pathogens, 4) Potential pathogens.

Nevertheless, the most used and generalized classification in insect pathology groups these pathogens in spore-forming and non spore-forming bacteria (Falcon, 1971).

ENTOMOPATHOGENIC SPORE-FORMING BACTERIA

The members of this group are Gram-positive; rod shaped cells, and belong to the *Bacillaceae* family. This is the most studied group that includes the entomopathogenic species of *Bacillus, Paenibacillus and Clostridium.* The sporulation process occurs when the nutrients in the medium are depleted. A sporangium structure is formed with two compartments consisting of a mother cell and the spore itself (Sneath, 1986). The spore-forming ability allows these species to be highly resistant to the environment, and they are considered highly pathogenic due to their toxin production.

GENUS *BACILLUS*

There are five fundamental species of importance in the control of pests and their distinctive characteristics as well as their biological control properties are following discussed.

One of the main distinctive characteristic is the form and position of spores inside the cell, which has been used as a criterion to classify the different species.

Bacillus thuringiensis (Bt).

It is a Gram-positive, spore forming bacterium, with peritrichous flagella from 3 to 5 μm of length by 1 to 1.2 μm of width. It can develop ellipsoidal resistance spores without sporangium swelling. It is a facultative anaerobic chemoorganotroph microorganism, with catalase activity. The different isolates from *Bt* generally present common biochemical characteristics. They are able to ferment glucose, fructose, trehalose, maltose and ribose, and to hydrolyze gelatin, starch, glycogen, sculling and N-acetyl-glucosamine. Nevertheless, the fundamental characteristic of *Bt* is that when sporulation occurs, simultaneously produces a protein nature body called δ-endotoxin, parasporal body, or insecticidal crystal protein (ICP), which is formed by one or several protein units. These proteins are highly toxic to numerous insect pests, without affecting beneficial insects and humans. These proteins, also denominated Cry (from Crystal) are the base of the most spread biological insecticide worldwide. The interest in this bacterium has remarkably been increased by virtue of the discovery of new diverse toxic activities against pests of different groups of organisms, which include mites and nematodes (Márquez, 2005; Federici, 1993).

Isolation, Identification and Classification

Bt is ubiquitous in the environment and can be isolated from soil, foliage, water and air, and other habitats due in great part to the spores dissemination (Ibarra *et al.*, 2003; Arrieta *et al.*, 2004). The coleopteran-active and lepidopteran-active *Bt* subspecies are primarily associated with the soil and phylloplane (leaf surfaces), whereas the dipteran-active *Bt* subspecies are commonly found in aquatic environments. In the environment, the spores persist and vegetative growth may occur when favorable conditions occur and nutrients are available.

The majority of the isolation procedures for *Bt* with entomopathogenic activity include a first selective stage consisting in a thermal shock process of the samples to eliminate vegetative flora without affecting the viability of the majority of spores. The samples in saline solution are warmed up to a temperature between 60-65°C (Bhalla *et al.*, 2005). The synthetic media usually provides the nutrients and energy needed for bacterial growth. These nutrients are assimilated by means of exoenzymes, which also maybe harmful or toxic for hosts (Aronson, 1993; De Maagd *et al.,* 2000). Other authors recommend the use of the polimixyn piruvate, egg yolk, mannitol and bromothymol blue agar (PEMBA). The selectivity of PEMBA medium is attributed to antibiotics that suppress growth of Gram-negative organisms and to the egg yolk, that biochemically identifies this species through lecithinase production.

The species of *Bt* is corroborated by the presence of the parasporal crystal by optical microscopy. *Bt* belongs to the group 1 of the *Bacillus* genus. It is a member of the *Bacillus cereus* group; the one which includes *B. anthracis*, *B. cereus*, *B. mycoides*, as well as the most recently described *B. pseudomycoides* (Nakamura, 1998) and *B. weihenstephanensis* (Lechner *et al.*, 1998). It is difficult to distinguish among these species as there are no sufficient differences in their morphological and biochemical characteristics, however, the Bergey´s *Manual of Determinative Bacteriology* recognizes the individuality of these species mainly being based on two different characteristics: the presence of the inclusion or parasporal crystal and the insecticidal properties (Sneath, 1986).

The classification of *Bt* subspecies based on the serological analysis of the flagellar (H) antigens was introduced in the early 1960s (de Barjac and Bonnefoi, 1962). This classification by serotype has been supplemented with morphological and biochemical criteria (de Barjac, 1981). Until 1977, only 13 *Bt* subspecies had been described, and at that time all subspecies were toxic to lepidopteran larvae only. The discovery of other subspecies toxic to Diptera (Goldberg and Margalit, 1977), Coleoptera (Krieg *et al.*, 1983) and apparently Nematoda (Narva *et al.*, 1991) enlarged the host range and markedly increased the number of subspecies. By the end of 1998 nearly 69 subspecies had been identified by the flagellar (H)-antigen (Lecadet *et al.*, 2001).

Höfte and Whiteley in 1989 proposed four main classes based only in the insecticidal activity spectrum. The four classes were: I specific against lepidopteran, II specific against lepidopteran and dipteran, III specific against coleopteran and IV specific against dipteran. The cytotoxic toxins Cyt were also included, and the classes V and VI, active against nematodes, were soon after considered by Feitelson *et al.*, (1992). This classification was unsuitable, because of the dual activity exhibited by one Cry protein against different insects. Further a new nomenclature based exclusively on the similarity of nucleotide sequence of the Cry proteins was created (Crickmore *et al.*, 1998).

Bt *Toxins*

This microorganism is characterized to produce a great variety of toxins with diverse properties. It is said that some of them constitute risks for the human health. Seven different toxins have been described, endotoxins (of great interest by their insecticidal power), exotoxins like phospholipase C, (the well-known α-exotoxin) and β-exotoxin, also known as thuringiensin which is a non-specific thermostable protein, identified as an rRNA synthesis inhibitor (Mackedonski and Hadjiolov, 1972), the *mouse factor* exotoxin which is toxic to mice and lepidopterans, and the vegetative insecticidal proteins (Vip proteins) of more recent isolation (Glare and O´Callaghan, 2000). However, *Bt* like other bacteria may produce during the vegetative growth and sporulation phase a variety of antibiotics, enzymes, metabolites and toxins that are biologically active and may have effects on both target and non-target organisms.

Mode of Action

The (Inclusion Crystalline Protein) ICP-spore complex of *Bt* is ingested by susceptible insect larvae. In the midgut the parasporal crystalline protein or protoxin is activated by gut proteases (Warren *et al.*, 1984; Jaquet *et al.*, 1987; Aronson *et al.*, 1993; Honée and Visser, 1993). Shortly afterwards, the gut becomes paralysed and the larva ceases to feed. The ICP structure and function have been reviewed in detail by Schnepf *et al.*, (1998). Binding of the ICP to putative receptors is a major determinant of ICP specificity and the formation of pores in the midgut epithelial cells is a major mechanism of toxicity.

The active toxin consists of three distinct domains (Höfte and Whiteley, 1989). The three domains interact in a complex manner, but experimental data suggest that the first and second domains of the toxin are involved in epithelial cell receptor binding and pore formation, while the third domain is primarily involved in recognition and pore formation (Huber *et al.*, 1981, Schnepf *et al.*, 1998; Dean *et al.*, 1996). Binding to specific receptors has demonstrated to be closely related to the insecticidal spectrum of the ICPs (Denolf *et al.*, 1997). Van Rie *et al.*, (1989) found the affinity of these toxins, in the tobacco budworm (*Heliothis virescens* Fabricius) (Lepidoptera: Noctuidae), and in the tomato hornworm (*Manduca sexta* Linnaeus) (Lepidoptera: Sphingidae), to bind to the brush border membrane vesicles. However, the number of binding sites was different, causing variability in the toxin affinity and consequently in the toxic activity.

Pore or ion channel formation occurs after the binding of the toxin to the receptor, and insertion of the N-terminal domain into the membrane, thus the regulation of the trans-membrane electric potential is disturbed. This can result in colloid-osmotic lysis of the cells, which is the main cytolytic mechanism that is common to all ICPs (Schnepf *et al.*, 1998). When the midgut epithelium of the larva is damaged, the hemolymph and gut contents can mix. This results in favorable conditions for the *Bt* spores to germinate. The resulting vegetative cells of *Bt* and the preexisting microorganisms in the gut proliferate in the haemocoel causing septicaemia, and finally death.

Production and Commercial Use

Conventional *Bt* products, which utilize naturally-occurring *Bt* strains, account for approximately 90% of the global market of microbial pest control agents. Most *Bt* products contain ICP and viable spores, but in some *Bt israelensis* (*Bti*)-based products the spores are inactivated.

Conventional *Bt* products have been targeted primarily against lepidopteran pests of agricultural and forestry crops; however in recent years, *Bt* strains active against coleopteran pests have also been marketed. Strains of *Bti* active against dipteran vectors of parasitic and viral diseases are being used in public health programs. Commercial *Bt* formulations may be applied on foliage, in soil, in aquatic environments or in food storage facilities. After the application of *Bt* to an ecosystem, the vegetative cells and spores may persist at concentrations that decrease gradually for weeks, months or years, becoming a component of the natural microflora (Boucias and Pendland, 1998).

The short half-life of *Bt* due to ultraviolet inactivation when topically applied, has stimulated considerable research into alternative delivery strategies. So far, the most controversial strategy is the use of insect-resistant transgenic crops expressing *Bt* delta-endotoxin genes.

In general, the most known commercial products are Bitoxibacillin™, Eksotoxin™, Agritol™, Bactospeine™, Bathurin™, Biospor™, Dipel™, Javelin™ and Sporeine™. Each year *Bt* biopesticides sales represent approximately USD $140 million, produced by aerobic fermentation technology. However, the global market of *Bt* biopesticides is about USD $250 million a year. More than the 50% of these products are stockpiled by the United States and Canada, while 18% by the East, 10% by China and 8% by South and Central America (Fernández and Juncosa, 2002). The industrial production of commercial products is distributed among a group of companies that produce agrochemicals, corresponding a 75% to large companies and the rest is distributed in a smaller group. There is a worldwide tendency to increase their production to be applied on crops for human consumption and for treatment of large agronomic areas (Ibarra and López-Meza, 2000).

Low Technology Systems and Industrial Production in Cuba

In 1988 the Ministry of Agriculture approved the National Program for Production of Biological Control Agents. To achieve this purpose the construction of plants for *B. thuringiensis* biopesticides production is being carried out making use of semi-industrial technologies. However, low technology systems are still in use in a network of productions centers named CREE (Entomopathogens and Entomophages Production Center). The first products were used in large scale against *H. virescens*, *Mocis latipes* (Lepidoptera: Noctuidae), *Plutella xylostella* Linnaeus (Lepidoptera: Plutellidae), *S. frugiperda* J.E. Smith (Lepidoptera:

Noctuidae), *Erinnys ello* Linnaeus (Lepidoptera: Sphingidae), and *Diaphania hyalinata* Linnaeus (Lepidoptera: Crambidae). Up to now, there are 195 CREE located throughout the country, that sale their productions to all agricultural organizations including small private growers. In addition, other four biopesticide plants produce in total 320 tons per year, and are located in the Western and Central regions of the country. The annual production of all production centers is 1,000 metric tons that are applied to an area of more than 120,000 ha (Fernández-Larrea, 2001).

The number of CREE and production plants is determined according to the necessity at that time, and to the region that will be covered, in correspondence with the crops and the main problems that they present. In this way, there is a relation between the type of the product elaborated and its use. The products are distributed by the CREE, avoiding to producers the transportation costs for long distances and storage.

Four lines of bioproducts with the generic name of THURISAV™ are obtained in fermentation plants (Table 1). They are watery concentrated fluids with a shelf-life of 6 months at 23-25°C. The successful results, sometimes, foster a greater demand that surpass production capacity of the existing plants.

Table 1. *Bacillus thuringiensis*-based products applied against different pests in Cuba

Products	***Pest***	***DosageL/ha***	***Crops***
Thurisav 1™	*Plutella xylostella*	5–10	Cabbage
	Mocis latipes	1–2	Pasture
Thurisav 13™	*Phyllocoptruta oleivora* Ashmead (Acari: Eriophyidae)	20	Citrus
	Polyphagotarsonemus latus Banks (Acari: Tarsonemidae)	3–5	Potato, Citrus
	Tetranychus tumidus Banks (Acari: Tetranichydae)	5–10	Banana
Thurisav 21™	*Heliothis virescens*	5–10	Tobbaco
	Plutella xylostella	1–5	Cabbage
Thurisav 24™	*Plutella xylostella, Trichoplusia ni* Hübner (Lepidoptera: Noctuidae) *Erinnys ello, Spodotera* spp., *Ascia monuste* (Lepidoptera: Pieridae) *Diaphania hyalinata*	4–5	Vegetables, roots, and tubers
Thurisav 25*	*Meloidogyne incognita* (Tylenchida: Heteroderidae)	20	Tomato and cucumber

* Product on development.

The *Bt* products are included in the Programs of Integrated Pest Management for different crops. These products are generally applied for each program along with preventive treatments, since the greater effectiveness is obtained in the first larval stages (Pérez and Vázquez, 2001).

The low technology system by means of static liquid cultures was first established in all the country. This technology involves the use of liquid media produced from agricultural or industrial by-products, mainly from the sugar industry. In this technology, crystal bottles are used as containers and culture medium is added in a 1:5 ratio (fifth volume bottle is filled with culture medium). After inoculation, bottles are settled at 28-30°C during 10-15 days depending on the strain and the culture medium. The product is harvested and a preservative is added to allow a longer storage period (up to 3 months at a temperature of 25°C or lower). Concentrations of 1-5 × 10^8 spores/ml corresponding to a potency of 15,000 IU compared to a National *Bt* pattern of *Spodoptera exigua* Hübner (Lepidoptera: Noctuidae) are obtained, with a cost of approximately U.S. $0.02 per liter and a production efficiency of 70% with respect to the installed capacity of the production centers.

The production on solid substrate is another low technology alternative to spread production in Cuba. The process has a first stage of bacterium propagation in liquid media constituted by different nutrients and salts. When the inoculum reaches the sporulation phase is used to inoculate the rice previously sterilized and distributed in plastic bottles. At this stage the moisture content should be enough to allow bacteria growth. After 5-7 days of incubation spores and crystals are formed, recovered and maintained at less than 20°C for 48-72 h. Finally the product is packaged into plastic bags and stored during 3 months at 20-25°C. The yield is around 6 × 10^8 and 1 × 10^9 spores/g of rice.

An industrial process produces higher concentrations of spores which are recovered by sedimentation. The process takes around 72 to 96 h, and the spores reach a concentration of 4-6 × 10^9 equivalent to 18000-22000 IU compared to a National *Bt* pattern of *S. exigua*. It is possible to store the product for 6 months at room temperature. This process has a production efficiency of 90% with respect to the installed capacity of the production centers, and the production cost is approximately USD $0.50-0.60/L. A third of the fermentation broth is recovered as a concentrate final product.

Bacillus sphaericus

It is a strictly aerobic bacterium, whose spore is totally spherical and terminal. During the sporulation process the most active strains produce a crystal protein with a high degree of activity against mosquito larvae. After *B. sphaericus* is ingested its toxins are released in the larval gut causing larvae to stop feeding and finally causing death. *B. sphaericus* is only effective against actively feeding larvae, and does not affect pupae or even adults. The growth conditions seem to have a remarkable influence on the insecticidal activity. It has been determine that the genera *Culex* sp. (Diptera: Culicidae), *Psorophora* sp. (Diptera: Culicidae), and *Anopheles* sp. (Diptera: Culicidae) are more sensitive to this bacterium than *Aedes* sp. (Diptera: Culicidae). *B. sphaericus* is typically applied to water with high organic content where mosquito larvae live. The areas where *B. sphaericus* is often the preferred choice for mosquitoes control include waste lagoons (where animal wastes are treated) and storm water catch basins. Although *B. sphaericus* has very few environmental risks associated with its use, its application is not permitted in reservoirs containing drinking water. Also it has been demonstrated that *B. sphaericus* is both non-toxic and non-pathogenic for a variety of species tested (Lacey, 1990). Environmental persistence of *B. sphaericus* varies depending on the formulation used and environmental conditions. Breakdown of *B. sphaericus* usually takes several weeks but residual levels have been shown to persist in some waters for up to nine months. Recycling of *B. sphaericus* from mosquito corpses can increase its persistence (Siegel and Shadduck, 1990). When used according to label rates, *B. sphaericus* does not appear to be harmful to mammals, birds, fish, or most non-target invertebrates (insects and worms).

These bacteria, have been formulated to be used against dipteran larvae which are vectors of important human diseases like dengue and yellow fever, causing epidemics in numerous countries of Latin America. The World Health Organization (WHO) in 1994 informed about the insecticide resistance development in 56 species of *Anopheles*, from those, 54 were resistant to DDT, 28 to organophosfates and 19 to carbamates and pyrethroids. For this reason in Brazil, Cuba, Nicaragua, Mexico, Colombia, Guatemala, Honduras, Peru, Ecuador and Dominican Republic efforts are combined to implement effective programs for eradication of different vectors, without environmental contamination and ecological risks. However, the limited data access generated by important corporations about the use of biological products reduces information availability in some countries.

The industrial products based on *Bt* and *B. sphaericus* used in the last decades need to be evaluated to improve formulations. This evaluation may include new ingredients to diminish disadvantages related to massive applications, high production costs, residual period time, and susceptibility to ultraviolet light. This is particularly important as it is already known that in China, India, France, the United States and Brazil have been demonstrated in laboratory and field, the development of resistance toward toxins of Bs 2362 in *C. quinquefasciatus* Say larvae, provoking the use of *Bt* var *isarelensis* as an alternative (Regis *et al.*, 2001).

Paenibacillus (Bacillus) Popilliae

The term "milky disease" comes from the pure white appearance of the grub when infected with *B. popilliae.* It was the first insect pathogen to be registered in the U.S. as a microbial control agent. This species is a Gram-negative spore-forming rod of 1.3 to 5.2 × 0.5 to 0.8 µ. This fastidious organism grows only on rich media containing yeast extract, casein hydrolysate, or an equivalent amino acid source, and sugars. Several amino acids are known to be required for growth, as well as the vitamins, thiamine and barbituric acid. Trehalose, the sugar found in insect hemolymph, is a favorable carbon source, although glucose can also be used. Japanese beetle is the exclusive host of *B. popilliae* which is sold commercially (Costilow and Coulter, 1971), However, other *B. popilliae* strains such as *B. lentimorbus,* (which is considered a strain of *B. popilliae* by some experts) have other scarab hosts and are specific to different beetles in the family Scarabaeidae, which also includes the Japanese beetle and the chafers (important pasture pests), as well as the beneficial dung beetles (Petterson *et al.*, 1999).

Spores which reside in the soil and have been ingested by beetle grubs, germinate in the larval gut within 2 days and the vegetative cells begin to proliferate, reaching maximum numbers within 3 to 5 days. By this time, some of the cells have penetrated the gut wall and have begun to grow in the hemolymph, where large numbers of cells develop from day 5 to 10. A few spores also are formed at this stage, but the main phase of sporulation occurs later and is completed in 14 to 21 days when the grub develops the typical milky appearance. In laboratory conditions the grub remains alive until this stage, and usually contains about 5×10^9 spores. In field conditions, however, there are reports that larvae sometimes die earlier, before the main phase of sporulation is completed. This is of concern because sporulation stops when the host dies and the grub ultimately releases fewer spores to maintain the level of infestation of a site (Redmon and Potter, 1995).

The advantages of using commercial preparations of *B. popilliae* include the very narrow host range (they are effective against Japanese beetles grubs, only), their complete safety for man and other vertebrates, their compatibility with other control agents including chemical insecticides and insect-pathogenic nematodes, and their persistence. The disadvantages include the high cost of production *in vivo,* the slow rate of action, the lack of effect on Japanese beetle adults, which also cause obvious and distressing damage, and the necessity for large areas to be treated. In addition, the narrow host range of *B. popilliae*, that seems to be environmentally desirable, is also a disadvantage, as other grubs present will not be attacked. So producers must accurately identify the infesting grub species to determine the presence of Japanese beetle (Zhang *et al.*, 1997).

The treatment is most effective when it is applied on a wide basis (or at least on relatively large areas) to reduce overall the levels of beetle infestation. It is less appropriate for using by small landowners, who may control the grubs in their own turf. Due to *B. popilliae* is an obligate dependent on its hosts for sporulation and some larvae may not ingest the spores (or not ingest enough to cause disease), a periodic resurgence and decline of the pest problem can be expected. The success of the control program must be judged not on this basis but by the fact that over a number of years the mean level of pest damage is lower than it would be in the absence of *B. popilliae* (Flexner and Benalvis, 2000).

The cause of death in insects infected with *B. popilliae* is not fully known. The growth of bacterial cells in the hemolymph seems to be the most likely explanation for physiological starvation. Fat reserves of diseased grubs have been shown to be much reduced compared with those of healthy grubs. Toxins also may be involved because they have been detected in bacterial culture filtrates and these have shown to be lethal by injection (Deacon, 1998). Recently, a crystal protein from sporulating cells of *B. popilliae* was found to have similarities to one of the Cry toxins of *Bt.* This protein also might contribute to pathogenic invasion through the gut wall.

Paenibacillus (Bacillus) larvae

This bacterium is the etiological agent of the American foulbrood (Genersch *et al.*, 2006), which is the most serious disease of the larval stage of domestic bees *Apis mellifera* Linnaeus (Hymenoptera: Apidae). Also it is highly contagious and is one of the few diseases able to kill a beehive. The control of this disease exposes serious problems because bacterial spores maintain their infectious capacity for long periods and can survive under adversely environmental conditions (Bailey and Ball,

1991). The clinical symptoms are more serious in young larvae. The affected cells in the nests (natural beehives) present the operculum sunk and darkened, with a humid appearance and irregular perforations. Inside the operculum, there is a dark yellowish or brownish viscous mass. Upon the course of the time, the resultant mass becomes a hard scale of that firmly adheres to the cells. Each scale contains 250 million of spores, which represent the infectious form. Spores can initiate the cycle of the disease once they are ingested by the larvae, along with the food (Shimanuki, 1990).

This species presents ellipsoidal spores from central to terminal position with sporangium deformation. Several methods have been developed for the isolation of viable spores of *P. larvae* from honey (Hansen, 1984); among them direct inoculation of honey sample in plate and concentration of spores by dialysis or centrifugation. The addition of different concentrations of antibiotics to basal media for the development of *P. larvae* allows a good recovery of spores from honeys with heterogeneous populations of aerobic sporulated bacteria and/or microaerophilic species (Hornitzky and Clark, 1991).

Paenibacillus (Bacillus) lentimorbus

It causes the same disease that *P. popilliae*. *Bacillus lentimorbus* is associated with type B milky disease (Petterson *et al.*, 1999). The latter is characterized by the appearance of brown clots that block circulation of hemolymph in the larva and lead it to gangrenous conditions in the affected parts (Stahly *et al.*, 1992). There are few physiological differences between the species, but growth in the presence of vancomycin or 2% NaCl is generally associated with strains of *B. popilliae* (Rippere *et al.*, 1998).

In natural form, both species infect larvae per *os* (by ingestion), although some beetles species do not become infected by oral route, they are susceptible through intrahaemocoel inoculation (Benintende and Márquez, 1996). This species is less spread than *P. popilliae*. In the midgut, the spore germinates, and the bacillus penetrates the midgut cells by phagocytosis, initiating the infection through the intestinal epithelium. Normally, the larva uses a known mechanism called "melanotic capsule", by its dark aspect. This is a layer composed by haemocytes that forms around the intestinal wall, like an immunological barrier to the subsequent infection of hemolymph. An acute infection is not stopped by this capsule, and the bacteria that invaded the interstice between intestinal epithelium and the basal layer, eventually pass to hemolymph, causing bacteraemia. Here the bacteria reproduce profusely developing spores, until appearance of the milky aspect of hemolymph. It is possible to find up to 10^9 spores/ml of hemolymph. The larva remains active until its death. The infection is generally slow, and hemolymph does not oxidize when it is exposed to the environment. The specific lethal cause of the infection is not known, but it could be due to the reduction of essential nutrients in hemolymph, or to the presence of thermostable toxins, or to the oxidative enzyme deterioration, or even to a combination of all these (Tanada and Kaya, 1993).

Because of both *P. lentimorbus* and *P. popilliae* are obligated pathogens, it is not possible to produce them industrially in artificial media. For that reason the commercial products are produced form larvae cadavers that were previously inoculated by injection. The cadavers are then macerated, filtered, homogenised and formulated. This process is time-consuming and increases production costs considerably. This is the technical base product for the Doom™ or the Japademic™ that are mainly used for the control of Japanese beetle *P. japonica* and for other beetle pests of grasses and fruit trees. One of the great advantages of these products is their high residuality, since spores can stay on the ground until 7 years, without necessity for being carried out another application, and their viability could be extended on the ground up to 21 years. This fact compensates the high cost of production. Although it has been possible to obtain a very limited vegetative growth in certain artificial media, no significant level of sporulation has been obtained (Ibarra, 2007).

Genus *Clostridium*

The members of this genus are strictly anaerobic, gram-positive, rod-shaped bacteria that produce oval or spherical endospores. Experimentally *C. brevifasciens* and *C. malacosomae* have caused death on the western tent caterpillar *Malacosoma californicum* (Packard) (Lepidoptera: Lasiocampidae) in laboratory tests. They are considered colonizing bacteria of the intestine where produce numerous toxins, enzymes and inhibitors with insecticidal properties.

These bacteria grow in the mesenteron and paralyze the larva, which dies by starvation and its body acquires a stretched appearance, in the form of telescope, called braquitosis (Tanada and Kaya, 1993). Due to most species are strictly anaerobic, the possibilities to develop bioinsecticides are low; nevertheless, some of their genes that express toxins with biological activity have already been cloned (Barloy *et al.*, 1996).

Entomopathogenic Non Spore-forming Bacteria

Serratia marcescens

It belongs to the Enterobacteriaceae family and was previously denominated *Bacillus prodigiosus*. It is facultatively anaerobic, easily recognize because of the brilliant red color colonies formed in agar plates. The color is due to the production of the prodigiosin pigment. It is a Gram-negative opportunistic microorganism with peritrichous flagella (Ruíz-Sánchez *et al.*, 2003). Its fast multiplication causes septicaemia and death of the insect between 24 and 72 h. *S. marcescens* is easily cultivated in laboratory, and is the only species among enterobacteria able to produce three enzymes, the DNAase, gelatinase and lipase, in addition, this bacterium can use the citrate as the only carbon source, and do not produces sulphur hydrogen. Some strains exhibit anti-tumour properties derived from the production of polypeptide fragments of serralisin (Cerrato-Soto, 2007). This finding has generated inventions in the field of the biotechnology and the pharmaceutical industry with their corresponding patents. Also *S. marcescens* is recognized as a model in molecular genetics laboratories for mutagenesis studies, for the development of mutations for antibiotic resistance, the regulation of gene expression and classical genetic studies based on analysis of mutant pigmentation (Haddix *et al.*, 2000). For insects reared under laboratory and greenhouse conditions, this microbial agent has been a persistent enemy, however, the expected effectiveness against insect pests is not the same under field conditions (Benintende and Márquez, 1996).

Serratia entomophila

It is a highly specific pathogen that causes the amber disease in larvae of *Costelytra zealandica* (White) (Coleoptera: Scarabaeidae). The bacterium was firstly applied as a liquid product named *Invade,* but recently *Bioshield* based on solid granulates was developed for application in field. The bacterium has been registered for sale in New Zealand after fulfilling required safety testings. *S. entomophila* can be used effectively as a biopesticide by introducing the bacterium into healthy insect pest populations. In order to use the bacterium as a biopesticide, a production method which produces a high concentrated fermentation broth was developed. This is effective when it is applied at the rate of 1 litre (4×10^{13} viable bacteria)/ha (Johnson *et al.*, 2001). Since 1992, approximately 15,000 ha of pasture have been treated with Invade™ in New Zealand.

The bacteria that cause this disease must be ingested by grass grub larvae to colonise the gut and eventually cause death. Ingestion of these cells triggers a reaction starting with feeding cessation, leading to gut clearance, and the switching off digestive enzyme production in the midgut epithelial cells (Jackson, 1995), which produces the resultant amber coloration. A long chronic disease period follows resulting in death of the infected hosts (Jackson *et al.*, 2001). An acute oral dose of as few as 3×10^4 cells constitutes the IC_{50} (concentration necessary for 50% infection) with most of these cells floating freely in the gut lumen. No cellular damage has been observed in the midgut epithelial cells, in spite of ingestion of high concentrations of bacteria. During the chronic stages of disease, bacteria are confined to the insect gut where they multiply. The chronic stages of disease (non-feeding and with a clear gut) cannot be reversed by elimination of bacteria through the administration of antibiotics. After application to the grass grubs, pathogenic *Serratia* cells stabilise at a level of about 10^3 cells/g. The level is maintained by multiplication in infected insects and subsequent release into the soil after insect death. If the host grass grub population decreases to low levels, the applied pathogenic strain will die (O'Callaghan and Jackson, 1993).

Xenorhabdus and *Photorhabdus*

These species are entomopathogenic bacteria with a wide host insect range. This species belong to the family Enterobacteriaceae. *Xenorhabdus* and *Photorhabdus* are species symbiotically associated with nematodes of the families Steinernematidae and Heterorhabditidae respectively. The factor(s) determining the symbiotic interaction between nematodes and bacteria are yet to be identified. *Xenorhabdus* and *Photorhabdus* species exist in two main phenotypic forms, a phenomenon known as phase variation (Boemare *et al.*, 1997).

The phase I (or primary form) differs from phase II (or secondary form) in certain physiological and morphological characteristics. There is no variation in the DNA integrity on phase I and phase II and this supports epigenetic regulatory mechanism in phase variation (Akhurst *et al.*, 1996).

Certain pathogenic determinants such as pili, lipopolysaccharides and toxins contribute to the pathogenicity of *Xenorhabdus* and *Photorhabdus* species, and both appear to be equally pathogenic to insects. The observed similarity in their virulence to host insects may reflect possibly an *in vivo* conversion from phase II to phase I, however the host cellular invasion and virulence is yet to be properly understood.

The virulence of *Xenorhabdus* variants varies among insects apparently due to different factors which include feeding habits. The molecular mechanism and biological significance of phase variation are presently unknown (Boemare, 2002).

The culture on solid substrate that uses crumbled polyurethane foam soaked with animal viscera supplemented media was described by Bedding (1981, 1984), as a method for the *in vitro* production of the nematode-bacteria complex. The symbiotic bacteria are first added to the culture medium, and after a growth period of 24 h the nematode infective stages that are previously sterilized on the surface, are inoculated. Two weeks later the nematodes are collected by centrifugation. To confirm that complex production has been successful, the bacteria are obtained by the method reported by Poinar (1990) or by hanging drop method. Polyurethane foam is up to date used with some variations made in relation to the culture media and methodologies used. The raw material costs are relatively low and there is no need of specialized labor. The nematode production can be done at small, medium or large scale, which is attractive for companies that commercialize entomopathogenic nematodes. In this kind of production, the raw material mixing, bacterial inoculation and juvenile 1 nematode stage production have needed some adjustments to improve yields in comparison with those obtained by *in vivo* production.

Another production system has been developed in submerged fermentation using tanks (airlift) of 3 to 10 m^3 capacity. Bacteria must be multiplied previously in fermentors of smaller capacity, with a strict quality control in order to avoid the dissociation to bacterial phase II. At present, the technology for obtaining nematodes by fermentation is the most novel and productive, and it is being developed in spite of some disadvantages concerning both the operation and design of the fermentors.

Several companies from developed countries (as Biosys, USA) use a fermentation broth that, although has a high initial cost, it has the advantage of producing large amounts of excellent product with a quality control of the raw materials. Although there are positive aspects, the quality control methods and performance are still unsatisfactory due to the high production cost (Ferraz *et al.*, 2008).

Generally, companies patent the entire production process for a specific nematode and its associated bacterium, so the composition of culture media used for growth remains under secret.

QUALITY CONTROL

An essential requirement for the production of any microbial control agent is an effective quality control system. In 1992, at the Meeting of the Active Group of Quality Control of the International Organization of Biological Control, several measurements were settled down to take control of the microbial products including all the operations, procedures, equipments and environmental conditions indicated (Guillon, 1997).

Production processes, downstream processing and extraction methods vary widely among products, so it is not possible to implement a single set of standardized procedures. In some countries registration product is enforced, as well as a complete description of the process.

Quinlan (1990) refers that the quality standards for *Bt* fermentation products accepted by International Union of Pure and Applied Chemistry (IUPAC) include limits on the concentration of microbial contaminants and metabolites as shown in the table below.

Table 2. Highest levels of microbial contamination allowed in bacterial insecticides according to IUPAC recommendation

Types of microorganisms	*Highest concentrations (g)*
Viable mesophiles	$< 1 \times 10^5$
Viable yeasts and moulds	< 100
Coliforms	< 10
Staphylococcus aureus	< 1
Salmonella	< 1/10
Lancefield Group D *Streptococci*	$< 1 \times 10^4$

(Quinlan, 1990).

In many countries, norms and procedures for quality control of products and processes have already been established; jointly with the registries since constitute the guarantee for the validation of the productions of biopesticides like *Bt*. For these reasons, in Cuba a quality control system has been established for being executed by the network of Plant Health Protection Province Laboratories (LAPROSAV). National quality control norms for all the bioproducts have been developed. The 2% of the daily production is sampled to evaluate purity, concentration, and viability, presence of crystals, virulence, and technical effectiveness in the field. The process to delivery certified stocks is done centrally.

The first point for quality control is to verify preservation methods, followed by verification of the subculture of the initial stock. In both cases, methods of preservation and maintenance must be established to guarantee the genetic stability of the isolate which should be verifiable through biochemical and molecular tests. The level of identification is particularly important as it provides a mechanism for tracking the downstream production process and the fate of the agent in the environment once released (for instance, a molecular fingerprinting), a validation check for the purity and accuracy of the formulations, and a standard reference that may be used to register or protect individuals isolates.

Is has been long accepted that serial sub culturing of *Bt* isolated on agar can result in the loss of certain characteristics of the original isolate. This can be avoided by using as many long-term storage methods as possible.

In a production process the registry of all the parameters such as temperature, moisture content, pH and monitoring must be kept by lot of production to facilitate the examination of all the critic steps from the beginning of the process to the final product (Traceability).

Elósegui *et al.* (2005, 2006) stated that contamination is a risk for microbial products and particularly for low technology systems-based products where it poses a much higher risk due to the poor automatization and more manual operations in the production environment. Anyway, the safe allowed number of CFU contaminants is independent of the technology used. Contamination can lead to a loss of efficiency of the active ingredient by reduction of the microorganisms due to the competence; In addition, an inhibitory effect of the toxic activity caused by the production of secondary metabolites may occur.

FINAL REMARKS

A limitation for a successful introduction of many potential entomopathogenic microorganisms to the field is due in great extent to the relative lack of knowledge about their ecology and fate. The mode of application, the persistence of introduced micro-organisms, their reproductive rate (multiplication), the gene transfer rate to indigenous organisms, dissemination from the site of application and the effects on the balance and functioning of the exposed ecosystem (safety, benefit and harm) are of major importance and must be assessed before a release can be considered (Travers *et al.*, 1987).

The studies concerning promissory entomopathogenic bacteria have lead to new multidisciplinary approaches. The progress achieved shows the great impact of these bacteria as biological control agents. The generalized adoption of concepts provided by integrated pest management with the integration of bioinsecticides is a well supported policy itself, which can contribute to a larger mass use of these microorganisms.

One of the main challenges is to reduce application costs in order to compete with chemical counterparts. The cost of toxicological tests, the registration demands as well as the time-consuming register process are discouraging factors for biopesticide producers.

Not only the cost is a determinant factor but also the quality product, the active ingredient strengths, the formulation stability in storage and in environmental constraints (solar radiation, rain wash-off) must be considered. Other factors such as mass production possibilities and biosafety must be taken into account.

The new trends should be addressed to finding new potential bacterial isolates with a wider activity spectrum to manufacture products able to play an efficient pest control in different ecological niches.

REFERENCES

Akhurst, R.J., Mourant, R.G., Baud, L. and Boemare, N. (1996). Phenotipic and DNA relatedness study between nematode symbiont and clinical strain of the genus *Photorhabdus* (Enterobactereaceae). *International Journal of Systematic Bacteriology*, **46**: 1034-1041.

Aronson, A.I. (1993). The Two Faces of *Bacillus thuringiensis*: Insecticidal protein and post exponential survival. *Molecular Microbiology*, **7**: 489-496.

Arrieta, G., Hernández, A. and Espinosa, A.M. (2004). Diversity of *B. thuringiensis* strains isolated from coffee plantations infested plantations infested with the coffee berry borer *Hypothenemus hampei*. *Journal of Biological Tropical*, **52(3)**: 757-764.

Bailey, L. and Ball, B.V. (1991). Bacteria. *In*: Bailey, L. and Ball, B.V. (*Eds.*), *Honey Bee Pathology*. Academic Press, London, pp. 35-52.

Barloy, P., Delécluse, A., Nicolas, L. and Lecadet, M.M. (1996). Cloning and expression of the first anaerobic toxin gene from *Clostridium bifermentans* subsp. *malaysia*, encoding a new mosquitocidal protein with homologies to *Bacillus thuringiensis* delta-endotoxins. *Journal of Bacteriology*, **178**: 3099-3105.

Bedding, R.A. (1981). Low cost *in vitro* mass production of *Neoplactana* and *Heterorhabditis* species (Nematoda) for field control of insect pest. *Nematologica*, **27**: 109-114.

Bedding, R.A. (1984). Large scale production, storage and transport of the insect parasitic nematodes *Neoplactana* spp. and *Heterorhabditis* spp. *Annals of Applied Biology*, **104**: 117-120.

Benintende, G. and Márquez, A. (1996). Bacterias entomopatógenas. *In*: Lecuona, R.E. (*Ed.*), *Microorganismos patógenos empleados en el control microbiano de insectos plagas*. Talleres Gráficos Mariano Mas. México, pp. 61-72.

Bhalla, R., Dalal, M., Panguluri, S., Jagadish, B., Mandaokar, A. and Singh, A. (2005). Isolation, characterization and expression of a novel vegetative insecticidal protein gene of *Bacillus thuringiensis*. *Microbiology Letters*, **243**: 467-72.

Boemare, N.E. (2002). *Biology, Taxonomy and Systematics of Photorhabdus and Xenorhabdus. Entomopathogenic Nematology.* Cabi Publishing, France, pp. 35-56.

Boemare, N.E., Akhurst, R.J. and Mourant, R.G. (1993). DNA relatedness between *Xenorhabdus* spp. (Enterobacteriaceae) simbiotic bacteria of entomopathogenic nematodes and a proposal to transfer *Xenorhabdus luminescens* to a new genus, *Photorhabdus* Gen. Nov. *International Journal of Systematic Bacteriology*, **43**: 249-255.

Boemare, N.E., Thaler, J.O. and Lanois, A. (1997). Simple bacteriological test for phenotypic characterization of *Xenorhabdus* and *Photorhabdus* phase variants. *Symbiosis*, **22**: 167-175.

Boucias, D.G. and Pendland, J.C. (1998). *Principles of Insect Pathology*. Klumer Academic Publishers, Norwell, Massachusetts, USA.

Bucher, G.E. (1960). Potential bacterial pathogens of insects and their characteristics. *Journal of Insect Pathology*, **2**: 172-195.

Carballo, M. Falguni, G. and López, J.A. (2004). *Control Biológico de Plagas Agrícolas*. Serie Técnica Manual Técnico no 53. CATIE, Managua, Nicaragua.

Castillo, P., Acosta, N. and Ciliezar, A. (1995). Control microbiológico de plagas artrópodas. *In*: EAP (*Ed.*), *Manual para la enseñanza del control biológico en América Latina*. Zamorano, Honduras, pp. 51-72.

Cerrato-Soto, V. (2007). Caracterización del efecto anticanceroso e identificación de dianas moleculares de principios activos procedentes de *Serratia marcescens*. Facultad de medicina-Campus de Bellvitge. Universidad de Barcelona, España.

Cloutier, C. and Cloutier, C. (1992). Les solutions biologiques de lutte pour la répression des insectes et acariens ravageurs des cultures. *In*: Vincent, C. and Coderre, D. (*Eds.*), *La Lutte Biologique*. Tec & Doc Lavoisier, Québec, Canada, pp. 33.

Costilow, R.N. and Coulter, W.H. (1971). Physiological studies of an oligosporogenous strain of *Bacillus popilliae*. *Applied Microbiology*, **22**: 1076-1084.

Crickmore, N., Zeigler, D.R., Feitelson, J., Schnepf, E., Van Rie, D., Lereclus. D., Baum, J. and Dean D.H. (1998). Revision of the nomenclature for *Bacillus thuringiensis* pesticidal crystal proteins. *Microbiology and Molecular Biology Reviews*, **62(3)**: 807-813.

de Barjac H. (1981). Insect pathogens in the genus *Bacillus*. *In*: Berkley, R.C.W. and Goodfellow, M. (*Eds.*), *The aerobic endospore-forming bacteria: classification and identification*. Academic Press Inc., New York, USA, pp. 241–250.

de Barjac, H. and Bonnefoi, A. (1962). Essai de classification biochimique et sérologique de 24 souches de *Bacillus* de type *thuringiensis*. *Entomophaga*, **7**: 5-31.

de Maagd, R.A., Bravo, A. and Crickmore, N. (2001). How *Bacillus thuringiensis* has evolved specific toxins to colonize the insect world. *Review Trends in Genetics*, **17(4)**: 193-199.

Deacon, J.W. (1998). Profiles of Microorganisms - Biological Control: *Bacillus popilliae*. Prepared for the course, Microbiology 3m, Biological Teaching Organisation, University of Edinburgh. URL. Available: http://helios.bto.ed.ac.uk/bto/microbes/control.htm. (Visited on 5 Sep 2009).

Dean, D.H., Rajamohan, F., Lee, M.K., Wu, S.J., Chen, X.J., Alcantara, E. and Hussain, S.R. (1996). Probing the mechanism of action of *Bacillus thuringiensis* insecticidal proteins by sitedirected mutagenesis – a minireview. *Gene*, **179**: 111-117.

Denolf, P., Hendrickx, K., Vandamme, J., Jansens, S., Peferoen, M., Degheele, D. and Van Rie, J. (1997). Cloning and characterization of *Manduca sexta* and *Plutella xylostella* midgut aminopeptidase N enzymes related to *Bacillus thuringiensis* toxin-binding proteins. *European Journal of Biochemistry*, **248**: 748-761.

Elósegui, O., Carr, A. and Fernández-Larrea, O. (2005). Influencia de la carga microbiana contaminante inicial del sustrato en la calidad final de biopreparados de *Trichoderma harzianum* Rifai y *Beauveria bassiana* Balsamo Vuillemin. *Fitosanidad*, **9(1)**: 51-55.

Elósegui, O., Fernandez-Larrea, O. and López, O. (2006). Control de la calidad de biopreparados de hongos. *In*: Carreras, B. (*Ed.*), *Memorias del Curso Internacional. La producción de microorganismos entomopatógenos y antagonistas para el control de plagas.* CIDISAV, La Habana, Cuba.

Falcon, L.A. (1971). Use of bacteria for microbial control. *In*: Burges, H.D and Huseey, N.W. (*Eds.*), *Microbial control of insects and mites*. Academic Press, New York, USA, pp. 67-95.

Federici, B.A. (1993). Insecticidal bacterial proteins identify the midgut ephitelium as a source of novel target sites for insect control. *Archives of Insect Biochemistry and Physiology*, **22**: 357-371.

Feitelson, J.S., Payne, J. and Kim, L. (1992). *Bacillus thuringiensis*: Insects and Beyond. *Bio/Technology*. **10**: 271-276.

Fernández, C. and Juncosa. R. (2002). Biopesticidas: ¿la agricultura del futuro?. *Phytoma*, **141**: 14-19.

Fernández-Larrea, O. (2001). Tecnologías de producción de *Bacillus thuringiensis. Manejo integrado de plagas y agroecología*, **64**: 110-15.

Ferraz, L.C., Leite, L.G., Lopes, R.G. and Moino, A. (2008). Uilização de nemátoides para o controle de pragas agrícolas e urbanas. *In*: Alves, S.B. and Lopes, R.B. (*Eds.*), *Controle microbiano de pragas na America Latina: avances e desafíos*. Piraciba, Brasil, pp. 414.

Flexner, J.L. and Benalvis, D.L. (2000). Microbial insecticides. *In*: Rechcigl, J.E. and Rechcigl, N.A (*Eds.*), *Biological and Biotechnological control of Insects Pest*. Lewis Publisher, Florida, USA, pp. 45-62.

Genersch, E., Forsgren, E., Pentikäinen, A., Ashiralieva, A., Rauch, S. and Kilwinski, J. (2006). Reclassification of *Paeni-bacillus larvae* subsp. *pulvifaciens* and *Paenibacillus larvae* subsp. *larvae* as *Paenibacillus larvae* without subspecies differentiation. *International Journal of Systematic and Evolutionary Microbiology*, **56**: 501-11.

Glare, T.R. and O'Callaghan, M. (2000). *Bacillus thuringiensis*: *Biology, Ecology and Safety.* John Wiley & Sons, LTD, New York, USA.

Goldberg, L.J. and Margalit, J. (1977). A bacterial spore demonstrating rapid larvicidal activity against *Anopheles sergentii*, *Uranotaenia unguiculata*, *Culex univitattus*, *Aedes aegypti* and *Culex pipiens*. *Mosquito News*, **37**: 355-358.

Guillon, M. (1997). Production of Biopesticides: Scale up and Quality Assurance. *In*: British Crop Protection Council (*Ed.*), Microbial Insecticides: Novelty or Necessity? Farham, UK; pp. 151-162.

Haddix, P.L., Paulsen, E.T. and Werner, T.E. (2000). Measurement of mutation to antibiotic resistance: ampicillin resistance in *Serratia marcescens*. *Journal of College Biology Teaching*, **26**: 17-21.

Hansen, H. (1984). Methods for determining the presence of the foulbrood bacterium *Bacillus larvae* in honey Dan. *Journal of Plant Soil Science*, **88**: 325-328.

Höfte, H. and Whiteley, H.R. (1989). Insecticidal Crystal Proteins of *Bacillus thuringiensis. Microbiology. Reviews*, **53(2)**: 242-255.

Honée, G. and Visser, B. (1993). The mode of action of *Bacillus thuringiensis* crystal proteins. *Entomologia Experimentalis et Applicata*, **69**: 145-155.

Hornitzky, M.A.Z. and Clark, S. (1991). Culture of *Bacillus larvae* from bulk honey samples for the detection of American Foulbrood. *Journal of Apicultural Research*, **30**: 13-16.

Huber, H.E, Lüthy, P., Ebersold, H.R. and Cordier, J.L. (1981). The subunits of the parasporal cristal of *Bacillus thuringiensis*: size, linkage and toxicity. *Archives Microbiology*, **129**: 14-18.

Ibarra, J.E. (1998). Bacterias entomopatógenas. *In*: Sociedad Mexicana de Control Biológico (*Ed.*). *Proceedings IX Curso Nacional de Control Biológico*. Río Bravo, Tamaulipas, México, pp. 76-89.

Ibarra, J.E. (2007). Uso de bacterias en el control biológico. *In*: Rodríguez-del Bosque, L.A. and Arredondo-Bernal, H.C. (*Eds.*), *Teoría y Aplicación del Control Biológico*. México, pp. 303.

Ibarra, J.E. and López Meza, J.E. (2000). Bacterias Entomopatógenas. *In*: Badii, M.H., Flores, A.E. and Galán Wong, L.J. (*Eds.*), *Fundamentos y Perspectivas de Control Biológico*. Universidad Autónoma de Nuevo León, México, pp. 462.

Ibarra, J.E., del Rincón, M.C., Orduz, S., Noriega, D., Benintende, G., Monnerat, R., Regis, L., de Olivera, C.M.F., Lanz, H., Rodríguez, M.H., Sánchez, J., Peña, G. and Bravo, A. (2003). Diversity of *Bacillus thuringiensis* strains from Latin America with insecticidal activity against different mosquito species. *Applied and Environmental Microbiology*, **69**: 5269-5274.

Jackson, T.A. (1995). Amber disease reduces trypsin activity in midgut of *Costelytra zealandica* (Coleoptera: Scarabaeidae) larvae. *Journal of Invertebrate Pathology*, **65**: 68-69.

Jackson, T.A., Boucias, D.G. and Thaler, J.O. (2001). Pathobiology of amber disease, caused by *Serratia* spp., in the New Zealand grass grub. *Journal of Inverterbrate Pathology*, **78**: 232-243.

Jaquet, F, Hütter, R. and Lüthy, P. (1987). Specificity of *Bacillus thuringiensis* delta-endotoxin. *Applied and Environmental Microbiology*, **53**: 500-504.

Johnson, V.W., Pearson, J.F. and Jackson, T.A. (2001). Formulation of *Serratia entomophila* for biological control of grass. *New Zealand Plant Protection*, **54**: 125-127.

Krieg, A., Huger, A.M., Langenbruch, G.A. and Schnetter, W. (1983). *Bacillus thuringiensis* var. *tenebrionis,* a new pathotype effective against larvae of Coleoptera. *Z Angew Entomol*ogy, **96(5)**: 500-508.

Lacey, L.A. (1990). Persistence and formulation of *Bacillus sphaericus* in bacterial control of mosquitoes and black flies. *In*: de Barjac, H. and Sutherland, D.J. (*Eds.*), *Biochemistry, Genetics and Applications of Bacillus thuringiensis israelensis and Bacillus sphaericus*. New Brunswick Rutgers, University Press, pp. 284-294.

Lecadet, M.-M., Frachon, E., Dumanoir, V.C., Ripouteau, H., Hamon, S., Laurent, P. and Thiéry, I. (1999). Updating the H-antigen classification of *Bacillus thuringiensis*. *Journal of Applied Microbiology*, **86(4)**: 660-672.

Lechner, S., Mayr, R., Francis, K., Pruss, B., Kaplan, T. and Wiessner-Gunkel, E. (1998). *Bacillus weihenstephanensis* sp. nov. is a new psychrotolerant species of the *Bacillus cereus* group. *International Journal of Systematic Bacteriology*, **48**: 1373-1382.

Márquez, M.E. (2005). Selección y evaluación tóxico-patogénicas de cepas cubanas de *Bacillus thuringiensis* con actividad nematicida. Universidad Agraria de La Habana. Cuba.

Mackedonski, V.V. and Hadjiolov, A.A. (1972). Preferential *in vivo* inhibition of ribosomal ribonucleic acid synthesis in mouse liver by the exotoxin of *Bacillus thuringiensis*. *Federation of European Biochemical Societies Letters*, **21**: 211-214.

Matheson, A. and Reid, M. (1992). Strategies for the prevention and control of American foulbrood. Part I. *American Bee Journal*, **132**: 399-402.

Narva, K.E, Payne, J.M., Schwab, G.E., Hickle, L.A., Galasan, T. and Sick, A.J. (1991). Novel *Bacillus thuringiensis* microbes active against nematodes, and genes encoding novel nematode active toxins cloned from *Bacillus thuringiensis* isolates: European patent application EP0462 721A2. Munich, Germany, European Patent Office.

Nakamura, L. (1998). *Bacillus pseudomycoides* sp. nov. *International Journal of Systematic Bacteriology*, **48**: 1031-1035.

O'Callaghan, M. and Jackson, T.A. (1993). Isolation and enumeration of *Serratia entomophila* –A bacterial pathogen of the New Zealand grass grub, *Costelytra zealandica*. *Journal of Applied Bacteriology*, **75**: 307-314.

Pérez, N. and Vázquez, L. (2001). Manejo ecológico de plagas. *In*: Funes, F., García, L., Bourque, M., Pérez, N. and Rosset, P. (*Eds.*), *Transformando el campo cubano. Avances de la Agricultura sostenible.* La Habana, Cuba, pp. 191-226.

Petterson, B., Rippere, K.E., Yousten, A.A. and Priest, F.G. (1999). Transfer of *Bacillus lentimorbus* and *Bacillus popilliae* to the genus *Paenibacillus* with emended descriptions of *Paenibacillus lentimorbus* comb. nov. and *Paenibacillus popilliae* comb. nov. *International Journal of Systematic Bacteriology*, **49**: 531-540.

Poinar, G.O. (1990). Biology and taxonomy of Steinernematidae and Heterorhabditidae. *In*: Gaugler, R. and Kaya, H.K. (*Eds.*), *Entomopathogenic nematodes in biological control*. CSR Press, Florida, USA, pp. 23-58.

Priest, F.G. (1992). Biological control of mosquitoes and other biting files by *Bacillus sphaericus* and *Bacillus thuringiensis*. *Journal of Applied Bacteriology*, **72**: 357-369.

Quinlan, R.J. (1990). Registration requirements and safety considerations for microbial pest control agents in the European Economic Community. *In*: Laird, M. Lacey, L.A. and Davidson, E.W. (*Eds.*), *Safety of microbial pesticides*. CRC Press; Florida, USA, pp. 11-18.

Redmond, C.T. and Potter, D.A. (1995). Lack of efficacy of *in vivo* and putatively *in vitro* produced *Bacillus popilliae* against field populations of Japanese beetle (Coleoptera: Scarabaeidae) grubs in Kentucky. *Journal of Economic Entomology*, **88**: 846-854.

Regis, L., Silva-Filha, M.H., Nielsen-Leroux, C. and Charles, J.F. (2001). Bacteriological larvicides of dipteran disease vectors. *Trends in Parasitology*, **17(8)**: 377-80.

Rippere, K.E., Tran, M.T., Yousten, A.A., Hilu, K.H. and Klein, M.G. (1998). *Bacillus popilliae* and *Bacillus lentimorbus,* bacteria causing milky disease in Japanese beetles and related scarab larvae. *International Journal of Systematic Bacteriology*, **48**: 395-402.

Ruíz-Sánchez, A., Cruz-Camarillo, R., Salcedo-Hernández, R. and Barboza-Corona, J.E. (2003). *Serratia marcescens*: de patógeno oportunista al control de insectos que afectan cultivos agrícolas. *Biotecnología*, **8**: 31-37.

Schnepf, E., Crickmore, N., Van Rie, J., Lereclus, D., Baum, J., Feitelson, J., Zeigler, D. and Dean, D.H. (1998). *Bacillus thuringiensis* and its pesticidal crystal proteins. *Microbiology and Molecular Biology Reviews*, **62**: 775-806.

Shimanuki, H. (1990). Bacteria. *In*: Morse, R.A. and Nowogrodzki. R. (*Eds.*), *Honey bee pests, predators and diseases*. Cornell University Press, Ithaca, N.Y., USA, pp. 27-47.

Siegel, J.P. and Shadduck, J.A. (1990). Mammalian safety of *Bacillus sphaericus* in bacterial control of mosquitoes and black flies. *In*: de Barjac, H. and Sutherland, D.J. (*Eds.*), *Biochemistry, Genetics and Applications of Bacillus thuringiensis israelensis and Bacillus sphaericus*. University Press, New Brunswick Rutgers, USA, pp. 202-217.

Sneath, P.H.A. (1986). Endospore-forming Gram-positive rods and cocci. *In*: Sneath, P.H.A., Mair, N.S., Sharpe, M.E. and Holt, J.G. (*Eds.*), *Bergey's Manual of Systematic Bacteriology*. Williams and Wilkins, Baltimore, USA, pp. 1104-1207.

Stahly, D.P., Andrews, R. and Yousten, A.A. (1992). The genus *Bacillus:* insect pathogens. *In*: Balows, A. Trüper, H.G. Dworkin, M. Harder W. and Schleifer, K.-H. (*Eds.*), *The Prokaryotes*. Springer, New York, pp.1697-1745.

Tanada, Y. and Kaya, H.K. (1993). *Insect Pathology*. Academic Press, Inc., New York, USA.

Travers, R.S., Martin, P.A.W. and Reichelderfer, C.F. (1987). Selective process for efficient isolation of soil *Bacillus* spp. *Applied and Enviromental Microbiology*, **53**: 1263-1266.

Van Rie, J., Jansens, S., Höfte, H., Degheele, D. and Van Mellaert, H. (1989). Specificity of *Bacillus thuringiensis* δ-endotoxins: Importance of specific receptors on the brush border membrane of the mid-gut of target insects. *European Journal of Biochemistry*, **186**: 239-247.

Vergara, R.R. and Varela, L.A.L. (1978). La importancia de la patología de insectos en el control biológico. Tunja, Universidad Pedagógica y Tecnológica de Colombia.

Warren, R.E., Rubenstein, D., Ellar, D.J, Kramer, J.M. and Gilbert, R.J. (1984). *Bacillus thuringiensis* var. *israelensis*: Protoxin activation and safety. *Lancet*, **24**: 678-679.

WHO (1994). La lucha mundial contra el Paludismo. *Boletín de la oficina Sanitaria Panamericana,* **116(6)**: 477-482.

Zhang, J., Hodgman, T.C., Krieger, L., Schnetter, W. and Schairer, H.U. (1997). Cloning and analysis of the first *cry* gene from *Bacillus popilliae. Journal of Bacteriology*, **179**: 4336-4341.

{2}

Entomopathogenic Viruses

Ma. Cristina Del Rincón-Castro and Jorge E. Ibarra

ABSTRACT

It is almost certain that there are no living forms in nature that escape infection by at least one kind of virus. Hence, a great variety of viruses attack and kill many insect species. These viruses are called entomopathogenic viruses and have been found in many insect orders. Some insect pests are also susceptible to viral infections and, therefore, these viruses can be or have been used as biological control agents. For this purpose, viruses are initially recovered from infected insects, which can later be produced and applied in the field as bioinsecticides. Although there is a great diversity of insect viruses, only a few are frequently observed in insect populations causing epizootics, such as baculoviruses, cypoviruses, entomopoxviruses, and iridoviruses. Undoubtedly, there are many more insect viruses still to be discovered; however, only few show potential to be used as control agents, especially in the case of baculoviruses. This chapter provides an abridged overview of insect viruses and their use in pest management, focusing its attention towards the description of the diverse groups of insect viruses, their characteristics, replication, genetics, use as bioinsecticides, genetic manipulation, production, application, and examples on their use as control agents.

INTRODUCTION

Viruses are well known by their devastating effects on human populations, causing epidemic diseases such as small pox, hepatitis, or AIDS. However, viral infections are not restricted to humans and their infection can be detected on plants, animals and microorganisms, some of which are also important to man. Research on animal viruses has proven to be important in the understanding of replication mechanisms, molecular biology, evolution, and host interaction. These studies have provided the bases for the understanding and development of the use of viruses as vaccines, such as the inoculation of attenuated virus particles to stimulate the immunological system that prevents potential infections by the production of protective antibodies.

Besides the negative effect of virus infections to humans, as well as to domestic animals and crop plants, viruses may have a positive effect when they attack insect pests or weeds. Some of these viruses are considered excellent candidates for insect pest control, mostly because: 1) they are highly specific, as they only infect one insect species or, at the most, some related species; 2) they are highly safe, as they show no negative effects on plants, domestic animals, birds, fish, or non-target insects; and 3) they have no secondary effect on the environment, as they are biodegradable and form an important part of the biotic factors in nature. Additionally, in recent years, viruses from the family Baculoviridae are extensively used in the overproduction of heterologous proteins, either for research, medical, or industrial purposes.

Insects have more than 400 million years of evolution and constitute the most diverse group of living things on the world. Today, few habitats are free of the insect presence, from steamy tropical jungles to freezing Polar Regions. Some insects have been associated to man throughout his evolution and play an important role since the beginning of the agriculture by becoming natural competitors for the food that man grows. To date, thousands of insect species damage our food and other agricultural products, destroy our possessions, attack our domestic animals, and transmit a number of human pathogens. Although many measures have been developed to control and eradicate them, only the use of chemical insecticides have proven to be the cheapest, fastest, and most practical technique to counteract their devastating effect. Unfortunately, the use of a single measure brought harmful consequences to the environment and the human health. Also, insects soon developed resistance to chemical insecticides and, because of its wide toxic range; beneficial insects are also destroyed as well as domestic animals and wildlife. Although still widely used, a more environmentally oriented society has made possible to reduce its application by about 1.5% annually. Additionally, more environmentally-friendly techniques, such as Biological Control, have been developed, which is based on the deliberate use and/or release of biotic factors to the environment with the purpose to reduce the pest populations. Biological control techniques may be cheaper and more efficient than the use of chemical insecticides, but is based on the basic knowledge of the environment, of the life history of the natural enemies, and their interactions with the insect pests.

Three major groups of biotic factors are used as biological control agents: 1) parasitoids, 2) predators, and 3) pathogens. The first two groups are mostly constituted by other insects, while the third group is constituted by infectious microorganisms, or entomopathogens, causing lethal or deleterious diseases, such as bacteria, fungi, protozoans, nematodes, and viruses. These microbes are called bioinsecticides when used massively to control insect pests. Entomopathogens are safe to man, beneficial insects, and any other non-target species and some can even be applied along with chemical insecticides. Viruses have demonstrated to be important biological control agents, in many cases. Nowadays, the market for bioinsecticides represents more than 2.5% of the total insecticide market and it is estimated to increase to 4.2% by 2010. Although bacterial bioinsecticides represent the greatest majority of them, viruses constitute an important component of this type of agents, especially those within the baculoviruses. This family constitutes the most diverse group of entomopathogenic viruses which is found almost exclusively attacking insects from the orders Lepidoptera, Hymenoptera, and Coleoptera, although some few have been found in crustaceans and spiders.

This chapter deals with the beneficial use of the entomopathogenic viruses, as many of their infections cause lethal diseases to susceptible individuals and, therefore, they can be important biotic factors to keep insect pest population densities under control. The following sections overview the basic biology of insect virus families, as well as their genetics, genetic manipulation, production as bioinsecticides, field application, and advantages and limitations. The goal of this chapter is to provide a summarized picture of the insect viruses and their uses as bioinsecticides.

GENERAL CHARACTERISTICS OF VIRUSES

Viruses constitute the simplest life forms, although their status as living things is still debatable. The simplest structure of a virus contains only a nucleic acid core, DNA o RNA, and a protein shell or capsid, which plays an important role in the host cell infection process. This nucleocapsid may also be surrounded by a lipid bilayer envelope, forming an infective unit called virion. Additionally, some viruses are occluded into a protein matrix. The matrix forms an occlusion body (OB). OBs are found in some virus families but appear to have evolved independently in each family. Nucleic acids in insect viruses may be double- or single-stranded DNA (dsDNA or ssDNA, respectively) or double- or single-stranded RNA (dsRNA or ssRNA, respectively), enveloped or non-enveloped, and occluded in OBs or non-occluded. Viruses lack self-movement and irritability, in contrast with the basic features that characterize all living things. Once a viral particle gains entry into a permissive (susceptible) cell, the DNA (o RNA) takes charge of the cell's metabolic system and profusely replicates into new virus particles, until the cell is normally depleted of all its content and dies. Viruses are considered obligate parasites, as they use the cell metabolic machinery to replicate. That is, viruses cannot replicate "*in vitro*" (*i.e.* on artificial media, where only organic material sources are provided). A great variety of viruses can be found in nature. Each group of viruses show some intrinsic features such as morphology, genome, infectivity, host range, chemical resistance, etc., which are distinctive of each group (Vaughn, 1992; Van Regenmortel *et al.*, 2000).

Viruses were unknown until the 1900s. By this time it was certain that viruses contain nucleic acids and performed functions separately, although dependent, of the host metabolism. This host dependency and their high specificity indicate that hosts evolved before their own viruses, and suggest that virus precursors were nucleic acid fragments which developed the ability of self-replication, similar to extra-chromosomal replicons known as plasmids. However, all viruses have viral genomes coated with a capsid and have the ability to invade cells and "capture" their metabolism. It has been suggested that primitive viruses may be similar to the actual viroids (short, non-coding, self-replicating, non-encapsulated, infective fragments of RNA) derived from chromosomal or transcriptional RNA from host cells. More complex virus groups may have evolved from DNA plasmids which acquired genes coding for capsid proteins and evolved to self-replicating infective particles. Complex virus groups may contain large and highly developed genomes as well as intricate structure as a consequence of a long and interrelated co-evolution with their hosts.

VIRUS CLASSIFICATION

Insect viruses are classified into 12 virus families and, as any other virus group; they are classified by the International Committee on Taxonomy of Viruses (ICTV) (Van Regenmortel *et al.*, 2000). Therefore, the ICTV follows the same criteria to classify the diversity of virus groups that attack insects, such as: type of genetic material (*i.e.* singe- or double-stranded DNA, or RNA, positive or negative strand), virion morphology and size (*i.e.* icosahedral, rod-shaped, etc.), presence of an envelope surrounding the virion, presence of an occlusion body engulfing the virions, host and host range, among many others. However, the ultimate criterion is the sequence of the genetic material which discriminates between viral species, and establishes the evolutionary relationship among viruses within the same group. Insect viruses are named by acronyms, according to their host and the viral group to which they belong to. For example, the *Autographa californica* multiple nucleopolyhedrovirus is named AcMNPV. Therefore, all nucleopolyhedroviruses are named NPV, just as the granuloviruses are named GV. In addition, lepidopteran baculoviruses were divided in NPVs (group I and II) and GVs by genomic sequence data (van Oers and Vlak, 2007). When genomic sequences from NPVs infecting sawfly and mosquito species became available, it was evident that lepidopteran NPVs and GVs were closer related to each other, than to NPVs from dipterans and hymenopteran hosts.

On the other hand, the entomopoxviruses are named EPV, the iridoviruses are IV, and the cytoplasmic polyhedrosis viruses (cypoviruses) are CPV. Insect viruses are highly diverse; however, only few groups are frequently found in insect populations and even fewer show potential to be used as biological control agents, highlighting the group of baculoviruses. The main features of the most important insect viruses are shown next.

Baculoviruses

Baculoviruses make up a family of insect viruses and are grouped into two main groups or genera: *Nucleopolyhedrovirus* or NPVs; and *Granulovirus* or GVs. Both groups contain circular double-stranded DNA genome of approximately 80-180 kbp, which is condensed within nucleocapsids and are predicted to encode about 90 to 180 genes (Okano *et al.*, 2006; van Oers and Vlak, 2007). Virions are enveloped rod-shaped nucleocapsids, which refers to the name baculovirus. These are occluded either within polyhedral or granular OBs (2–15 μm in size) mainly formed by the protein polyhedrin (Figs. 2.1a, 2.1c). Although there have been more than 600 isolates reported from a variety of insect species, 90% of these have been isolated from lepidopteran hosts (Faulkner, 1981; Adams and McClintock, 1991; Vlak, 1992). NPVs are also divided into two groups: multiple nucleopolyhedroviruses or MNPVs, with several virions per envelope (Fig. 2.1b); and single nucleopolyhedroviruses or SNPVs, with only one virion per envelope (Fig. 2.1d). NPV virions replicate only in the nuclei of susceptible cells and their OBs fluctuate between one and 15 μm in size (Tinsley and Kelly, 1985). On the other hand, GV virions are always single within the envelope and the OBs are very small (0.2 to 0.5 μm) as compared with NPV polyhedra, but this is because there is only one virion per OB. GV virions replicate in the cytoplasm of susceptible cells.

Entomopoxviruses

The Poxviridae family is separated into two subfamilies: the Entomopoxvirinae, which comprises poxviruses of insects; and the Chordopoxvirinae, which comprises poxviruses of vertebrates (Goodwin *et al.*, 1991). The first subfamily or Entomopoxvirus (EPV) comprises three genera based on insect host and virion morphology. Genera are designated as *Entomopoxvirus* A, (infects only coleopteran), *Entomopoxvirus* B (infects lepidopteran and orthopteran), and *Entomopoxvirus* C (infects only dipteran) (Arif and Kurstak, 1991). They show allantoid- to brick-shaped virions, occluded within ovoid OBs called spheroids (Tinsley and Kelly, 1985; Adams, 1991). Virions size up to 400 nm long and 250 nm width, contain dsDNA ranging from 270 to 320 kbp, and replicate in the cytoplasm of susceptible cells. Entomopoxviruses have been isolated from 27 species of orthopterans, lepidopterans, dipterans, and coleopterans.

Cypoviruses

The family Reoviridae includes 12 genera of segmented dsRNA viruses, and some of them infect mammals (Evans and Entwistle, 1987). The viruses isolated from insects are called cytoplasmic polyhedrosis viruses or cypoviruses (CPVs). CPVs are commonly isolated from insects and have only one genus: *Cypovirus*. These viruses have linear dsRNA genomes divided into 10–12 segments of about 12 to 32 kbp in total. Size and number of fragments depend on the species. The non-enveloped icosahedral virions, with a diameter of 60-80 nm, show 12 lateral projections and are occluded within large isometric OBs named polyhedra of up to 10 μm in size (Hukuhara and Bonami, 1991). Virions replicate in the cytoplasm of the insect midgut epithelial cells (Hukuhara, 1985), and they are mainly isolated from Lepidoptera, and occasionally from Diptera or Hymenoptera, and rarely from Coleoptera or Neuroptera. The *Bombyx mori* cytoplasmic polyhedrosis virus or BmCPV, was the first occluded cytoplasmic insect virus reported (Xeros, 1954).

Iridoviruses

Iridoviruses are large icosahedral viral particles of 120 to 300 nm in diameter and consist of a central core of nucleic acid and proteins. Virions bud from the plasma membrane, and are not occluded in a protein matrix. Iridoviruses have been isolated from Diptera, Hemiptera, Lepidotera, Coleoptera and Hymenoptera, and belong to two genera: *Iridovirus*, whose viral particles fluctuate between 120 to 130 nm in size and the type species is the rice borer *Chilo suppressalis* (Walker) (Lepidoptera: Crambidae) (Devauchelle *et al.*, 1985); and *Chloriridovirus*, with a larger viral particle (180 nm) and the type species was isolated from *Aedes taeniorhynchus* Wiedemann (Diptera: Culicidae) mosquito larvae. Genomic DNA of iridoviruses is a linear molecule fluctuating from 140 to 303 Kbp (Goorha and Murti, 1982). The replication of iridoviruses includes nuclear and cytoplasmic stages, but virion assembly occurs exclusively in the cytoplasm (Goorha, 1982). The most distinctive feature of this family is a particular iridescence of the infected tissues that vary in color according to the host species. All host species of iridoviruses are associated with aquatic environments.

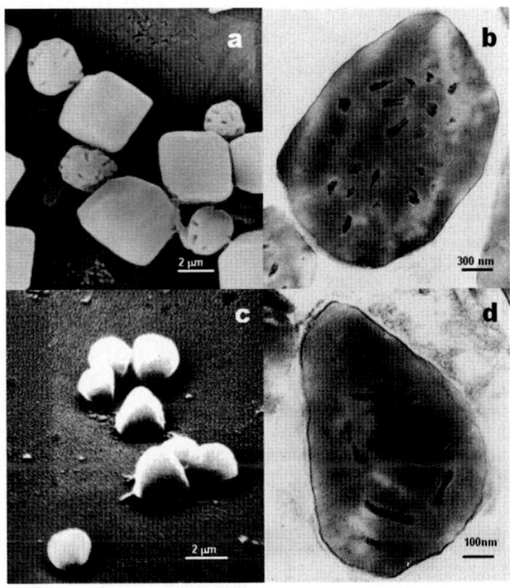

Fig. 2.1. Baculovirus occlusion bodies. a) Scanning electron microscopy of AcMNPV; b) Transmission electron microscopy of AcMNPV; c) Scanning electron microscopy of TnSNPV and; d) Transmission electron microscopy of TnSNPV

VIRUS REPLICATION

When insects inadvertently consume virus-contaminated food the infection cycle starts. Virus gains entry into a permissive host cell, using a variety of infective mechanisms: endocytosis, phagocytosis, pinocytosis, membrane fusion, etc. Subsequently, the viral genome is dissociated from the capsid and released into the host cell in a great variety of mechanisms. Once in the cell (either in the nucleus or in the cytoplasm) the viral genome replicates itself, frequently using the host cell's enzymatic machinery, creating a great number of identical copies. Also, structural, functional, and auxiliary viral genes are transcribed and translated into proteins, which will be used to assemble, part by part, new viral particles. The new copies of the viral genome are packed within the new capsids and the new viral progeny is released by a variety of procedures, which are distinctive for each viral group. Some viruses are released by inverted pinocytosis, which engulfs the viral particle within a membrane vesicle, creating what is known as "enveloped viruses".

Replication in Insect Cell Lines

Baculoviruses have been widely used for the production of numerous recombinant proteins in insect cells (Condreay and Kost, 2007). The most popular insect cell lines are Sf-9 and Sf-21, derived from the ovarian tissue of *Spodoptera frugiperda* (J.E. Smith) (Lepidoptera: Noctuidae), and BTI-TN-5B1-4, derived from *Trichoplusia ni* (Hübner) (Lepidoptera: Noctuidae) (also called Hi-Five) (Ikonomou *et al.*, 2003). In the past, culture media that allowed the *in vitro* growth of cells were merely saline solutions supplemented with 5 to 50% insect hemolymph. Nowadays, media are prepared with chemical substitutes of the hemolymph components, such as protein hydrolysates and fetal bovine serum (FBS) (Cameron *et al.,* 1989). Later on, a variety of insect cell lines supported the efficient replication of some MNPVs. More recently, at the beginning of the 1990s, some cell lines were developed to allow the replication of some SNPVs (Granados *et al.,* 1994). So far, the establishment of insect cell lines able to support replication of GVs has not been well standardized, in spite of some reports indicating their feasibility. The optimal growth of NPVs in cell lines is greatly influenced by many factors, such as temperature, culture media composition, quality of inoculum, cell density, cell division rate, among others (Bilimoria, 1991).

Insect cell cultures are excellent systems of gene expression, with respect to the pattern and capacity of posttranslational modifications. However, their cultivation is more complicated and costly, and usually yields lower product titers. One of the most important advances in insect virus technology was the development of infection and assembly of virus in tissue cells. The development of tissue culture media, the establishment of continuous lepidopteran cell cultures, the plaque purification assays and the ability to propagate baculovirus in tissue cultures, contributed to understand the *in vitro* replication of some baculoviruses, CPVs, IVs and EPVs.

In the insect cells, baculovirus infection starts with the attachment of the so-called extracellular virions (EVs) or budded non-occluded virions (BV), which naturally spread the infection systemically throughout the insect body. Nucleocapsids penetrate the cell by viropexis or fusion (Adams *et al.,* 1977), and once in the cytoplasm they move towards the nucleus, where nucleocapsids discharge the DNA, either by injecting the nucleic acid through a nuclear pore or by penetrating the nucleus and discharging the DNA once inside (Hirumi, 1975; Granados, 1980). The cytopathic effect of the infection is detected until the hypertrophy of the nucleus (Fig. 2.2) the degradation of the compact chromatin, and the formation of the virogenic stroma, which appears like a dark area in the nucleus, and becomes darker as replication proceeds. New nucleocapsides are assembled and some leave the cell as new EVs, which spread the infection throughout the insect or the cell culture. The rest are occluded in new OBs, that increase in number until nuclear and cell membranes burst and OBs are released into the hemolymph or the culture medium (Volkman *et al.,* 1976).

The EPV replication occurs successfully in a cell line derived from *Estigmene acraea* (Drury) (Lepidoptera: Arctiidae) (BTIEAA). This replication process is very similar to that observed in the living insect. Virion penetrates the cell by fusion between the viral envelope and the cell membrane. Once in the cytoplasm, replication starts by the formation of a virogenic stroma (Arif and Kurstak, 1991). High titers of new virions are produced per cell. On the other hand, CPVs have also been replicated in insect cell lines. *In vitro* replication of CPVs is identical to that observed in the living insect. A virogenic stroma is formed in the cytoplasm and new assembled virions are efficiently occluded by polyhedra. However, the morphology of OBs produced *in vitro* by CPVs is very different from that produced in the living insect (Hukuhara and Bonami, 1991). The Chilo iridiscent virus or IIV-6 strain can be easily grown and propagated in *Choristoneura fumiferana* cells (cell line CF-124) (Jakob and Darai, 2002).

Replication in insects

All baculovirus infections are initiated *per os* through ingestion of food previously contaminated with OBs. Alternative routes of infection occur when adult females transmit the infection to their progeny (vertical

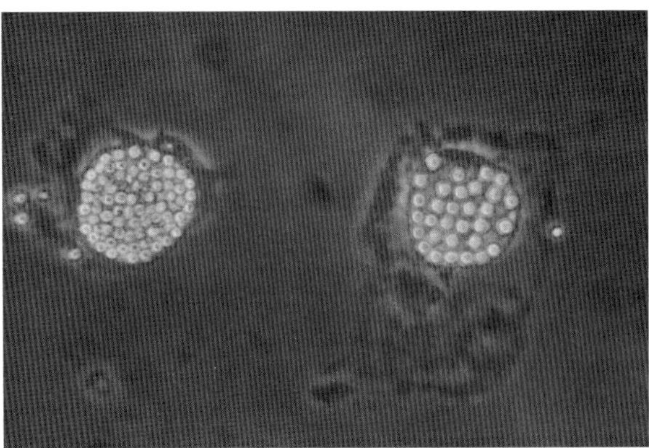

Fig. 2.2. *Trichoplusia ni* cells infected with AcMNPV. Hypertrophied nuclei appear filled with newly formed OBs. Cytoplasmic membrane is about to be degraded

transmission) by inoculation of the eggs either superficially or internally (transovarial transmission) (Granados, 1980; Mazzone, 1985). In addition, if a parasitoid lays eggs into an infected individual, the parasitoid becomes able to transmit the virus to healthy individuals, during subsequent ovipositions (horizontal transmission). In a *per os* infection by baculoviruses, once the inoculum (polyhedra or granules) are ingested, they reach the midgut (mesenteron) lumen, where pH is normally high (9.5 to 11.5). Under these conditions and perhaps the joint action of proteolytic enzymes, OBs are degraded and the enveloped virions are released into the lumen (Granados, 1980). Virion envelopes possess high affinity to the membrane of the midgut epithelial cells (especially those of the microvilli) and attach to them, releasing naked virions (nucleocapsids). Nucleocapsids then move towards the cell nucleus were they replicate (Mazzone, 1985; Granados and Williams, 1986). Recently, an excellent review about the role of the actin cytoskeleton in the baculovirus infectivity elucidated that actin cytoskeleton is a key factor in NPV infectivity (Volkman, 2007).

A baculovirus infection (Fig. 2.3) is typically a biphasic process in which different virus forms produced are genotypically identical, but phenotypically different. In the infection caused by baculoviruses, intracellular occlusion-derived virus (ODV) can be produced as well as extracellular viral progeny (budded virus or BV). The ODVs transmit infections from insect to insect, whereas the BVs spread the infection from cell to cell in an infected insect (Fig. 2.3b) (Granados, 1980). The indicators of a baculovirus infection are the swollen nucleus and the formation of virogenic stroma (Arif and Kurstak, 1991). Further changes leading to the formation of OB are observed as they are formed exclusively in the nucleus in NPV infections (Fig. 2.3b), while in GV infections, capsules may form in the nucleus and in the cytoplasm. Dipteran and hymenopteran NPVs are thought to replicate only within midgut cells. Only the lepidopteran NPVs provoke systemic infections in their hosts (Volkman, 2007).

On the other hand, the primary infection with entomopoxviruses is very similar to that of baculoviruses with the exception that, once OBs are degraded, virions gain entry to the epithelial cells via phagocytosis. Then, virions are released into the cytoplasm were they replicate (Moss, 1996). The CPV OBs are degraded in the midgut lumen where naked (non-enveloped) virions are released. The virions attached to the microvilli release their genome into the cytoplasm for replication. At a late stage of infection, polyhedra are produced in the cytoplasm, and many virus particles are occluded in polyhedra, which provide some protection to virions from environmental factors, such as UV light, desiccation, a wide range of pH, and degradation by microorganisms (Hukuhara and Bonami, 1991). Additionally, oral and transovarian transmission of IIV-3 have been documented for mosquito larvae (Hall *et al.,* 1985). Early mosquito larval stages are most susceptible to IIV-3 infection and naked virions are directly transferred into the epithelial cells via pinocytosis. Virons, encapsulated within a vacuole, are transported to the inner cytoplasm and released, where they replicate (Ward and Kalmakoff, 1991; Tanada and Kaya, 1993).

The development of external symptoms (signs) of entomopathogenic virus infection is very different in each viral group. Most infections become apparent only until the infection is widespread throughout the host body. The first detectable signs of a viral disease appear when insect becomes sluggish, stops feeding, and growth ceases. Insects infected by baculoviruses become whitish because of the massive infection of the fat body, visible through a more translucent integument (exoskeleton), which turns thinner as the infection advances until it ruptures. A grayish to creamy liquid is released, containing billions of suspended OBs. By this time the larva has crawled up and hanged head down from its crochets in an inverted "V" position. This facilitates the spread of the inoculum in the field (Fig. 2.3a) (Mazzone 1985; Granados and Williams, 1986).

The insects infected with entomopoxviruses show similar symptoms to those of baculovirus infections, except that the whitish coloration of the fat body is located at the posterior part of the abdomen, which becomes swollen due to the accumulation of spheroids. Some coleopterans show a swollen rectal sac, which becomes whitish and irregular. The infection develops slowly in all cases, and rarely causes epizootics in insect populations. However, the infection is common in nature. The insects infected with cipoviruses are only affected at the midgut epithelial tissue; in this case larvae develop diarrhea and vomit, and stop feeding and growing. Although larvae lose color, the integument keeps its typical consistency. Due to the great amounts of OBs produced by the epithelial cells, the midgut becomes white-creamy colored. Most diseases caused by CPVs infections are chronic, and mortality levels are normally low in insect populations. Conversely, infections caused by IVs are normally lethal in the immature stages. Usually, symptoms are obvious at the end of the larval stages and death overcomes just before pupation. However, IV infections can be detected earlier in the form of iridescent spots in legs, prolegs, antennae, and thorax (Fig. 2.3c) (Tanada and Kaya, 1993). In advanced stages of infection, some tissues show iridescent spots (Fig. 2.3d). Interestingly, no behavioral changes are observed until larvae stop feeding and die.

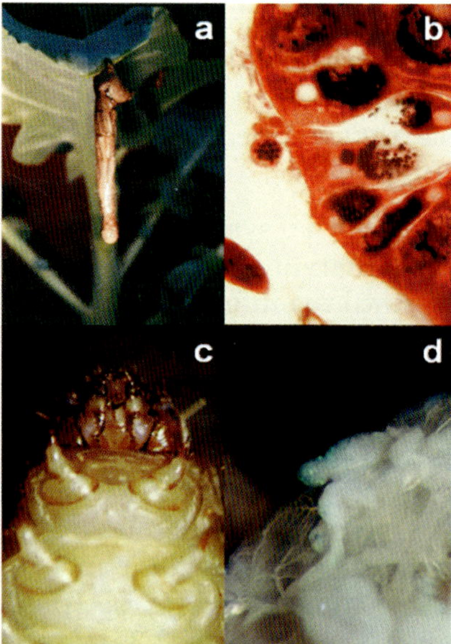

Fig. 2.3. Signs and symptoms in insects infected with viruses. a) Gross pathology of *T. ni* larvae infected with TnSNPV; b) Micrograph of *T. ni* fat body infected with AcMNPV. c) *Galleria mellonella* (Linnaeus) (Lepidoptera: Pyralidae) larvae infected with iridovirus; d) *G. mellonella* fat body infected with iridovirus

APPLICATION OF VIRUSES

Production and Formulation

Production of baculoviruses as bioinsecticides has been carried out exclusively *in larva*, by massively infecting susceptible individuals. The production strategy requires a highly reliable technique to maintain a massive rearing of an insect colony, in such a way that it can provide massive numbers of susceptible individuals in a continuous and synchronized manner. Mechanization of the rearing process is highly recommended, both to standardize the production and to reduce costs. This can be achieved by using artificial or semi-artificial diets, which not only expedite the production process but also allows keeping an acceptable level of control of undesirable infectious agents (Vanderzant *et al.,* 1962). The use of artificial diets also helps to obtain large individuals which will maximize the yield of OBs per larva. Although the infection of larvae may be achieved by injection of EVs, the most practical method is based on the contamination of the insect diet with the OBs (Tanada and Kaya, 1993). In general terms, it is suggested that 90% of the individuals in a colony is used for the production of baculoviruses, while the resting 10% is used for the colony maintenance. The optimum inoculum concentration and the most adequate larval stage to be infected are determined experimentally, in order to ensure the maximum production of OBs per larva. However, it is common to use 1 to 5 million OBs per diet container with 10-third instar larvae. Infection at this larval stage allows the development of the highest larval biomass the insect can accumulate at the moment of maximum infection of tissues.

Once dead, infected insects are collected and immediately processed, or stored under freezing conditions (-20°C or lower). The most frequent processing of larvae consists of the disintegration of the infected tissues, usually by blending the dead larvae, followed by sieving through a wide mesh which will eliminate large pieces of unblended tissues (normally, uninfected tissues) and large pieces of integument. The obtained OB concentrate should then be quantified and formulated, or could be stored under freezing conditions. Because of baculoviruses are highly virulent and infective, there is a high possibility that the whole insect colony get infected with it. That is why the colony maintenance and virus production (two-phase production) must be physically isolated and personnel working in phase one, must be different from that working in phase two, along with other safety measures. The insect colony used to produce the virus is usually of the same species than that which will be controlled in the field. However, in some situations, the viral strain shows a somewhat extended host range. Therefore, alterative hosts may be used in the production of the virus, if production results practical and cheaper.

The selection of inert materials such as talc or diatomaceous earth that will carry the OB concentrate is a very important step in the formulation process of viral products. For this purpose, some additives, such as dispersants, adherents, UV-protectants and feeding stimulants, can also be included in the formulation. Minerals such as bentonite, attapulgite, silica sand, and others, as well as organic materials such as wheat bran, milled nut shells, and crushed corn cobs may be added to formulations as inert carrying material (Williams and Cisneros, 2001). A wide variety of surfactants such as Adsee, Agral NN, Chevron, Rhoplex B60A, Tween 80, etc., contribute to a more efficient and well dispersed application of formulated viral products. Also, it is important to consider the persistence of an application on the leaf surface. Therefore, some chemical adhesive can be added to the formulations, such as Agral NN polyvinyl alcohol, Hyvis 150, skim milk, etc. Because the viral product must be ingested by the insect, the addition of baits and phagostimulants such as sugar, cotton seed oil, glycerin, Gustol, etc., will improve the efficacy of the product. Also, UV protectants such as Tinopal LPW, Phorwite AR, Leucophor BS, activated carbon, folic acid, etc. may be added, in order to avoid exposure of viruses to UV rays (Williams and Cisneros, 2001).

All these materials and ingredients improve the application and persistence of the formulation in the field. At the end of the formulation process, the products can be presented as wettable powders, granules, suspensions, etc. Most of these formulations can be easily sprayed in the field, using the same equipment employed for the application of chemical insecticides. Nowadays, formulation technology has evolved to very high levels of efficiency and complexity, and some formulations have been developed recently, using micro-encapsulation. The micro-encapsulation process combines a mixture of baculoviruses with some polymers such as gelatin, pectin, chitin, and calcium alginate, or starch which is the most common and useful encapsulating agent. A micro-encapsulated viral product should exhibit a shelf-life of at least 18 months.

Quality control measures should take into account the reduction to a minimum of contaminant microorganisms in the product, especially if the microbial content includes some potential vertebrate pathogens. Quality tests may also include resistance to abrasive materials, susceptibility to UV rays from sunlight, possible synergistic or antagonistic factors, rain effect, etc. More detailed tests may include DNA tests to prove the genome integrity, serological tests of the capsid proteins, and safety tests to vertebrates. However, most of these tests are rather required during the registration process of the product. Each lot of a viral product must be tested for its efficiency by performing quantitative bioassays on a susceptible pest. LC_{50} values indicate the virulence level of each lot of production and adjustments must be done during formulation, in order to standardize every lot to a certain level of virulence. Potency of the product is usually measured by the concentration of OBs per weight unit of the product (*e.g.* 3×10^9 OB gr^{-1}). This is important information required to recommend the application doses in the field and for product comparison, as there is a variety of factors than can vary among the different manufacturers.

Some viruses can be produced *in vitro*. The *in vitro* production (in insect cells) of viral insecticides constitutes a very attractive alternative as it would substitute the laborious and expensive maintenance of the insect colonies, for a highly homogeneous, and accurately controlled cell culture maintained in large fermentors or bioreactors. In this technique the sterile culture medium in the fermentor is inoculated with a cell line highly susceptible to the virus to be produced and, most importantly, adapted to the *in vitro* massive production. Adequate conditions (*i.e.* temperature, agitation, aeration, pH, etc.) in the fermentor will ensure the optimum proliferation of the cell line, up to a density that is appropriate to optimize the infection. At this point, inoculation should be done with EVs extracted from previously inoculated cell cultures, at a density predetermined to achieve an effective and rapid infection of the cells, as OBs must be produced as fast as possible. Once cells burst and release the OBs to the medium, these are collected and concentrated, usually by centrifugation, and the concentrate is formulated as described above. Other advantages of the *in vitro* production are the elimination of insect parts, tissue residues, other microorganisms, and any other contaminant, usually found in the *in larva* production. Unfortunately, the *in vitro* production still faces important limitations that hamper the use of this technique at an industrial level. However, insect cell culture systems are currently under development, mostly for other uses, and may eventually become a practical technique to produce viral pesticides at more affordable costs.

Strategies for Virus Application in the Field

The classical biological control technique has proven to be a successful pest control strategy for some baculoviruses. The introduction of a new virus into an insect population, may accomplish the establishment and persistence of the virus into the insect population without any further application. This is mostly due to the important role of viruses as biotic factors in the ecosystem, regulating insect population densities in nature (Thomas *et al.,* 1972; Podgwaite 1985; Hostetter *et al.,* 1985).

Augmentation is also a good strategy for virus application, mainly when they are found in low levels in the insect populations, because either they constitute a minor natural factor of mortality or because they are at the beginning of an epizootic cycle. Either in an inoculative or an augmentative strategy, there is a variety of methods to spread the pathogen, for example, the release of parasitoids inoculated with the virus or the release of inoculated birds that may spread out the virus through their feces. In any case, the application of the virus in the insect population increases the natural level of inoculum triggering an epizootic that would cause the reduction of the pest population. The augmentation technique will always be more successful if the insect population is in low densities, and therefore could be used as a preventive measure (Podgwaite 1985). Its use in forest ecosystems has proven very successful, as the virus is well preserved in forest soils. On the other hand, the inundative strategy would be similar to the repeated release of other biological control agents, such as parasitic wasps or predatory ladybirds. However, the application technique and the equipment used are more similar to the application of a chemical insecticide. Another method used for the release of insect viruses employs baits mixed with the virus, which can be complemented with a variety of attractants and phagostimulants.

GENETICS OF ENTOMOPATHOGENIC VIRUSES

Recent advances in the molecular biology of insect viruses have been achieved through the use of the insect cell lines. Comparatively with other viruses, genetics of the entomopathogenic viruses are intricate and complex because most virus genomes may be as large as 300 kbp. A total of 52 complete genomes of baculovirus species have been sequenced to date (Table 2.1) (Jehle *et al.,* 2006; Velasco de Castro Oliveira *et al.,* 2006; van Oers and Flak 2007). Baculovirus genomes contain from 100 to 160 open reading frames (ORFs), but more than 800 different genes have been identified in completely sequenced genomes (Herniou and Jehle, 2007). Among the numerous baculovirus species known to date, *Autographa californica* multiple NPV (AcMNPV) is the most well studied and most extensively used. AcMNPV has a circular double-stranded DNA genome of 133,894 bp (Ayres *et al.,* 1994). This species is the only baculovirus species showing a wide host range, with more than 30 lepidopteran species reported to be susceptible to this virus (Groner, 1986; Adams and McClintock, 1991). Its genome was the first one to be sequenced among baculoviruses (Ayres *et al.,* 1994). A detailed analysis of this genome elucidated the function of a great proportion of genes, including the characterization of gene promoters and enhancer elements, structure of mRNAs, late mRNA transcription initiation, elongation and termination, enzyme complex formation, function of mRNA biosynthetic genes, and dependency of late gene transcription on viral DNA replication. Additionally, two genomes from EPVs have been obtained (Afonso *et al.,* 1999; Bawden *et al.,* 2000), as well as two from IVs, one from the genus *Iridovirus* (Jakob *et al.,* 2001) and another from the genus *Chloriridovirus* (Delhon *et al.,* 2006); and three from ascoviruses: *Trichoplusia ni* ascovirus 2c (TnAV-2c), *S. frugiperda* ascovirus1a (SfAV-1a), and *H. virescens* ascovirus 3e (HvAV-3e) (Cui *et al.,* 2007).

Baculovirus genomes show a complex regulation system that sets the bases for their gene classification. There are four classes of genes classified according to their expression activity: immediate early genes (IE), delayed-early genes (DE), late genes (L), and very late genes (VL). The organized temporarily-regulated cascade of gene regulations is dependent on the systematic expression of genes; genes expressed in the previous phase are necessary to activate genes in subsequent phases (Friesen and Miller, 1986). Thus, the IE and DE genes are expressed before the replication of viral DNA. However, the expression of IE genes depends solely on the host factors. The expression of DE genes depends on the expression products of the IE genes. The late genes are expressed after the viral DNA replication took place, and are related to the expression of capsid proteins and all the structural proteins required to assemble new virions, especially the EVs. A fourth class of genes is found almost exclusively in baculoviruses (Miller, 1988). These are called very late genes and their expression products are related to the assembly of all the structural proteins of the OBs, mainly the genes expressing the polyhedrin which is expressed at very high levels.

At least one late gene is required to activate very late promoters. One of the most important genes of this phase is the IE-1 gene which acts as a transcriptional activator of other early genes, operating in "trans". Transitory expression assays support the role of this gene. Also, there are six homologous regions scattered in the AcMNPV genome, highly rich in EcoRI sites, which contains repeated sequences of 60 bp fragments with a 26 bp imperfect palindromic sequence. These regions, along with the IE-1 gene, act like "enhancers", analogous to those found in other DNA viruses and may function as sites for replication origin (Guarino and Summers, 1986). Late genes are expressed from 6 to 18 hours post-infection (hpi). During this phase, intensive DNA replication occurs, and new EVs are formed. All nucleocapsid proteins are synthesized during this phase, mainly the major capsid protein and the DNA binding protein. Also, a 64 kDa glycoprotein is abundantly synthesized during this phase. This protein is located at the EV envelope, and is involved in the recognition of cell receptors during the invasion process of a new host cell (Volkman, 1986).

AcMNPV produces an average of 70-100 OBs per nucleus 3 days post-infection, and about this time polyhedrin may constitute 25 to 50% of the total protein (Miller, 1988). The polyhedrin gene and the rest of the very late genes are expressed from 20 to 72 hpi, this process is also known as the inclusion phase. Along with the polyhedrin, other proteins are also over expressed during the last infection phase, such as the so-called p10 protein. This is a 10 kDa protein which forms fiber-like structures throughout the OB and provides stability to the closely attached molecules of polyhedrin (Van der Wilk *et al.,* 1987). Due to the remarkably high levels of expression of these two proteins, their promoters have been extensively studied, and constitute the basis for the use of baculoviruses as expression vectors for heterologous proteins. Conveniently for this purpose, both proteins are not required in the virus replication process, as both are involved in the formation of OBs, once virions are fully assembled. That is, recombinant EVs are totally infective to insect cells, even if they lack the ability to form OBs.

Five genomes of CPVs have been sequenced so far: BmCPV-1, LdCPV-1, DpCPV-1, LdCPV-14, and TnCPV-15 (Mertens and Bamford, 2009). The genomes of these CPVs include an average of 10 dsRNA fragments varying in length from 0.4 to 4 Kb. That is, fragments are separated and visualized by agarose gel electrophoresis producing particular banding patterns, called electropherotypes. Different electropherotypes from cypoviruses were identified by differences in the migration patterns of their genome segments. The sum of all these fragments resulted in a total genome length of 19 to 25 Kb. To date, 14 different electropherotypes have been identified by this method (Mertens *et al.,* 1989). Due to these studies, it is known that the smallest fragment contains the coding gene for the CPVs polyhedrin, which is the major component of the OBs.

The molecular biology of entomopoxviruses has been limited by the lack of information on EPV genomics. EPV genomic organization and molecular mechanisms of replication, pathogenesis, and host range are largely unknown. Only two genomes of EPVs are sequenced so far: *Melanoplus sanguinipes* EPV (MsEPV) (Afonso *et al.*, 1999) and *Amsacta moorei* EPV (AmEPV) (Bawden *et al.*, 2000). The 236, 120 bp MsEPV genome contains a subset of genes shared among all poxvirus sequences analyzed so far, which reinforce the idea of a common genetic core of poxvirus genes. This poxvirus gene core includes many genes associated to RNA transcription, posttranscriptional modification, DNA replication, and structural proteins. EPV genomes are constituted of a single dsDNA molecule of 130 to 375 kbp in length. Such an extensive genome may have the potential to code for as many as 150 to more than 300 genes (Arif, 1984). A peculiarity of these genomes is the presence of isometric DNA sequences at the terminal ends of the molecule, constituted by inverted palindromic sequences. Similar to baculovirus genes, the EPV genes are classified according to phases of expression. These are: a) early genes, b) intermediate genes, and c) late genes (Moss, 1996). Early genes are expressed before the genomic DNA is replicated. Some of these genes are expressed even before the DNA is totally released from the capsid. It is known that proteins expressed at the early events of infection inhibit the synthesis of macromolecules of the host cell. These early genes code for non-structural proteins, including enzymes involved in the DNA replication, the modification of DNA and RNA, ass well as those implicated in the inactivation of the host defense mechanisms. Additionally, these genes code for transcriptional factors used by the intermediate genes. During the intermediate phase, viral DNA replicates and intermediate genes code for transcriptional factors used by the late genes. Finally, the expression of the late genes occurs until the viral DNA has been replicated and has coded mostly for structural proteins, involved in the assembling of the capsid and spheroid formation. Paradoxically, some of these genes also code for transcriptional factors used by early genes (Moss, 1996).

Only two entomopathogenic IV genomes have been sequenced to date: *Chilo* IV or IIV-6 of 212, 482 bp (Jakob *et al.*, 2001) and *Aedes taeniorhynchus* IV or IIV-3 of 191,100 bp (Delhon *et al.*, 2006). The IV genome is a dsDNA molecule circularly permuted and terminally redundant (Goorha and Murti, 1982), containing 96 to 234 largely non-overlapping ORFs. The G-C content range from 27 to 55%, and the complex repeat sequences are mostly located between coding regions. In addition, these genomes exhibit little or no co-linearity among genera (Delhon *et al.*, 2006). The replication process of IV viral genomes is very complex. Replication starts early after the infection. DNA replicates in the nucleus, producing short copies. As the replication continues in the nucleus, some copies are transported to the cytoplasm where replication goes on and recombination among DNA strands occurs. Recombinant copies build a complex of concatamers (Ward and Kalmakoff, 1991). An enzyme called DNA integrase-recombinase may be involved in the recombination of small pieces of DNA as well as in the resolution of the concatamer, just before the DNA is packed within the capsids (Jakob and Darai, 2002).

Table 2.1. Updated list of sequenced genomes of baculoviruses

Virus species	***Lenght (bp)***	***Proteins***
Alphabaculovirus		
Adoxophyes honmai NPV	113220	125
Adoxophyes orana NPV	111724	121
Agrotis ipsilon MNPV	155122	163
Agrotis segetum NPV	147544	153
Antheraea pernyi NPV	126629	147
Anticarsia gemmatalis NPV	132239	152
Autographa californica NPV	133894	156
Bombyx mandarina NPV	126770	141
Bombyx mori NPV	128413	143
Choristoneura fumiferana DEF MNPV	131160	149
Choristoneura fumiferana MNPV	129593	146
Chrysodeixis chalcites NPV	149622	151
Clanis bilineata NPV	135454	129
Ecotropis obliqua NPV	131204	126
Epiphyas postvittana NPV	118584	136
Euproctis pseudoconspersa NPV	141291	139
Helicoverpa armigera NPV	130759	137
Helicoverpa armigera NPV NNg1	132425	143
Helicoverpa armigera MNPV	154196	162
Helicoverpa armigera NPV G4	131405	135
Helicoverpa zea SNPV	130869	139
Hyphantria cunea NPV	132959	148
Leucania separata NPV	168041	169
Lymantria dispar MNPV	161046	164
Mamestra configurata NPV-A	155060	169
Mamestra configurata NPV-B	158482	168
Maruca vitrata MNPV	111953	126
Orgyia leucostigma NPV	156179	135
Orgyia pseudotsugata MNPV	131995	152
Plutella xylostella multiple NPV	134417	152
Rachiplusia ou MNPV	131526	149
Spodoptera exigua MNPV	135611	139
Spodoptera frugiperda MNPV	131330	142
Spodoptera litura NPV	139342	141
Spodoptera litura NPV II	148634	147
Trichoplusia ni SNPV	134394	145
Betabaculovirus		
Adoxophyes orana GV	99657	119
Agrotis segetum GV	131680	132

Table 2.1. (*Contd...*)

Table 2.1. (*Contd...*)

Virus species	*Lenght (bp)*	*Proteins*
Choristoneura occidentalis GV	104710	116
Cryptophlebia leucotreta GV	110907	128
Cydia pomonella GV	123500	143
Helicoverpa armigera GV	169794	179
Phthorimaea operculella GV	119217	130
Pieris rapae GV	108592	120
Plutella xylostella GV	100999	120
Pseudaletia unipuncta GV	176677	183
Spodoptera litura GV	124121	136
Xestia c-nigrum GV	178733	181
Deltabaculovirus		
Culex nigripalpus NPV	108252	109
Gammabaculovirus		
Neodiprion abietis NPV	84264	93
Neodiprion lecontii NPV	81755	89
Neodiprion sertifer NPV	86462	90

RECOMBINANT VIRUSES AS PESTICIDES

The development of new insect cell lines, permissible to the *in vitro* virus infection, and the development of baculoviruses as expression vectors of heterologous proteins are both factors that have a positive influence on the development of recombinant baculoviruses (Miller, 1988; Cameron *et al.,* 1989). At present, a variety of insecticidal factors have been integrated into the genomes of AcNPV and the BmNPV, mostly due to the availability of permissible insect cell lines for these two viruses. The lack of new cell lines permissible to the infection of other baculoviruses or even other insect viruses, has restricted the development of other recombinant viruses, including other NPVs and GVs. Recently, Obregón-Barboza and collaborators, applied the microprojectile bombardment technique on *T. ni* larvae, not only to infect with virions from TnGV or AcMNPV, or transfect the larvae with genomic DNA from both virus, but also to develop recombinant granuloviruses (TnGV) by co-transfection with genomic DNA and a transfer vector carrying the GFP gene (Obregón-Barboza *et al.,* 2007).

The insect cell systems for production of recombinant proteins are based on the unessential role of the polyhedrin or the p10 gene in the replication cycle of the virus, as well as on the significantly high transcriptional activity of these gene's promoters (Miller, 1988). Both proteins are expressed at the end of the replication cycle and at levels that might reach 25 to 50% of the total proteins (Wood, 1996). In fact, the virus genome is turned into a gene delivery system by replacement of the polyhedrin gene coding sequence with that of the transgene to be expressed, so the polyhedrin gene promoter will drive the expression of the transgene (Summers and Smith, 1985). To construct a transfer vector, the recombinant gene expression cassette must be inserted into the viral genome. Transfer vectors normally contain an *E. coli* origin of replication, an antibiotic resistance gene, and a baculovirus genome segment, which contains either the polyhedrin or the p10 promoters or both, as well as a non-essential sequence. This last sequence is important to integrate the transgene into the baculovirus genome by means of homologous recombination. In general, a recombinant baculovirus is obtained by co-transfection on insect cell monolayers, with a DNA mixture from the transfer vector, containing the gene of interest, and the baculovirus genome (Summers and Smith, 1987). Once new viruses are formed and released to the culture medium from the infected cells, recombinant viruses are selected using plaque purification assays (Brown and Faulkner, 1977) and molecular techniques such as PCR, sequencing, etc. (Summers and Smith, 1987).

The genomes of some baculoviruses, especially those of AcNPV, BmNPV, and HearSNPV, have been modified in order to obtain viral strains with enhanced insecticidal properties (Table 2.2). For this purpose, genes coding for hormones, enzymes, and toxins have been used to modify baculoviruses genomes (Hughes *et al.,* 1997). In regard to the use of insecticidal toxins, a toxin obtained from the scorpion *Buthus eupeus* was tested, but the recombinant AcNPV obtained exhibited no effect on the paralysis or death of larvae (Carbonell *et al.,* 1988). Additionally, genes coding for venoms from three arthropods have been introduced into the AcNPV. One is the neurotoxin AaHIT from the scorpion *Androctonus australis*, which causes paralysis and death to a great variety of insects. Recombinant AcNPV virions expressing this toxin caused a mean killing time (LT_{50}) reduction of 25% (Stewart *et al.,* 1991), and the recombinant *Helicoverpa armigera* SNPV expressing the same gene reduced the mean survival time (ST_{50}) 17–34% (Sun *et al.,* 2004). Recombinant viruses that carry the neurotoxin TxP-1 from the mite *Pyemotes tritici* infected the greater wax moth larvae, *G. mellonella*. Paralysis was observed two days after infection; however, mortality was actually caused by the viral infection rather than by the toxin, and the killing time was similar to that observed with the parental strain (Tomalski and Miller, 1991).

Three more toxins from animal origin were tested when their respective genes were integrated into the AcNPV genome (Prikhod'ko *et al.*, 1996). These are: the µ-Aga-IV venom from the spider *Agelenopsis aperta*, and the As II and Sh1 toxins from the sea anemones *Anemonia sulcata* and *Stichodactyla helianthus*, respectively. All three toxins cause neurotoxic effects on insects. Although the efficiency of each recombinant virus varied between each other and among the different insect species tested, the µ-Aga-IV venom showed the highest activity on *S. frugiperda* larvae, with a 37% reduction in the LT_{50}, (Prikhod'ko *et al.,* 1996). The As II and Sh1 toxins were more efficient against *T. ni* larvae, with LT_{50} reductions of 38.4 and 36%, respectively (Prikhod'ko *et al.,* 1996).

On the other hand, the gene coding for the diuretic hormone from the tobacco hornworm *Manduca sexta* (Linnaeus) (Lepidoptera: Sphingidae) was integrated into the BmNPV genome. Infection of the recombinant virus on silkworms significantly increased their water loss rate and reduced the LT_{50} by 20%, as compared with that of the parental strain (Eldridge *et al.,* 1991). In another example, the gene coding for the juvenile hormone esterase was integrated into the AcNPV genome. This enzyme hydrolyzes the hormone which causes the triggering of a molt in the insect, as well as the stop of feeding. When recombinant viruses expressing this gene infected *T. ni* larvae, although feeding was reduced, no effect was detected on the LT_{50} (Hammock *et al.,* 1990). Genes that express toxins from other sources, such as the δ-endotoxin gene from *Bacillus thuringiensis* (*B. t.*), have been integrated into the AcNPV genome (Martens *et al.,* 1990). In the same way, genes expressing protoxins from *B. t aizawai* and *B. t. kurstaki* were also successfully expressed in recombinant viruses. Even, the typical *B. t.* crystals were observed in the infected *in vitro* cell cultures. However, when recombinant viruses were tested against lepidopteran larvae, no difference was observed neither in the LT_{50}, nor the LC_{50} (mean lethal concentration) of those obtained with the parental strain (Merryweather *et al.*, 1990).

Additionally, a gene expressing the toxin URF13 obtained from maize was cloned in AcNPV, showing a significant 40% reduction in the LT_{50}, when tested against *T. ni* larvae (Korth and Levings, 1993). A recombinant AcMNPV, AcMNPV-enMP2, that expresses the MacoNPV enhancin gene under the control of its native promoter, was developed and characterized (Li *et al.,* 2003). In *T. ni* larvae, the LD_{50} of the AcMNPV-enMP2 was 4.4 times lower than that of the parental strain. Finally, the AcMNPV was engineered to express the insect-selective toxin IT2 from the scorpion *Leiurus quinquestriatus*. This recombinant virus killed faster the *H. virescens* larvae than the wild-type virus (van Beek *et al.,* 2003). Interestingly, a GV gene was introduced into the AcNPV genome, that is, a gene from one baculovirus into another baculovirus. This is the *vef* gene from the TnGV which codes for the enzyme called "enhancin". This enzyme degrades the peritrophic membrane of the insect, facilitating the contact of the virions with the midgut epithelial cells. When the recombinant virus was tested against *T. ni* larvae, a reduction of 22% in the LT_{50} was observed (Del Rincón-Castro and Ibarra, 2005).

Recently, the potential adverse effects of a recombinant AcMNPV (AcAaIT) were studied in rabbits and fish. In the first case, no relevant changes were observed in the production of some enzymes. Immunohistochemical observation of tissues from the stomach, intestine, liver, kidney, brain, spleen, and lung only revealed slight changes. In the second case, no mortality was detected in treated (AcAaIT) or untreated fish during the experimental period (Ashour *et al.*, 2007). Some tests using recombinant baculoviruses have been focused mainly on their prevalence in the field, rather than on their performance as a bioinsecticide. This is because of the concerns related to dissemination of the recombinant virus and the possibility of horizontal transmission of the transgene. That is why strict regulatory safety requirements, developed in many countries, must be followed before the release of any genetically modified organisms, including viruses.

Table 2.2. Recombinant baculoviruses as bioinsecticides

Gene	***Virus***	***Effect***	***Reference(s)***
B. eupeus toxin I	AcMNPV	No effect	Carbonell *et al.*, 1988
Diuretic hormone	BmNPV	20% reduction on LT_{50}	Maeda, 1989
B. thuringiensis δ-endotoxin	AcMNPV	Crystal formation, no effect	Merryweather *et al.*, 1990
Juvenile hormone esterase	AcMNPV	Reducing insect feeding	Hammock *et al.*, 1990
A. australis neurotoxin AaIT	AcMNPV	Reduction of time required to kill the host	Stewart *et al.*, 1991
Pyemotes tritici neurotoxin gene TxP-I	AcMNPV	Paralysis of larvae in two days	Tomalski and Miller, 1991
Deletion of *egt* gene	AcMNPV	Feeding cessation, wandering and spining	O'Reilly and Miller, 1991
Manduca sexta eclosion hormone gene	AcMNPV	No effect	Eldridge *et al.*, 1991
Juvenile hormone esterase	AcMNPV	No effect	Eldridge *et al.*, 1992
Corn Protein URF-13	AcMNPV	40% reduction on LT_{50}	Korth and Levings, 1993
PTTH Hormone	AcMNPV	No effect	O'Reilly *et al.*, 1995

Table 2.2. (*Contd...*)

Table 2.2. (*Contd...*)

Gene	*Virus*	*Effect*	*Reference(s)*
Agelenopsis aperta toxin μ-Aga-IV	AcMNPV	37% reduction on LT_{50}	Prikhod'ko *et al.*, 1996
Anemonia sulcata toxin As II	AcMNPV	38.4% reduction on LT_{50}	Prikhod'ko *et al.*, 1996
S. helianthus toxin Sh1	AcMNPV	36% reduction on LT_{50}	Prikhod'ko *et al.*, 1996
Digetia canities neurotoxin	AcMNPV	Paralysis in 2 days	Hughes *et al.*, 1997
Algal virus pyrimidine dimer-specific glycosylase (av-PDG)	AcMNPV	Protection against UV in budded virus	Petrik *et al.*, 2003
Leiurus quinquestriatus scorpion toxin LqhIT2	AcMNPV	Elicited response significantly faster	van Beek *et al.*, 2003
Enhancin gene from MacoNPV	AcMNPV	4.4.-fold reduction on LD_{50}	Li *et al.*, 2003
A. australis neurotoxin AaIT	HearSNPV	17–34% reduction on ST_{50}	Sun *et al.*, 2004
TnGV enhancin gene	AcMNPV	22% reduction on LT_{50}	Del Rincón-Castro and Ibarra, 2005

AcMNPV: *Autographa californica* multiple nucleopolyhedrovirus, LT_{50}: estimated lethal time that kills 50% of the tested population, LC_{50}: estimated concentration that kills 50% of the tested population. LD_{50}: mean lethal doses. LT_{50}: estimated survival time that kills 50% of the tested population.

USE OF INSECT VIRUSES IN THE FIELD

For many years, baculoviruses have been applied as specific biocontrol agents against forest and agriculture pests. Many products are already available, developed and/or commercialized in many countries of the world. Despite this widespread interest and intrinsic benefits of their use, their application is still limited. However, in some Latin-American countries, baculoviruses have been incorporated into a variety of insect control programs. One of the most useful examples on the use of baculovirus in agricultural systems is the control of the velvet bean caterpillar *Anticarsia gemmatalis* (Hübner) (Lepidoptera: Noctuidae) by an NPV. This is an important pest of soybeans in the Americas. Once an AgNPV was isolated from this pest, it has been widely used to control it. The virus is highly virulent against *A. gemmatalis* and usually only one application per crop cycle is needed. AgNPV application is well established in Brazilian pest management programs in more than one million hectares of soybeans. These programs were initiated 15 years ago with excellent results (Moscardi and Sosa-Gómez, 1993; Moscardi, 1999). Applications of AgNPV practically eliminated the use of chemical insecticides against this pest. In experimental tests, virus preparations are applied at 1.5×10^{11} occlusion bodies per hectare (about 20 g or 50 larval equivalents), achieving control efficiencies ranging from 60 to 70% (Moscardi and Sosa-Gómez, 1993; Moscardi, 1999). Also, another soybean pest, the green clover worm *Plathypena scabra* (Fabricius) (Lepidoptera: Noctuidae), has been efficiently controlled with its own GV (PsGV). PsGV tested at 2.47×10^{11} OB per hectare showed no significant differences in the control levels obtained when compared with a *B. t.* product and a chemical insecticide, after six days of treatment.

The potato tuber worm *Phthorimaea operculella* (Zeller) (Lepidoptera: Gelechiidae) was also efficiently controlled with its own GV (PoGV) in Peru (Sporleder *et al.,* 2007). This pest is very important, mostly during the tuber storage. So, farmers dust the potatoes just after harvest and before storage with a dry formulation of PoGV, provided mainly by government agencies. This method proved to be practical and efficient to control this pest. In contrast, in Southern Mexico, the SfNPV from *S. frugiperda* has been experimentally used against the same pest in maize. Results indicated moderate control levels, similar to those obtained in Northern Mexico with the use of AgNPV against the velvet bean caterpillar (Avila Valdéz and Rodríguez del Bosque, 2003; Cisneros *et al.,* 2003). In the same country, TnNPV and AcNPV have been experimentally tested against *T. ni* in broccoli. Tests carried out in 1993 in the central Mexico, indicated that both viruses were able to achieve the same levels of control as those obtained with the chemical insecticide and a *B. t.* product, after two applications (Ibarra and Aguilar, 1993).

Moreover, the MbNPV isolated from the European cabbage moth *Mamestra brassicae* (Linnaeus) (Lepidoptera: Noctuidae) has also been used to successfully control this pest in Europe. Spray-dry formulations of this virus were used by the late 1970s, containing about 6×10^9 OB g^{-1}. After six applications, *M. brassicae* populations were reduced from 54 to 97% in cauliflower. Similarly, the PrGV isolated from the imported cabbageworm *Pieris (Artogeia) rapae* (Linnaeus) (Lepidoptera: Pieridae) has been used for the control of this pest in broccoli and cauliflower, with good level of success. In an experiment carried out in the United States using this baculovirus, control levels ranged between 87 and 97% nine days after application, at doses between 2.4×10^{11} and 2.4×10^{13} OB (granules) ha^{-1} (Huber, 1986).

On the other hand, one NPV isolated from the cotton pest *H. armigera* (HearNPV) has been used to control this pest in cotton.

HearNPV has been produced to be used against the cotton bollworm in China. Some farmers processed about 1,600 tons of infected insects with this virus. In America, tests carried out in 1972 and 1973 at 190 larval equivalents (LEs) per hectare showed efficient control levels on this pest. One LE is the highest amount of OBs produced by one larva. In the case of *H. armigera* that is the amount produced when a third-instar larva is infected, producing an average of 3×10^9 OBs per larva. Therefore, 190 LE ha^{-1} are about 5.7×10^{11} OB ha^{-1} (Tanada and Kaya, 1993). An important example on the use of baculovirus as bioinsecticide is the NPV of *H. zea*. This strain has been used since the 1970s to control *Heliothis virescens* (Fabricius) (Lepidoptera: Noctuidae) and *Helicoverpa zea* (Boddie) (Lepidoptera: Noctuidae) in cotton grown in the United States, as well as in other countries. The *Heliothis* / *Helicoverpa* complex can damage a variety of important crops worldwide, such as cotton and soybean (Huber, 1986). A commercial product called Elcar was developed, which demonstrated to be very efficient in controlling *H. virescens* in cotton, until pyrethroids were approved to be used on cotton (Ignoffo and Couch, 1981).

Other lepidopteran pests of cotton, such as *S. exigua* (Hübner) (Lepidoptera: Noctuidae), *T. ni*, *Pectinophora gossypiella* (Saunders) (Lepidoptera: Gelechiidae) as well as the *Heliothis* / *Helicoverpa* complex have been controlled with AcNPV. The experimental product is called SAN 404 I WP, and has shown significant levels of control. In soybeans, *H. zea* has also been efficiently controlled with Elcar, as shown in an experimental work where doses from 2.5 to 247 LE ha^{-1} were tested. Pest populations and crop damage were reduced significantly (Ignoffo and Couch, 1981). On the other hand, the *Pseudoplusia includens* NPV (PiNPV) was isolated from field collected larvae in Guatemala. In the early 1970s this virus was tested experimentally in soybean fields at 49.4, 98.8, and 197.6 LE ha^{-1}, causing mortality levels of 52.8, 70.9, and 76.2%, respectively. By 1975, PiNPV was tested in four and eight hectares of soybean fields by airplane application at 247 LE ha^{-1}, reducing the pest population by 60.4 and 75.6% during the first and second week after treatment, respectively. In soybean fields, populations of the beet armyworm, *S. exigua* were reduced up to 55% when 25, 50, and 100 LE ha^{-1} of the SeNPV were tested (Yaerian and Young, 1982).

AcNPV has been successfully used in the control of the cabbage looper, *T. ni* in broccoli and cauliflower. This baculovirus causes natural epizootics in *T. ni* populations, but if the natural inoculum is increased by applications in the field, populations can be reduced from 75 to 98% in broccoli, cauliflower and cabbage. In lettuce, when used at concentrations of 4.5×10^9, 2.3×10^{10} and 4.5×10^{10} OB ha^{-1}, control levels raised up to 100% (Carter, 1984). Other cases have shown high pest control levels when AcNPV is applied along with chemical insecticides. This may be due to the sub-lethal, weakening effect of chemicals, which make the insect more susceptible to develop the infection. AcNPV has been applied both on foliage and soil. When four applications of 1.5×10^{12} OB ha^{-1} were made on foliage, control achieved was similar to that obtained with the application of chemical insecticides. Similarly, when 7.5×10^{12} OB ha^{-1} were applied in soil, an adequate control of *T. ni* was achieved in cauliflower. Also, when AcNPV has been applied either alone or in combination with TnNPV, control levels were similar to those obtained with chemical insecticides.

The use of HzNPV against *H. virescens* attacking tobacco was very successful, as tests of 4.9×10^8 OB per plant achieved up to 100% mortality. Likewise, HzNPV has also been used to control *H. zea* in sorghum. When this virus was applied at 2.7×10^{12} OB ha^{-1}, pest populations were reduced up to 88% (Huber, 1986). The same virus, formulated as Elcar and as the experimental product Sandoz 240 WP-78, were applied against the same pest and the same crop at a dose of 98.8 LE ha^{-1}, with a significant reduction of pest populations, similar to that obtained with chemical insecticides. Comparable results were observed against the same pest in peanuts, using Elcar. In another crop, CpNPV was tested to control populations of *Colias philodice* Godart (Lepidoptera: Pieridae) on alfalfa. Applications of concentrated suspensions (5×10^7 and 1×10^8 OB ml^{-1}) started an early epizootic development in a population that normally shows the epizootics until the pest has caused significant damage to the crop.

Also, the codling moth *Cydia pomonella* (Linnaeus) (Lepidoptera: Tortricidae), a pest of pear and apple crops, has developed resistance to several chemical insecticides. One baculvirus isolated of this pest, CpGV, is highly virulent to codling moth with LD_{50}s as low as 1.2-5 occlusion bodies per insect. CpGV was originally isolated from *C. pomonella* in Mexico, 40 years ago (Tanada, 1964). Commercial formulation of OBs of this virus are called Granusal™, Granupom™, Carpovirusine™, Virin-CyAP, Cyd-X, Virosoft CP4, and Madex™. The virus has been successfully used in the United States, Canada, and Europe as bioinsecticide on thousands of hectares.

On the other hand, the use of baculoviruses in forest ecosystems has been very successful and well studied. One of these examples is the use of the OpNPV from the Douglas-fir tussock moth, *Orgyia pseudotsugata* (McDunnough) (Lepidoptera. Lymantriidae). This pest has seriously devastated wide forest areas in the United States. As a result, the US Forest Service started a large program on the production of this baculovirus which has been extensively applied mostly in Northern United States, showing control levels of up to 100% efficient in some areas. Another very important forest pest in Northern United States, Central and Eastern Europe, the Mediterranean region, and Japan is the gypsy moth, *Lymantria dispar* (Linnaeus) (Lepidoptera: Lymantriidae), as it defoliates a wide variety of trees (*e.g. Quercus*, *Populus*, etc.). Only in the United States, gypsy moth has devastated more than 800,000 hectares of forests. The baculovirus LdNPV has been developed to control this pest and has been registered as Gypcheck, which has been highly effective to control this pest.

Recently, the baculovirus NaNPV isolated from *Neodiprion abietis* (Harris) (Hymenoptera: Diprionidae) was registered. This virus causes a lethal disease on larvae of the balsam fir sawfly, *N. abietis*. The formulated product called Abietiv FBI is applied to balsam fir forests to decrease the pest populations and reduce trees damage. Another important forest pest in pine trees is the European pine sawfly, *Neodiprion sertifer* (Geoffroy) (Hymenoptera: Diprionidae). Like the NaNPV, the NsNPV is a virus from a non-lepidopteran pest which was also developed as bioinsecticide, called Neocheck. A major advantage of this virus is its high infectivity rate and long persistence in forest soils.

ADVANTAGES AND LIMITATIONS OF ENTOMOPATHOGENIC VIRUSES

For many years, baculoviruses have been successfully applied as biological control agents against a variety of pests. This viruses show a series of advantages that make them important alternative measures within pest management programs. First, baculoviruses only infect arthropods, with the overwhelming majority within the insect group. This feature facilitates the registry of virus-based bioinsecticides, as possible safety restrictions may be faced by other entomopathogenic virus families (*i.e.* EPVs, CPVs) containing species pathogens to vertebrates. Second, baculoviruses are not only restricted to arthropod hosts but the host specificity of each baculovirus species is limited to one or a few highly related insect species. This feature, shared by many insect viruses, makes them even safer to use, as non-target insects are not threatened by any eventual infection. Third, baculovirus applications are innocuous to the environment as residual accumulation and contamination does not occur. OBs are rapidly degraded in the field, mostly by UV light that causes inactivation and proteolysis of polyhedra.

Fourth, no cases of resistance to baculoviruses have been recorded except for some examples of field-collected individuals that survived to the application and subsequently were subjected to strong selection under laboratory conditions. Fifth, baculoviruses can be applied along with many other pesticides, because they tolerate this interaction and standard spray equipment can be used for the application of these bioinsecticides.

On the other hand, baculoviruses also show some characteristics that limit their use as biological control agents. An important limitation is their slow activity, as many baculoviruses may take 4 to 14 days to kill their host (Gröner, 1986). Unfortunately, infected larvae keep feeding on plants during this period and, even worst, frequently a baculovirus infection induces the larva to grow bigger than usual, so it consumes even more plant material. This is a problem that genetic engineering of baculoviruses is trying to solve by the development of recombinant baculoviruses with added virulence factors. Another limitation of the baculovirus is their high specificity, which, from the environmental point of view is an advantage. This is mostly due to the narrow markets of a very specific viral bioinsecticides. From a commercial point of view, specificity may be a major constraint to invest in the development of such products. This is the main reason why many viral bioinsecticides are developed and produced by non-profit organizations, such as government agencies.

Another shortcoming of baculoviruses is their restriction to infect only larval stages and, even more, their use as a control measure should target the first instars, as larger larvae require higher doses to be killed. Additionally, production of viral bioinsecticides is expensive due to high costs required to maintain a mass rearing colony, as it is a labor-intensive work. This is one of the reasons why *in vitro* production is gaining great interest in industry as it would solve this important limitation. Interestingly, this limitation may be a reason to promote the production of baculoviruses *in larva* in developing countries, where labor is cheaper than in industrialized countries.

REFERENCES

Adams, J.R., Goodwin, R.H. and Wilcox, T.A. (1977). Electron microscope investigation on invasion and replication of insect baculoviruses *in vivo* and *in vitro*. *Biology of the Cell*, **28**: 261-268.

Adams, J.R. (1991). Introduction and classification of viruses of invertebrates. *In*: Adams, J.R and Bonami, J.R. (*Eds.*), *Atlas of Invertebrate Viruses*. CRC Press. Boca Raton, Florida. pp. 1-8.

Adams, J.R. and McClintock, J.T. (1991). Baculoviridae. Nuclear polyhedrosis viruses. *In*: Adams, J.R and Bonami, J.R. (*Eds.*), *Atlas of Invertebrate Viruses*. CRC Press. Boca Raton, Florida, pp. 87-204.

Afonso, C.L., Tulman, E.R., Lu Z., Oma, E., Kutish, G.F. and Rock, D.L. (1999). The genome of *Melanoplus sanguinipes* entomopoxvirus. *Journal of Virology*, **73**: 53-552.

Arif, B.M. (1984). The entomopoxviruses. *Advances in Virus Research*, **29**: 195-201.

Arif B.M. and Kurstak, E. (1991). Entomopoxviruses. *In*: Kurstak, E. (*Ed.*), *Viruses of invertebrates*. Marcel Dekker Inc., New York, pp. 179-195.

Ashour, M.B., Ragheb D.A., El-Sheikh, E.A., Gomaa, E.A., Kamita, S.G. and Hammock, B.D. (2007). Biosafety of recombinant and wild type nucleopolyhedroviruses as bioinsecticides. *International Journal of Environmental Research and Public Health*, 4(2): 111-125.

Avila Valdez, J. and Rodríguez del Bosque, L.A. (2003). Uso de los nucleopoliedrovirus de *Anticarsia gemmatalis* como principal estrategia del MIP en soya de la región sur de Tamaulipas. *In: Memorias del XXVI Congreso de Control Biológico*. Guadalajara, Jal. México. pp. 327-330.

Ayres, M. D., Howard, S.C., Kuzio, J., Lopez-Ferber, M. and Possee, R.D. (1994). The complete DNA sequence of *Autographa californica* nuclear polyhedrosis virus. *Virology*, **202**: 586-605.

Devauchelle, G., Attias, J., Monnier, C., Barray, S., Cerutti, M., Guerillon, J. and Orange-Balange, N. (1985). *Chilo* iridescent virus. *Current Topics in Microbiology and Immunology*, **116**: 37-48.

Bawden, A.L., Glassberg, K.J., Diggans, J., Shaw, R., Farmerie, W. and Moyer, R.W. (2000). Complete genomic sequence of the *Amsacta moorei* entomopoxvirus: analysis and comparison with other Poxviruses. *Virology*. **274**: 120-139.

Bilimoria, S.L. (1991). The biology of nuclear polyhedrosis viruses. *In*: Kurstak, L. (Ed.), *Viruses of Invertebrates*. Marcel Dekker, Inc., New York, N.Y., pp. 1-72.

Brown, M. and Faulkner, P. (1977). A plaque assay for nuclear polyhedrosis viruses using a solid overlay. *Journal of General Virology*, 36: 361-364.

Cameron, I.R., Possee, R.D. and Bishop, D.H. (1989). Insect cell culture technology in baculovirus expression systems. *Trends in Biotechnology*, **7**: 66.

Carbonell, L.F., Hodge, M.R., Tomalski, M.D. and Miller, L.K. (1988). Synthesis of a gene coding for an insect-specific scorpion neurotoxin and attempts to express it using baculovirus vectors. *Gene*, **73**: 409-418.

Carter, J.B. (1984). Viruses as Pest-Control Agents. *In*: Rusell, G.E. (*Ed.*), *Biotechnology and Genetic Engineering Reviews*. Intercept, Newcastle-upon-Tyne, United Kingdom.

Cisneros, J., Penagos, D.I. and Williams, T. (2003). Potencial de un nucleopoliedrovirus para el control de *Spodoptera frugiperda* (Lepidoptera: Noctuidae) en el maíz. *In: Memorias del XXVI Congreso de Control Biológico*, Guadalajara, Jal. pp. 315-318.

Condreay, J.P. and Kost, T.A. (2007). Baculovirus expression vectors for insect and mammalian cells. *Current Drug Targets*, **8**: 1126-1131.

Cui, L., Cheng, X., Li. L. and Li, J. (2007). Identification of *Trichoplusia ni* ascovirus 2c virion structural proteins. *Journal of General Virology*, **88**: 2194-2197.

Del Rincon-Castro, M.C. and Ibarra, J.E. (2005). Effect of nucleopolyhedrovirus of *Autographa californica* expressing the enhancin gene of TnGV on *Trichoplusia ni* larvae. *Biocontrol Science and Technology*, **15(7)**: 701-710.

Delhon, G., Tulman, D.R., Afonso, C.L., Lu, Z., Becnel, J.J., Moser, B.A., Kutish, G.F. and Rock, D.L. (2006). Genome of invertebrate iridescent virus type 3 (mosquito iridescent virus). *Journal of Virology*, **80(17)**: 8439-8449.

Eldridge, R.F., Horodyski, M., Morton, D.B., O'Reilly, D.B., Truman, J.W., Riddiford, L.M. and Miller, L.K. (1991). Expression of an eclosion hormone gene in insect cells using a baculovirus vectors. *Insect Biochemistry and Molecular Biology*, **21**: 341-351.

Eldridge, R., O'Reilly, D.R., Hammock, B.D. and Miller, L.K. (1992). Insecticidal properties of genetically engineered baculoviruses expressing an insect juvenile hormone esterase gene. *Applied and Environmental Microbiology*, **58**: 1583-1591.

Evans, H.F. and Entwistle, P. (1987). Viral Diseases. *In*: Fuxa, J.R. and Tanada, Y. (*Eds.*), *Epizootiology of insect diseases*. J. Wiley & Sons, New York, pp. 257-322.

Faulkner, P. (1981). *Baculovirus*. *In*: Davidson, E.W. (*Ed.*), *Pathogenesis of invertebrate microbial diseases*. Allanheld, Osmun CO Publishers. Inc., Totowa, New Jersey, pp. 3-37.

Friesen, P.D. and Miller, L.K. (1986). The regulation of baculovirus gene expression. *In*: Doerfler, W. and Boehm, P. (*Eds.*), *The molecular biology of baculoviruses*, Springer-Verlag, Berlin, pp. 31-49.

Goodwin, R.H., Milner, R.J. and Beaton, C.D. (1991). *Entomopoxvirinae*. *In*: Adams, J.R and Bonami, J.R. (*Eds.*), *Atlas of invertebrate viruses*. CRC Press, Boca Raton, Florida, pp. 259-285.

Goorha, R. (1982). Frog virus 3 DNA replication occurs in two stages. *Journal of Virology*, **43**: 519-528.

Goorha, R. and Murti, K.G. (1982). The genome of frog virus 3, an animal DNA virus, is circularly permuted and terminally redundant. *Proceedings of the National Academy of Sciences of the United States of America*. **79**: 248-252.

Granados, R.R. (1980). Infectivity and mode of action of baculoviruses. *Biotechnology and Bioengineering*. **23**: 1377-1405.

Granados, R.R. and Williams, K.A. (1986). *In Vivo* infection and Replication of Baculoviruses. *In*: Granados, R.R. and Federici, B.A. (*Eds.*), *The biology of baculoviruses*. CRC Press, Boca Raton, Florida, pp. 89-108.

Granados, R.R., Guoxun, L., Derksen, A.C.G. and McKenna, K.A. (1994). A new insect cell line from *Trichoplusia ni* (BTI-Tn-5B1-4) susceptible to *Trichoplusia ni* single enveloped nuclear polyhedrosis virus. *Journal of Invertebrate Pathology*, **64**: 260-266.

Gröner, A. (1986). Specificity and safety of baculoviruses. *In*: Granados, R.R. and Federici, B.A. (*Eds.*), *The biology of baculoviruses*. Vol. I. CRC Press, Boca Raton, Florida, pp. 177-202.

Guarino, L.A. and Summers, M.D. (1986). Interspersed homologous DNA of *Autographa californica* nuclear polyhedrosis virus enhances delayed-early gene expression. *Journal of Virology*, **60**: 215-219.

Hall, D.W. (1985). Pathobiology of invertebrate icosahedral cytoplasmic deoxyriboviruses (Iridoviridae). *In*: Maramorosh, K. and Sherman, K.E. (*Eds.*), *Viral insecticides for biological control*. Academic Press, Orlando, Florida, pp. 163-196.

Hammock, B.D., Bonning, B.C., Possee, R.D. Hanzlik, T.N. and Maeda, S. (1990). Expression and effects of the juvenile hormone esterase in a baculovirus vector. *Nature*, **344**: 458-461.

Herniou, E.A. and Jehle, J.A. (2007). Baculovirus phylogeny and evolution. *Current Drug Targets*, **8(10)**: 1-8.

Hirumi, H., Hirumi, K. and McIntosh, A.H. (1975). Morphogenesis of a nuclear polyhedrosis virus of the alfalfa looper in a continuos cabbage looper cell line. *Annals of the New York Academy of Sciences*, **226**: 302-324.

Hostetter, D.L. and Bell, M.R. (1985). Natural dispersal of baculovirus in the environment. *In*: Maramorosch, K. and Sherman, K.E. (*Eds.*), *Viral Insecticides for Biological Control*. Academic Press, Orlando, Florida. pp. 249-284.

Huber, J. (1986). Use of baculoviruses in pest managenment programs. *In*: Granados, R.R. and Federici, B.A. (*Eds.*), *The biology of baculoviruses*. Vol. II. CRC Press, Inc., Boca Raton, Florida, pp. 181-202.

Hughes, P.R., Wood, H.A., Breen, J.P., Simpson, S.F., Duggan, J.K. and Dybas, J.A. (1997). Enhanced bioactivity of recombinant baculoviruses expressing insect-specific spider toxins in lepidopteran crop pests. *Journal of Invertebrate Pathology,* **69**: 112-118.

Hukuhara, T. (1985). Pathology associated with cytoplasmic polyhedrosis viruses. *In*: Maramorosh, K. and Sherman, K.E. (*Eds.*), *Viral insecticides for biological control*. Academic Press, Orlando, Florida, pp. 121-162.

Hukuhara, T. and Bonami, J.R. (1991). Reoviridae. *In*: Adams, J.R. and Bonami, J.R. (*Eds.*), *Atlas of invertebrate viruses*. CRC Press, Inc., Boca Raton, Florida, pp. 393-434.

Ibarra, J.E. and Aguilar, L. (1993). Pruebas de dos virus de la poliedrosis nuclear, para el control del falso medidor de la col, *Trichoplusia ni*, en brócoli. *In: XV Congreso Nacional de Control Biológico. Sociedad Mexicana de Control Biológico*. Monterrey, NL. México. pp. 98-103.

Ignoffo, C.M. and Couch, T.L. (1981). The nucleopolyhedrosis virus of *Heliothis* species as a microbial insecticide. *In*: Burges, H.D. (*Ed.*), *Microbial control of pests and plant diseases 1970-1980*. Academic Press, New York. pp. 330-361.

Ikonomou, L., Schneider, Y.J. and Agathos, S.N. (2003). Insect cell culture for industrial production of recombinant proteins. *Applied and Environmental Microbiology*, **62**: 1-20.

Jakob, N.J., Muller, K., Bahr, U. and Darai, G. (2001). Analysis of the first complete DNA sequence of an invertebrate iridovirus: coding strategy of the genome of *Chilo* iridescent virus. *Virology*, **286**: 182-196.

Jakob, N.J. and Darai, G. (2002). Molecular anatomy of *Chilo* iridescent virus genome and the evolution of viral genes. *Virus Gene*, **25(3)**: 299-316.

Jehle, J.A., Blissard, G.W., Bonning, B.C., Cory, J.S., Herniou, E.A., Rohrmann, G.F., Theilmann, D.A., Thiem, S.M. and Vlak, J.M. (2006). On the classification and nomenclature of baculoviruses: A proposal for revision. *Archives of Virology*, **151**: 1257-1266.

Korth, K.L. and Levings, C.S. (1993). Baculovirus expresion of maize mitochondrial protein URF13 confers insecticidal activity in cell cultures and larvae. *Proceedings of the National Academy of Sciences of the United States of America,* **90**: 3388-3392.

Li, Q., Li, L., Moore, K., Donly, C., Theilmann, D.A. and Erlandson, M. (2003). Characterization of *Mamestra configurata* nucleopolyhedrovirus enhancin and its functional analysis via expression in an *Autographa californica* M nucleopolyhedrovirus recombinant. *Journal of General Virology*, **84**: 123-132.

Maeda, S. (1989). Increased insecticidal effect by a recombinant baculovirus carrying a synthetic diuretic hormone gene. *Biochemical and Biophysical Research Communications*. **165**: 1177-1183.

Martens, J.W., Honee, G., Zuidema, Van Lent, D.J.W., Visser, M.B. and Vlak, J.M. (1990). Insecticidal activity of a bacterial crystal protein expressed by a recombinant baculovirus in insect cells. *Applied and Environmental Microbiology*, **56(9)**: 2764-2770.

Mazzone, H.M. (1985). Pathology associated with baculovirus infection. *In*: Maramorosh, K. and Sherman, K.E. (*Eds.*), *Viral insecticides for biological control*. Academic Press, Orlando, Florida, pp. 81-120.

Mertens, P.P.C. and Bamford, D.H. (2009). The RNAs and Proteins of dsRNA Viruses. Available: http://www.reoviridae.org/dsRNA_virus_proteins/. Visited on 18 Dec 2009.

Mertens, P.P.C., Crook, N.E., Rubinstein, R., Pedley, S. and Payne, C.C. (1989). Cytoplasmic polyhedrosis virus classification by electropherotype: validation by sereological analyses and agarose gel electrophoresis. *Journal of General Virology*, **70**: 173-185.

Merryweather, A.T., Weyer, U., Harris, M.P.G., Hirst, M., Booth, T. and Possee, R.D. (1990). Construction of genetically engineered baculovirus inscticides containing the *Bacillus thuringiensis* subsp. *kurstaki* HD-73 delta endotoxin. *Journal of General Virology,* **71**: 1535-1544.

Miller, L.K. (1988). Baculovirus as gene expression vectors. *Annual Review of Microbiology*, **42**: 177-199.

Moscardi, F. and Sosa-Gómez, D.R. (1993). A case study in biological control: soybean defoliating caterpillars in Brazil. *In*: Buxton, D.R., Shibles, R., Forsberg, R.A., Blad, B.L., Asay, K.H., Paulsen, G.M. and Wilson, R.F. (*Eds.*), *International Crop Science*. Crop Science Society of America, Inc. Madison, Wisconsin, pp. 115-119.

Moscardi, F. (1999). Assessment of the application of baculoviruses for control of Lepidoptera. *Annual Review of Microbiology*, **44**: 257-289.

Moss, B. (1996). Poxviridae: The viruses and their replication. *In*: Fields, R.R., Knipe, D.M. and Howley, P.M. (*Eds.*), *Virology*. Vol. 2. Rayen Press, New York, pp. 2637-2671.

Obregón-Barboza, V., Del Rincón-Castro, M.C., Cabrera-Ponce, J.L. and Ibarra, J.E. (2007). Infection, transfection, and co-transfection of baculoviruses by microprojectile bombardment of larvae. *Journal of Virological Methods*, **140**: 124-131.

Okano K., Vanarsdall, A.L., Mikhailov, V.S. and Rohrmann, G.F. (2006). Conserved molecular systems of the Baculoviridae. *Virology,* **344**: 77-87.

O'Reilly, D.R., Kelly, T.J., Master, E.P., Thyagaraja, B.S., Robson, R.M., Shaw, T.C. and Miller, L.K. (1995). Overexpression of *Bombyx mori* prothoracicotropic hormone using baculovirus vectors. *Insect Biochemistry and Molecular Biology,* **25**: 475-485.

O'Reilly, D.R. and Miller, L.K. (1991). Improvement of a baculovirus pesticide by deletion of the EGT gene. *Biotechnology*, **9**: 1086-1089.

Petrik, D.T., Montelone, I.A., Van Etten, B.A. and Clem, R.J. (2003). Improving baculovirus resistance to UV inactivation: increased virulence resulting from expression of a DNA repair enzyme. *Journal of Invertebrate Pathology*, **82**: 50-56.

Podgwaite, J.D. (1985). Strategies for field use of baculoviruses. *In*: Maramorosch, K. and Sherman, K.E. (*Eds.*), *Viral insecticides for biological control*. Academic Press, Orlando, Florida, pp. 775-797.

Prikhod'ko, G.G., Robson, M., Warmke, J.W., Cohen, C.J., Smith, M.M., Wang, P., Warren, V., Kaczorowski, V., Van der Ploeg, L.H.T. and Miller, L.K. (1996). Properties of three baculovirus-expressing genes that encode insect-selective toxins: m-Aga-IV, As II and Sh I. *Biological Control*, **7**: 236-244.

Sporleder, M., Rodriguez, E.M., Huber, J. and Kroschel, J. (2007). Susceptibility of *Phthorimaea operculella* Zeller (Lepidoptera; Gelechiidae) to its granulovirus *Po*GV with larval age. *Agricultural and Forest Entomology,* **9 (4)**: 271-278.

Stewart, L.M.D., Hirst, M., López Ferber, M., Merryweather, A.T., Cayley, P.J. and Possee, R.D. (1991). Construction of an improved baculovirus insecticide containing an insect-specific toxin gene. *Nature,* **352**: 85-88.

Summers, M. and Smith, G.E. (1985). Genetic engineering of the genome of the *Autographa californica* nuclear polyhedrosis virus. *In*: Fields, B., Martini, M.A. and Kamely, D. (*Eds.*), *Genetically altered viruses and the environment*. Cold Spring Harbor Laboratory, New York, N.Y., pp. 319-329.

Summers, M.D. and Smith, G.E. (1987). A manual of methods for baculovirus vectors and insect cell culture procedures. *Texas Agricultural Experiment Station*. Bulletin No. 1555.

Sun, X., Wang, H., Sun, X., Chen, X., Peng, C., Pan, D., Jehle, J.A., van der Werf, W., Vlak, J.M. and Hu, Z. (2004). Biological activity and field efficacy of a genetically modified *Helicoverpa armigera* single-nucleocapsid nucleopolyhedrovirus expressing an insect-selective toxin from a chimeric promoter. *Biological Control,* **2**: 124-137.

Tanada Y. (1964). A granulosis virus of the codling moth, *Carpocapsae pomonella* (Linnaeus) (Olethreutidae, Lepidoptera). *Journal of Insect Pathology,* **6**: 78-80.

Tanada, Y. and Kaya, H.K. (1993). *Insect pathology*. Academic Press, San Diego, California,

Thomas, D.E., Reichelderfer, C.F. and Heimpel, A.M. (1972). Accumulation and persistence of a nuclear polyhedrosis virus of the cabbage looper in the field. *Journal of Inverebrate Pathology,* **20**: 157-164.

Tinsley, T.W. and Kelly, D.C. (1985). Taxonomy and Nomenclature of Insect Pathogenic Viruses. *In*: Maramorosh, K. and Sherman, K.E. (*Eds.*), *Viral insecticides for biological control.* Academic Press, Orlando, Florida, pp. 3-25.

Tomalski, M.D. and Miller., L.K. (1991). Insect paralysis by baculovirus-mediated expression of a mite neurotoxin gene. *Nature*, **352**: 82-85.

van der Wilk, F., Lent, J.W.M. and Vlak, J.M. (1987). Immunogold detection pf polyhedrin, p10 and virion antigens in *Autographa californica* nulcear polyhedrosis virus-infected *Spodoptera frugiperda* cells. *Journal of General Virology*, **68**: 2615-2623.

van Beek, N., Lu, A., Presnail, J., Davis, D., Greenamoyer, C., Joraski, K., Moore, L., Pierson, M., Herrmann, R., Flexner, L., Foster, J., Van, A., Wong, J., Jarvis, D., Hollingshaus, G. and McCutchen, B. (2003). Effect of signal sequence and promoter on the speed of action of a genetically modified *Autographa californica* nucleopolyhedrovirus expressing the scorpion toxin LqhIT2. *Biological Control*, **27**: 53-64.

van Oers, M.M. and Vlak, J.M. 2007. Baculovirus Genomics. *Current Drug Targets*, **8**: 1051-1068.

van Regenmortel, M.H.V., Fauquet, C.M., Bishop, D.H.L., Cartens, E.B., Estes, M.K., Lemon, S.M., Maniloff, J., Mayo, M.A., McGeoch, D.J., Pringle, X:R. and Wickner, R.B. (2000). *Virus taxonomy.* Seventh report of the international committee of taxonomy of viruses. Academic Press, San Diego.

Vanderzant, E.S., Pool, M.C. and Richardson, C.D. (1962). The role of the ascorbic acid in the nutrition of three cotton insects. *Journal of Insect Physiology,* **8**: 287-297.

Vaughn, J.L. (1992). Virus and rickettsia diseases. *In*: Cantwell, G.E. (*Ed.*), *Insect diseases*. Vol.1. VMI Out of Print Books on demand. Univeristy of Michigan Ann Arbor, USA. pp. 49-85.

Velasco de Castro Oliveira, J., Wolff, J.L.C., Garcia-Maruniak, A., Ribeiro, B.M., Batista de Castro, M.E., Lobo de Souza, M., Moscardi, F., Maruniak, J.E. and de Andrade Zanotto, P.M. (2006). Genome of the most widely used viral biopesticide: *Anticarsia gemmatalis* multiple nucleopolyhedrovirus. *Journal of General Virology*, **87**: 3233-3250.

Vlak, J.M. (1992). The biology of baculovirus *in vivo* and in cultured insect cells. *In*: Vlak, J.M., Schlaeger, E.J. and Bernard, A.R. (*Eds.*), *Baculovirus and recombinant protein production processes.* Editiones Roche, Interlakend, Switzerland, pp. 2-10.

Volkman, L.E. (1986). The 64K envelope protein of budded *Autographa californica* nuclear polyhedrosis virus. *In*: Doerfler, W. and Bohm, P. (*Eds.*), *The molecular biology of baculoviruses*. Springer-Verlag, New York N.Y. pp. 103.

Volkman, L.E. (2007). Baculovirus Infectivity and the Actin Cytoskeleton. *Current Drug Targets*, **8**: 1075-1083.

Volkman, L.E., Summers, M.D. and Hsieh, C.M. (1976). Occluded and nonoccluded nuclear polyhedrosis virus grown in *Trichoplusia ni* comparative neutralization, comparative infectivity, and *in vitro* growth studies. *Journal of Virology*, **19**: 820-822.

Ward, V.K. and Kalmakoff, J. (1991). Invertebrate Iridoviridae. *In*: Kurstak, E. (*Ed.*), *Viruses of invertebrates*. Marcel Dekker, Inc., New York, pp. 197-226.

Williams, T. and Cisneros, J. (2001). Formulación y aplicación de los baculovirus bioinsecticidas. *In*: Caballero, P., López-Ferber, M. and Williams, T. (*Eds.*), *Los baculovirus y sus aplicaciones como bioinsecticidas en el control biológico de plagas*. PHYTOMA-Universidad Pública de Navarra, España, pp. 313-372.

Wood, H.A. (1996). Genetically enhanced baculovirus insecticides. *In*: Gunasekaran, M. and Weber, D.J. (*Eds.*), *Molecular biology of the biological control of pests and diseases of plants*. CRC Press Inc., Boca Raton, Florida, pp. 91-104.

Xeros, N. (1954). A second virus disease of the leather jacket *Tipula paludosa*. *Nature*, **174**: 562-564.

Yearian, W.C. and Young, S.Y. (1982). Control of insect pests of agricultural importance by viral insecticides. *In*: Kurstak, E. (*Ed.*), *Microbial and viral pesticides*. Marcel Dekker, Inc., New York, N.Y., pp. 387-423.

{3}

Entomopathogenic Fungi

Ninfa M. Rosas-García, Alejandro Sánchez-Varela, Jesús M. Villegas Mendoza and Jesús D. Quiroz-Velázquez

ABSTRACT

There are more than 700 species of entomopathogenic fungi belonging to approximately 90 genera, however few species have been considered as biological control agents. They are able to kill any insect developmental stage and are wide spectrum pathogens. The general fungal taxonomy is now based on phylogenetically studies and includes Zygomycota, Ascomycota, Deuteromycota and Chytridiomycota. The mode of action of entomopathogenic fungi is a complex process that includes enzymes production to penetrate insect cuticle. Among these enzymes proteases called Pr play one of the most important roles followed by chitinases. The insects exhibit some defense mechanisms against fungal infection which include the presence of inhibitory compounds, but also possess cellular defense reactions which act when fungus invade the hemocele. Other cytotoxic metabolites are produced by entompathogenic fungi such as destruxins by Metarhizium anisopliae and beauvericin by Beauveria bassiana, however their role in pathogenicity is not completely clear. Some important genera are described in this chapter.

INTRODUCTION

Among the biological control agents, the fungi play an important role for insect control. Many pathogenic fungi are natural control agents of many insect pests and other arthropods, frequently cause epizooties and significantly reduce host populations. Although entomopathogenic fungi have long been recognized to exhibit a strong potential in controlling insect pests, attempts to manipulate fungi as real control agents began in the late ninetieth century with little or moderate success. But until the late sixties the interest in the use of fungi as biological control agents increased due to problems caused by intensive use of chemical control.

There are more than 700 species of entomopathogenic fungi belonging to approximately 90 genera (Alves, 1998) however few species have been considered as biological control agents. Some aspects such as host insect and interaction mechanisms between fungus and insect should be considered if fungi are being studied as potential control agents. This chapter provides an overview of the interaction between insect and pathogen, and the description of the fungal pathogenesis. In addition, many entomopathogenic fungi are described including those which are not considered but offer a potential alternative to insect control. Entomopathogenic fungi are wide spectrum pathogens, and this is the reason why they may attack a number of insect pests living in different environments. As an example, they attack from insects living in aquatic environments to phytophagous insects living in the aerial parts of the plant. Similarly they are able to kill any insect developmental stage such as eggs, larvae, pupae and adults (Alves, 1998). An interesting example is observed in *Anopheles*, *Culex* and *Aedes*, when conidia of *Beauveria bassiana* kill *Anopheles* and *Culex* larvae, and produce 100% mortality in adults of the three genera (Clark *et al.*, 1968).

FUNGAL TAXONOMY AND NOMENCLATURE

The basic rank in taxonomy is the species and different concepts have been used to define fungal species. These concepts include morphology, combination of characters (polythetic concept), and ecology among others. According to biological concept a species is defined as an interbreeding population isolated by intrinsic reproductive barriers. As any other organism, fungi must be named according international rules governed by the International Code of Botanical Nomenclature (ICBN). The taxonomy of some pathogenic fungi is unstable and controversial so far, and the modality of fungal propagation such as the sexual state (teleomorph) or the sexual state (anamorph), has allowed to group fungi in some way, but the correct classification should be studied more extensively (Guarro *et al.*, 1999).

General reclassification of fungi is now based on phylogenetically studies. The phylum Zygomycota, subphylum Zygomycotina, and class Zygomycetes have disappeared and replaced by the phylum Glomeromycota. The Entomophthorales have been reclassified in the subphylum Entomophthoromycotina Humber, and other new subphyla in this group are Murmomycotina, Zoopagomycotina, and Kicxellomycotina. Although these concepts are abandoned it is recommended to continue to use these discontinued names until correct classificiation become accepted (ARSEF 2009).

Among entomopathogen Ascomycetes, a major phylogenetic reclassification of the Clavicipitaceae based on analyses consisting of five to seven loci, including the nuclear ribosomal small and large subunits (*nrSSU* and *nrLSU*), the elongation factor 1α (*tef1*), the largest and the second largest subunits of RNA polymerase II (*rpb1* and *rpb2*), α-tubulin (*tub*), and mitochondrial ATP6 (*atp6*), has split this family into three families; Clavicipitceae, Cordycipitaceae, and Ophiocordycipitaceae (Sung *et al.*, 2007) Available at *http://www.studiesinmycology.org*)

A general diagram is shown in Figs. 3.1, 3.2 and 3.3. The figures only show classification in which entomopathogenic fungi are included (Hawksworth, 2002; Catalog of life, 2010), no other pathogenic or non pathogenic fungi are considered here. The Table 3.1 shows entomopathogenic fungi reported up to date, classified by class and order.

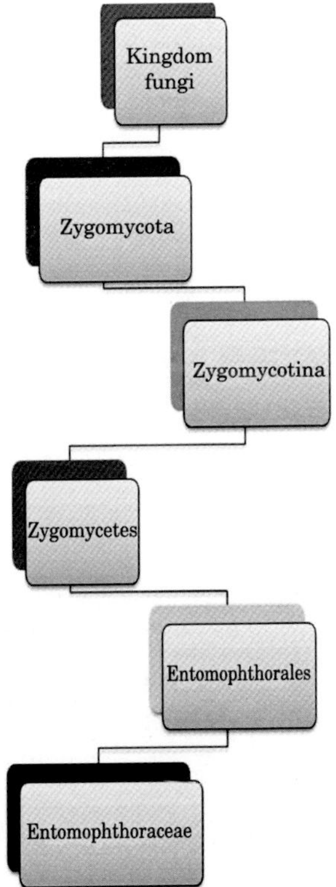

Fig. 3.1. Taxonomic classification of entomopathogenic fungi

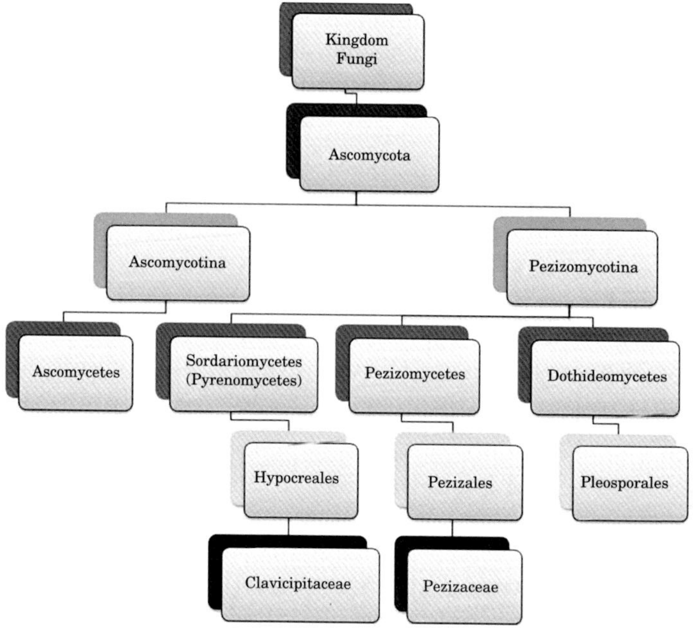

Fig. 3.2. Taxonomic classification of entomopathogenic fungi

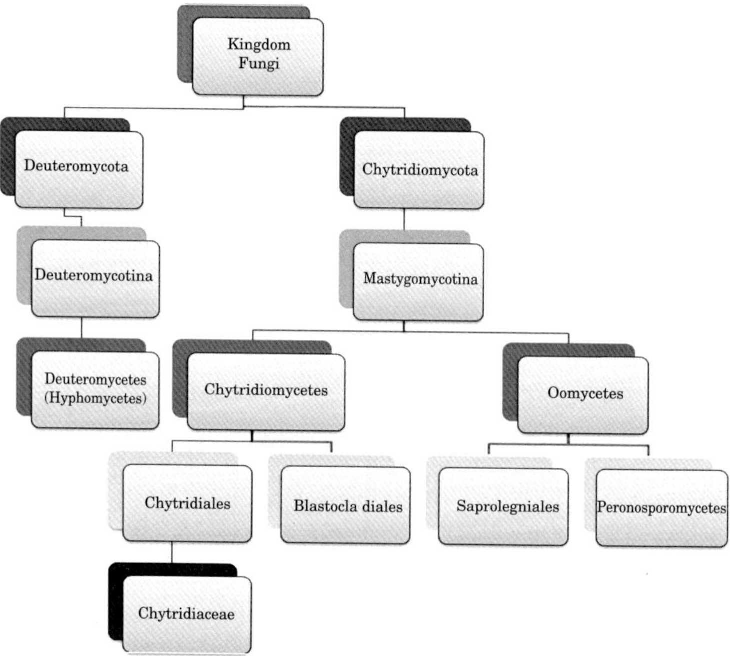

Fig. 3.3. Taxonomic classification of entomopathogenic fungi

Table 3.1. Classification of fungi with entomopathogenic activity reported by the ARSEF

Class: Order	***Fungus***
Sordariomycetes: Hipocreales	*Acremonium* sp.; *Acremonium alternatum*; *Acremonium berkeleyanum*; *Acremonium furcatum*; *Acremonium psammosporum*; *Acremonium roseum*; *Akanthomyces aculeatus*; *Akanthomyces clavatus*; *Akanthomyces gracilis*; *Aphanocladium* sp.; *Aphanocladium album*; *Aschersonia* sp.; *Aschersonia andropogonis*; *Aschersonia basicystis*; *Aschersonia cubensis*; *Aschersonia goldiana*; *Aschersonia incrassata*; *Aschersonia insperata*; *Aschersonia napoleonae*; *Aschersonia placenta*; *Aschersonia rhombispora*; *Aschersonia turbinata*; *Aschersonia viridans*; *Ascopolyporus philodendrus*; *Ascopolyporus villosus*; *Atricordyceps harposporifera*; *Balansia* sp.; *Beauveria amorpha*; *Beauveria bassiana*; *Beauveria caledonica*; *Beauveria malawiensis*; *Beauveria pseudobassiana*; *Beauveria vermiconia*; *Clavicipitaceae anamorph*; *Clonostachys rosea*; *Cordyceps agriota*; *Cordyceps bassiana*; *Cordyceps bifusispora*; *Cordyceps brongniartii*; *Cordyceps cardinalis*; *Cordyceps gunnii*; *Cordyceps isarioides*; *Cordyceps militaris*; *Cordyceps pruinosa*; *Cordyceps pseudoinsignis*; *Cordyceps roseostromata*; *Cordyceps scarabaeicola*; *Cordyceps spegazzinii*; *Cordyceps staphylinidaecola*; *Culicinomyces bisporalis*; *Culicinomyces clavisporus*; *Cylindrocarpon* sp.; *Drechmeria coniospora*; *Elaphocordyceps capitata*; *Elaphocordyceps ophioglossoides*; *Escovopsis aspergilloides*; *Escovopsis* sp.; *Escovopsis weberi*; *Evlachovaea* sp.; *Evlachovaea kintrischica*; *Fusarium* sp.; *Fusarium acuminatum*; *Fusarium avenaceum*; *Fusarium coccophilum*; *Fusarium lateritium*; *Fusarium merismoides*; *Fusarium moniliforme*; *Fusarium moniliforme* var. *Intermedium*; *Fusarium nygamai*; *Fusarium oxysporum*; *Fusarium oxysporum*; *Fusarium polyphialidicum*; *Fusarium semitectum*; *Fusarium solani*; *Fusarium subglutinans*; *Fusarium torulosum*; *Gibellula clavulifera* var. *alba*; *Gibellula pulchra*; *Gibellula* sp.; *Gliocladium* sp.; *Gliocladium roseum*; *Gliocladium* sp.; *Haptocillium balanoides*; *Haptocillium sinense*; *Harposporium anguillulae*; *Harposporium arcuatum*; *Harposporium bysmatosporum*; *Harposporium cerberi*; *Harposporium cicloides*; *Harposporium dicorymbum*; *Harposporium helicoides*; *Harposporium janus*; *Harposporium leptospira*; *Harposporium lilliputanum*; *Harposporium microsporum*; *Harposporium oxycoracum*; *Harposporium* sp.; *Harposporium subuliforme*; *Hirsutella* sp.; *Hirsutella bogoriensis*; *Hirsutella brownorum*; *Hirsutella citriformis*; *Hirsutella cryptosclerotium*; *Hirsutella fusiformis*; *Hirsutella gigantea*; *Hirsutella guyana*; *Hirsutella haptospora*; *Hirsutella illustris*; *Hirsutella kirchneri*; *Hirsutella longicolla* var. *cornuta*; *Hirsutella minnesotensi*; *Hirsutella necatrix*;

Table 3.1 *(Contd...)*

Table 3.1 *(Contd...)*

Class: Order	*Fungus*
	Hirsutella nodulosa; *Hirsutella nr. Haptospora*; *Hirsutella radiata*; *Hirsutella repens*; *Hirsutella rhossiliensis*; *Hirsutella sinensis*; *Hirsutella strigosa*; *Hirsutella subulata*; *Hirsutella thompsonii*; *Hirsutella thompsonii* var. *synnematosa*; *Hirsutella thompsonii* var. *vinacea*; *Hirsutella versicolor*; *Hirsutella verticillioides*; *Hymenostilbe sphecophila*; *Hymenostilbe verrucosa*; *Hyperdermium* sp.; *Hypocrea rufa*; *Hypocrella* sp.; *Hypocrella Africana*; *Hypocrella amazonica*; *Hypocrella andropogonis*; *Hypocrella discoidea*; *Hypocrella javanica*; *Hypocrella macrostroma*; *Hypocrella panamensis*; *Hypocrella phyllogena*; *Hypocrella raciborskii*; *Isaria* sp.; *Isaria amoenerosea*; *Isaria andoi*; *Isaria cateniannulata*; *Isaria catenioblíqua*; *Isaria cicadae*; *Isaria farinose*; *Isaria fumosorosea*; *Isaria guaensis*; *Isaria javanica*; *Isaria poprawskii*; *Isaria tenuipes*; *Lecanicillium* sp.; *Lecanicillium aphanocladii*; *Lecanicillium aphanocladii*; *Lecanicillium flavidum*; *Lecanicillium fusisporum*; *Lecanicillium lecanii*; *Lecanicillium longisporum*; *Lecanicillium muscarium*; *Lecanicillium muscarium*; *Lecanicillium pissodis*; *Lecanicillium psalliotae*; *Leptolegnia* sp.; *Leptolegnia chapmanii*; *Mariannaea elegans*; *Mariannaea pruinosa*; *Metacordyceps yongmunensis*; *Metarhiziopsis microspora*; *Metarhizium* sp.; *Metarhizium acridum*; *Metarhizium album*; *Metarhizium anisopliae* var. *acridum*; *Metarhizium anisopliae* var. *anisopliae*; *Metarhizium anisopliae* var. *lepidiotae*; *Metarhizium anisopliae* var. *majus*; *Metarhizium anisopliae sensu lato*; *Metarhizium anisopliae sensu stricto*; *Metarhizium brunneum*; *Metarhizium flavoviride*; *Metarhizium flavoviride* var. *flavoviride*; *Metarhizium flavoviride* var. *minus*; *Metarhizium flavoviride* var. *novazealandicum*; *Metarhizium flavoviride* var. *pemphigum*; *Metarhizium flavoviride* 'Type E'; *Metarhizium frigidum*; *Metarhizium globosum*; *Metarhizium guizhouense*; *Metarhizium lepidiotae*; *Metarhizium majus*; *Metarhizium pingshaense*; *Metarhizium robertsii*; *Metarhizium taii*; *Microhilum oncoperae*; *Myrothecium* sp.; *Myrothecium verrucaria*; *Neobarya parasitica*; *Nomuraea* sp.; *Nomuraea anemonoides*; *Nomuraea atypicola*; *Nomuraea cylindrospora*; *Nomuraea rileyi*; *Nomuraea viridulus*; *Ophiocordyceps aphodii*; *Ophiocordyceps bisporus*; *Ophiocordyceps gracilioides*; *Ophiocordyceps heteropoda*; *Ophiocordyceps myrmecophila*; *Ophiocordyceps nutans*; *Ophiocordyceps robertsii*; *Ophiocordyceps variabilis*; *Paecilomyces* sp.; *Paecilomyces crustaceus*; *Paecilomyces lilacinus*; *Paecilomyces marquandii*; *Paecilomyces reniformis*; *Paecilomyces sinensis*; *Paecilomyces viridis*; *Plesiospora globosa*; *Pochonia bulbillosa*; *Pochonia chlamydosporia*; *Pochonia chlamydosporia*

Table 3.1 *(Contd...)*

Table 3.1 *(Contd...)*

Class: Order	***Fungus***
	var. *chlamydosporia*; *Pochonia microbactrospora*; *Pochonia suchlasporia* var. *catenulate*; *Polycephalomyces ramosus*; *Polycephalomyces tomentosus*; *Pseudogibellula formicarum*; *Regiocrella camerunensis*; *Rotiferophthora amamiensis*; *Rotiferophthora angustispora*; *Rotiferophthora denticulospora*; *Rotiferophthora japonica*; *Rotiferophthora lacrima*; *Rotiferophthora minutispora*; *Scopulariopsis* sp; *Simplicillium* sp.; *Simplicillium lamellicola*; *Simplicillium lanosoniveum*; *Sorosporella* sp.; *Sorosporella uvella*; *Spicellum ovalispora*; *Sporothrix* sp.; *Sporothrix insectorum*; *Stilbella buquetii*; *Synglicocladium* sp.; *Synglicocladium tetanopsis*; *Synnematium jonesii*; *Syspastospora parasitica*; *Tolypocladium* sp.; *Tolypocladium cylindrosporum*; *Tolypocladium extinguens*; *Tolypocladium geodes*; *Tolypocladium inflatum*; *Tolypocladium lignicola*; *Tolypocladium microsporum*; *Tolypocladium nubicola*; *Tolypocladium parasiticum*; *Tolypocladium tundrense*; *Torrubiella* sp.; *Torrubiella homopterorum*; *Torrubiella piperis*; *Torrubiella ratticaudata*; *Trichoderma* sp.; *Trichoderma lignorum*; *Trichoderma viride*; *Trichothecium* sp.; *Trichothecium roseum*; *Verticillium* sp.; *Verticillium cinnamomeum*; *Verticillium epiphytum*; *Verticillium insectorum*; *Verticillium lecanii*; *Verticillium pseudohemipterigenum*
Entomophthoromycotina: Entomphthorales	*Basidiobolus* sp.; *Basidiobolus haptosporus*; *Basidiobolus haptosporus* var. *minor*; *Basidiobolus heterosporus*; *Basidiobolus magnus*; *Basidiobolus meristosporus*; *Basidiobolus microspores*; *Basidiobolus ranarum*; *Batkoa* sp.; *Batkoa apiculata*; *Batkoa gigantean*; *Batkoa major*; *Batkoa papillata*; *Conidiobolus* sp.; *Conidiobolus adiaeretus*; *Conidiobolus antarcticus*; *Conidiobolus bangalorensis*; *Conidiobolus brefeldianus*; *Conidiobolus coronatus*; *Conidiobolus firmipilleus*; *Conidiobolus heterosporus*; *Conidiobolus iuxtagenitus*; *Conidiobolus lamprauges*; *Conidiobolus megalotocus*; *Conidiobolus obscurus*; *Conidiobolus osmodes*; *Conidiobolus paulus*; *Conidiobolus pseudapiculatus*; *Conidiobolus pumilus*; *Conidiobolus rhysosporus*; *Conidiobolus stromoideus*; *Conidiobolus thromboides*; *Entomophaga* sp.; *Entomophaga asiatica*; *Entomophaga aulicae*; *Entomophaga australiensis*; *Entomophaga calopteni*; *Entomophaga conglomerata*; *Entomophaga macleodii*; *Entomophaga maimaiga*; *Entomophaga praxibuli*; *Entomophthora chromaphidis*; *Entomophthora culicis*; *Entomophthora muscae*; *Entomophthora planchoniana*; *Entomophthora scatophagae*; *Entomophthora schizophorae*; *Entomophthora syrphi*; *Entomophthora thripidum*; *Erynia aquatic*; *Erynia conica*; *Erynia curvispora*; *Erynia ovispora*; *Erynia rhizospora*; *Erynia*

Table 3.1 *(Contd...)*

Table 3.1 *(Contd...)*

Class: Order	***Fungus***
	sciarae; *Eryniopsis caroliniana*; *Eryniopsis ptychopterae*; *Furia* sp.; *Furia Americana*; *Furia gastropachae*; *Furia ithacensis*; *Furia neopyralidarum*; *Furia pieris*; *Furia virescens*; *Macrobiotophthora vermicola*; *Massospora cicadina*; *Neozygites* sp.; *Neozygites floridana*; *Neozygites parvispora*; *Neozygites tetranychi*; *Pandora* sp.; *Pandora blunckii*; *Pandora delphacis*; *Pandora dipterigena*; *Pandora kondoiensis*; *Pandora neoaphidis*; *Pandora nouryi*; *Schizangiella* sp.; *Schizangiella serpentis*; *Strongwellsea castrans*; *Zoophthora* sp.; *Zoophthora anglica*; *Zoophthora aphidis*; *Zoophthora lanceolata*; *Zoophthora occidentalis*; *Zoophthora phalloides*; *Zoophthora phytonomi*; *Zoophthora radicans*
Orbiliomycetes: Orbiliales	*Arthrobotrys amerospora*; *Arthrobotrys brochopaga*; *Arthrobotrys cladodes* var. *macroides*; *Arthrobotrys conoides*; *Arthrobotrys haptotyla*; *Arthrobotrys musiformis*; *Arthrobotrys oligospora*; *Arthrobotrys oligospora* var. *oligospora*; *Arthrobotrys superba*; *Arthrobotrys thaumasia*; *Dactylaria brochopaga*; *Dactylaria candida*; *Dactylella* sp.; *Dactylella aphrobrocha*; *Dactylella doedycoides*; *Dactylella heterospora*; *Dactylella leptospora*; *Dactylella oxyspora*; *Drechslerella dactyloides*; *Duddingtonia flagrans*; *Gamsylella gephyropaga*; *Gamsylella lobata*; *Geniculifera paucispora*; *Monacrosporium cionopagum*; *Monacrosporium doedycoides*; *Monacrosporium elegans*; *Monacrosporium ellipsosporum*; *Monacrosporium eudematum*; *Monacrosporium parvicolle*; *Monacrosporium psychrophilum*; *Monacrosporium thaumasium*
Blastocladiomycetes: Blastocladiales	*Catenaria anguillulae*; *Coelomycidium simulii*
Dothideomycetes: Capnodiales	*Cladosporium* sp. *Mycosphaerellaceae*; *Cladosporium cladosporioides*; *Cladosporium herbarum*; *Cladosporium sphaerospermum*
Dothideomycetes: Dothideales	*Cladosporium cladosporioides*; *Lasiodiplodia theobromae*; *Podonectria coccicola*
Dothideomycetes: Pleosporales	*Alternaria* sp.; *Bipolaris* sp.; *Curvularia* sp.; *Phoma cava Schulzer*
Sordariomycetes: Phyllachorales	*Colletotrichum gloeosporoides*
Sordariomycetes: Microascales	*Scopulariopsis* sp.; *Acremonium furcatum*
Kickxellomycotina: Kickxellales	*Coemansia* sp.; *Ramicandelaber longisporus*
Kickxellomycotina: Harpellales	*Smittium culicis*; *Smittium culisetae*; *Smittium megazygosporum*
Peronosporomycetes: Myzocytiopsidales	*Crypticola clavulifera*
Peronosporomycetes: Lagenidiales	*Lagenidium caudatum*; *Lagenidium giganteum*
Peronosporomycetes: Pythiales	*Pythium* sp.

Table 3.1 *(Contd...)*

Table 3.1 *(Contd...)*

Class: Order	***Fungus***
Hyphochytriomycetes: Hyphochytriales	*Hyphochytrium catenoides*
Pezizomycetes: Pezizales	*Ostracoderma* sp.
Chytridiomycetes: Chytridiales	*Batrachochytrium dendrobatidis*
Chytridiomycetes: Spizellomycetales	*Powellomyces variabilis*; *Rhizophlyctis rosea*; *Spizellomyces kniepii*
Chytridiomycetes: Rhizophydiales	*Rhizophydium sphaerotheca*
Pucciniomycetes: Septobasidiales	*Septobasidium apiculatum*; *Septobasidium burtii*; *Septobasidium carestianum*
Peronosporomycetes: lagenidiales	*Lagenidium caudatum*; *Lagenidium giganteum*
Mucormycotina: Mucorales	*Mortierella histoplasmatoides*; *Mortierella hialina*; *Mucor circinelloides*; *Mucor* sp.; *Phycomyces blakesleeanus*; *Sporodiniella umbellata*; *Syncephalastrum racemosum*
Eurotiomycetes: Onygenales	*Myriodontium keratinophilum*
Agaricomycetes: Agaricales	*Agaricus brunnescens*; *Volvariella volvácea*; *Nematoctonus* sp.; *Nematoctonus concurrens*; *Nematoctonus leiosporus*; *Nematoctonus leptosporus*; *Conidiobolus coronatus*; *Armillaria Mellea*; *Entoloma* sp.
Eurotiomycetes: Ascosphaerales	*Ascosphaera* sp.; *Ascosphaera aggregata*; *Ascosphaera apis*; *Ascosphaera arienoides*; *Ascosphaera atra*; *Ascosphaera duoformis*; *Ascosphaera flava*; *Ascosphaera larvis*; *Ascosphaera major*; *Ascosphaera proliperda*; *Ascosphaera solina*; *Ascosphaera subcuticulata*
Oomycetes: Saprolegniales	*Aphanomyces* sp.; *Leptolegnia* sp.; *Leptolegnia chapmanii*; *Saprolegnia* sp.
Euromycetes: Eurotiales	*Aspergillus* sp.; *Aspergillus clavatus*; *Aspergillus flavus*; *Aspergillus flavus*; *Aspergillus nidulans*; *Aspergillus niger*; *Aspergillus niger*; *Aspergillus ochraceus*; *Aspergillus ochraceus*; *Aspergillus parasiticus*; *Aspergillus sydowii*; *Aspergillus tamari*; *Aspergillus tamari*; *Paecilomyces crustaceus*; *Penicillium* sp.; *Penicillium sumatrense*; *Penicillium thomii*; *Penicillium vulpinum*; *Elaphomyces* sp.
Hyphomycetes	*Cephalosporium-artige*; *Beniowskia* sp.; *Calcarisporium arbuscula*; *Calcarisporium ovalisporum*; *Calcarisporium parasiticum*; *Engyodontium album*; *Engyodontium aranearum*; *Humicola* sp.; *Phialophora* sp.; *Phialophora malorum*; *Stachybotrys* sp.; *Tilachlidium* sp.; *Tritirachium isariae*
Dothideomycetes: Pleosporales	*Alternaria* sp.; *Bipolaris* sp.; *Curvularia* sp.; *Phoma cava*
Saccharomycetes: Saccharomycetales	*Candida* sp.; *Candida cf. pseudoglaebosa*; *Geotrichum* sp.; (ARSEF, 2009)

Table 3.1 *(Contd...)*

ZYGOMYCOTA

The Zygomycota are a group of lower fungi, which are easily recognized in host tissue by their randomly aseptate hyphae. The sexual spore is called zygospore (Guarro *et al.*, 1999). It includes the order Entomophthorales which comprises several entomopathogenic species (Alves, 1998):

(1) *Entomophthora* which attacks hemipterans, diperans, orthopterans and lepidopterans. It is considered an excellent pathogen of houseflies.
(2) *Zoophthora*, particularly *Z. radicans* is highly epizootic and predominates over other species of Entomophthorales. It is highly virulent toward *Empoasca kraemeri* Ross & Moore (Hemiptera: Cicadellidae), a common bean pest, and *E. fabae* (Harris) (Hemiptera: Cicadellidae), which damages potatoes.
(3) *Entomophaga* forms ovoid conidia and spores called azigospores. This genus involves more than 20 species, being the most common for lepidopterans *E. maimaga*, *E. tenthredinis*, and for dipterans *E. caroliniana* and *E. culicis*.
(4) *Neozygites* forms simple conidiophores and primary conidia are rounded or ovoid. *N. adjarica* attacks aphids of the genus *Aphis*, and *N. parvispora* attacks *F. occidentalis*.
(5) *Conidiolobus* presents branched mycelia, and primary conidia are pyriforms. The most important species are: *C. obscures*, *C. apiculatus*, *C. coronatus* and *C. major*, which are virulent to aphids, coleopterans and dipterans.

CHYTRIDIOMYCOTA

These are the most primitive fungi. The zoospores and gamets are mobil by flagella. These fungi are able to produce epizootics and are found in water or wet soil. It comprises several important species:

(1) *Coelomomyces* is the most important genus and produces epizootics in dipterans such as *Aedes*, *Anopheles*, *Culex*, *Culiceta*, and *Chironomus*.
(2) *Myiophagus* is a genus that can be found parasitizing mealybugs.
(3) *Oomycetes* are a distinct group of organisms which form filamentous hyphae during vegetative growth. They produce motile, biflagellated zoospores during asexual sporulation. Many of them

are aquatic species, and some other are terrestrial (Hardham, 2001). They are virulent against *Anopheles*, *Culiceta* and *Culex*.

DEUTEROMYCOTA

All the fungi that do not have any connection with any ascomycetous or basidiomycetous taxa are included in the broad artificial group of deuteromycetes. Numerous other terms such as "fungi imperfecti", "asexual fungi", or "conidial fungi" are used to name these fungi (Guarro *et al.*, 1999). They produce mycelium with conidia and are able to live in aquatic habitats, but the majority of them are soil-borne. Numerous species are included:

Metarhizium, *Beauveria*, *Nomurea*, *Verticillium*, *Hirsutella*, *Aschersonia*, *Aspergillus*, *Paecilomyces*, *Fusarium*, *Sorosporella*, *Tolypocladium*, *Sporothrix*, *Tetracrium*, *Acremonium*.

All these species are not described individually as they have numerous host insect, and many of them are described throughout the chapter.

ASCOMYCOTA

This is the largest phylum of Fungi. Their basic characteristic is the presence of asci inside the ascomata and they possess bilayered hyphal walls. Important genera belong to this phylum:

(1) *Cordyceps* is a genus that comprises more than 150 species being the majority insect parasites. Their asexual state is represented by the genera *Beauveria*, *Hirsutella*, *Hymenostilbe*, *Nomuraea*, *Verticillium*, etc., some of them previously mention in Deuteromycota.
(2) The orders Diptera, Hymenoptera, Coleoptera, Lepidoptera, Hemiptera, Isoptera, and Orthoptera include many susceptible species to *Cordyceps*.
(3) *Hypocrella* is virulent to mealybugs and white flies being *Aschersonia* its asexual state.
(4) *Torrubiella*, *Myriangium*, *Nectria*, *Podonectria* can attack mealybugs.

In the Table 3.2 are shown some of the most common entomopathogenic fungi, and their susceptible insects.

Table 3.2. Entomopathogenic fungi and their host insects

Entomopathogenic fungus	***Order: Family Susceptible insect***	***Common name***	***Reference*(s)**
	Hymenoptera: Formicidae		
B. bassiana	*Solenopsis reichteri* Forel	Black imported fire ant	Broom *et al.*, 1976
B. bassiana	*Solenopsis invicta* Buren	Red imported fire ant	Siebeneicher *et al.*, 1992
	Diptera: Culicidae		
B. bassiana	*C. tarsalis* Coquillet *C. pipiens* (Linnaeus) *Ae. aegypti* (Linnaeus) *Ae. sierrensis* (Ludlow) *Ae. nigromaculis* (Ludlow) *An. albimanus* (Wiedemann)	Western encephalitis mosquito Northern house mosquito Yellow fever mosquito Western tree hole mosquito Irrigated pasture mosquito New world malaria mosquito	Clark *et al.*, 1968
	Diptera: Calliphoridae		
Conidiobolus coronatus	*Calliphora vicina* Jean-Baptiste Robineau-Desvoidy	Blow-fly	Bogús *et al.*, 2007
	Diptera: Tephritoidea		
M. anisopliae	*Tetanops myopaeformis* (Roder)	Sugar beet root maggot	Jackson and Jaronski, 2009
	Orthoptera: Acrididae		
B. bassiana *B. brongniartii* *B. cylindrospora* *B. densa; B. nivea* *P. farinosus; V. lecanii*	*Melanoplus sanguinipes* (Fabricus)	Migratory grasshopper	Khachatourians,1992 Bidochka *et al.*, 1999, Bidochka and Khachatourians, 1994
Metarhizium spp.	*Schistocerca gregaria* Forsskål	Desert locust	Gillespie *et al.*, 1998

Table 3.2 *(Contd...)*

Table 3.2 *(Contd...)*

Entomopathogenic fungus	***Order: Family Susceptible insect***	***Common name***	***Reference*(s)**
	Lepidoptera: Noctuidae		
B. bassiana	*Helicoverpa zea* (Boddie)	Bollworm	Cheung and Grula, 1982
B. bassiana	*Trichoplusia ni* (Hübner)	Cabbage looper	Gupta *et al.*, 1994
B. bassiana	*Spodoptera exigua* (Hübner)	Beet armyworm	Hung and Boucias, 1992
	Lepidoptera: Sphingidae		
M. anisopliae	*Manduca sexta* (Linnaeus)	Tobacco hornworm	St. Leger *et al.*, 1988
	Lepidoptera: Tortricidae		
B. bassiana	*Cydia pomonella* (Linnaeus)	Codling moth	García-Gutiérrez *et al.*, 2004
	Lepidoptera: Lasiocampidae		
Conidiobolus coronatus	*Dendrolimus pini* (Linaeus)	Pine-tree Lappet	Bogús *et al.*, 2007
	Lepidoptera: Pyralidae		
Conidiobolus coronatus, *B. bassiana*	*Galleria mellonella* (Linnaeus)	Greater wax moth	Bogús *et al.*, 2007; Gupta *et al.*, 1994
	Coleoptera: Curculionidae		
Acremonium Beauveria Cladosporium; Clonostachys Paecilomyces	*Hypothenemus hampei* (Ferrari)	Coffee berry borer	Vega *et al.*, 2008
	Coleoptera: Scarabaeidae		
M. anisopliae	*Hoplia philanthus* (Füessly)	Garden chafer	Ansari *et al.*, 2004
B. brongniartii	*Melolontha melolontha* (Linnaeus)	European cockchafer	Enkerli *et al.*, 2004
	Coleoptera: Scolytidae		
B. bassiana	*Scolytus scolytus* (Fabricius)	Large elm bark beetle	Barson, 1977
	Coleoptera: Cerambycidae		
B. brongniartii	*Anoplophora glabripennis* (Motschulski)	Asian longhorned beetle	Dubois *et al.*, 2004

Table 3.2 *(Contd...)*

Table 3.2 *(Contd...)*

Entomopathogenic fungus	***Order: Family Susceptible insect***	***Common name***	***Reference*(s)**
	Coleoptera: Bostrichidae		
B. bassiana	*Rhyzopertha dominica* (Fabricius)	Lesser grain borer	Lord, 2001a
	Coleoptera: Silvanidae		
B. bassiana	*Oryzaephilus surinamensis* (Linnaeus)	Sawtoothed grain beetle	Lord, 2001b
	Hemiptera: Cicadellidae		
B. bassiana	*Empoasca vitis* (Göthe)	False-eye leafhopper	Fang *et al.*, 2004
	Hemiptera: Reduviidae		
B. bassiana	*Triatoma sordida* (Stäl)	No common name	Luz *et al.*, 2004
	Hemiptera: Aphididae		
Lecanicillium longisporum and *L muscarium*	*Myzus persicae* (Sulzer)	Green peach aphid	Down *et al.*, 2009
	Hemiptera: Aleyrodidae		
Lecanicillium longisporum and *L. muscarium*	*Bemicia tabaci* (Gennadius)	Sweetpotato whitefly	Down *et al.*, 2009
B. bassiana *Lecanicillium lecanii*	*Trialeurodes vaporariorum* (Westwoll)	Greenhouse whitefly	Fargues *et al.*, 2003
	Hemiptera: Scutelleridae		
B. bassiana *Beauveria* sp.; *Verticillium lecanii*; *V. psalliotae* *Verticillium* sp. *Paecilomyces farinosus*	*Eurygaster integriceps* Puton	Sunn pest	Parker *et al.*, 2003
	Thysanoptera: Thripidae		
Lecanicillium longisporum and *L. muscarium, B. bassiana*	*Frankliniella occidentalis* (Pergande)	Western flower trips	Down *et al.*, 2009, Ungine *et al.*, 2005

Mode of Action

The mode of action is similar for all entomopathogenic fungi, so that a general description is given below.

Asexual fungal conidia are generally responsible for infection, and passively disperse through the environment. They attach to the cuticle of a suitable insect host in a nonspecific way (Boucias *et al.*, 1988). Through a hydrophobic interaction between insect cuticle and aerial conidia (Holder *et al.*, 2007) the germination process begins (Shah and Pell, 2003). To overcome host resistance mechanisms fungal metabolites enable the pathogen for the penetration process, e.g. cuticle degrading enzymes, inhibition of selective process of the host and interference with the regulatory systems of the host (Hajek and St. Leger, 1994).

Entomopathogens are able to secrete proteases capable of degrading elastin or mucin (St. Leger *et al.*, 1997), and are probably adapted to an ecological niche, as they produce only the necessary enzymes to colonize an insect, such as proteases and chitinases, which degrade the insect cuticle, but not polysaccharidases (St. Leger *et al.*, 1997).

Fungal cuticle degrading enzymes presents a considerable affinity for insect cuticle. However adsorption is necessary for cuticle degradation by the major proteases, such as Pr1, as has been previously demonstrated in *Metarhizium anisopliae*. St. Leger *et al.*, (1986a). The mechanism of cuticle degradation by Pr1 initiates when enzyme is adsorbed via nonspecific electrostatic bonds, which occur between positively charged groups on the basic protease molecule and carboxyl groups on the cuticle. This process does not involve the production of any adhesive materials, so therefore the hydrophobicity is the responsible mechanism for conidial attachment. This process has been reported to occur in the entomopathogenic deuteromycetes, *B. bassiana*, *Nomurea rileyi*, and *M. anisopliae*. The attachment step involves elastase activity. The active site of this enzyme comes into contact with any part of the cuticle and splits its peptide bonds, and solubilized proteins are further degraded.

The first enzymatic activities that appeared during the initiation of the pathogenesis are those of the proteolytic complex (protease, aminopepetidase, and carboxypeptidase), and esterases, which appear within the first 24 h. These processes provide nutrients to the fungus enabling the onset of cuticle penetration; subsequently chitinases appear as a result of induction as chitin becomes available after proteins cuticle degrade. Lately lipase production occurs when conidia are germinating as some lipids of the epicuticle become present (St. Leger *et al.*, 1986b).

Once penetration has occurred, conidia are involved in a favorable environment with suitable temperature, pH, and moisture, so the germination process can begin. At the apical end of the germ tube, hyphae begin expanding forming a structure called appressorium. During this phase there is a migration of the citoplasmic content towards the germ tube, transforming this structure in a high metabolic activity center. The hyphae lower apical ends differentiate into a thin and protruding structure (Alves, 1998). Once the hyphae have invaded the insect, hemolymph components suffer some changes after infection. These changes include decline of some nutrients, increase of total amount of basic amino acids, and sugar reduction. Physical presence of the fungus or toxin liberation into the alimentary tract tissue leads to malfunction of the digestive tract with subsequent feeding cessation ensuing death. At the end of the infection process, the mycelium generates conidiophores that grow out of the insect body through spiracles, mouth parts, and intersegmental membranes. The emerging hyphae begin to cover the insect cadaver (Kershaw *et al.*, 1999) which can become completely mycosed (Figs. 3.4 and 3.5). Outgrowing mycelia form primary condia at their tips (Federici, 1999), and these can be dispersed to the environment.

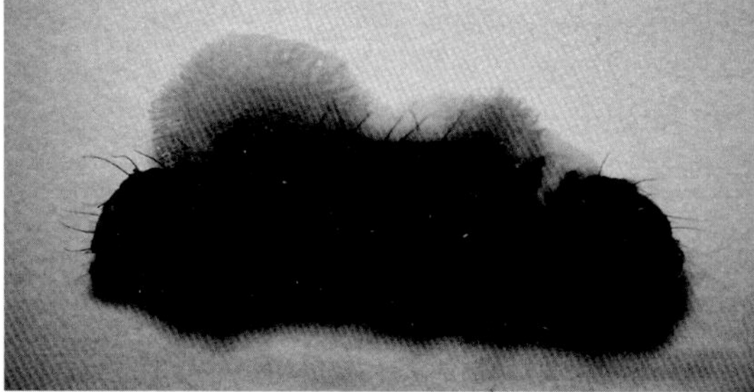

Fig. 3.4. Lateral view of a dead insect larva with hyphae emerging from insect body

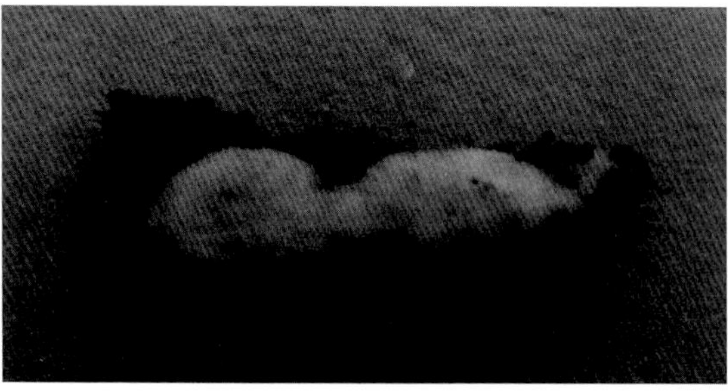

Fig. 3.5. Dorsal view of a dead insect larva with hyphae emerging from insect body

Enzymatic Activity

The most studied enzymes as virulence factors in entomopathogenic fungi are proteases. A basic protease designated as Pr1 with chymoelastase activity was found as an important pathogenic determinant to extensively degrade cuticle. A second enzyme was designated as a trypsin-like protease Pr2 (St. Leger *et al.*, 1987).

The first protease with a chymotrypsin primary specificity was reported in *M. anisopliae*. As this was the first chymotrypsin it is thought that its corresponding gene was transferred laterally from an actinomycete bacterium (Screen and St. Leger, 2000).

The first fungal metallocarboxypeptidase was also reported from *M. anisopliae*, and showed closest amino acid identity to human carboxypeptidases. Some analyses in *M. sexta* indicated that this enzyme together with other endoproteases participate in procuring nutrients (Joshi and St. Leger, 1999).

Pr1 is an up regulated enzyme and is coded by the *cag* genes. This protease facilitates hyphal breaching of the cuticle allowing the fungus to emerge from the insect cadaver (Small and Bidochka, 2005). It is supposed that subtilisin-related proteases and trypsin-related enzymes play important roles in pathogenesis, as these enzymes are not secreted by saprophytes (St. Leger *et al.*, 1992).

Chitinase activity was found only in the presence of chitin, indicating that its synthesis is regulated by an inducer-repressor mechanism (St. Leger *et al.*, 1986c). This means that release of chitinase is dependent on the accessibility of its substrate (St. Leger *et al.*, 1996a). However, the regulation of enzymes such as subtilisin-like proteases (Pr1a and Pr1b), trypsin-like proteases (Pr2), aspartyl proteases, metalloproteases, aminopeptidases and chitinases are expressed accordingly to the ambient pH, at which they function effectively. Interestingly they are able to elevate pH in the surrounding medium to produce basic compounds (St. Leger *et al.*, 1999), but it has been observed that chitinase production also occurs under nonoptimal conditions supporting the evidence that it plays a secondary role in infection process (St. Leger *et al.*, 1998).

The most important protease in *M. anisopliae* is Pr1; however proteinase 2 Pr2 has tryptic specificity for basic residues. Two isoforms of Pr2 (4.4-Pr2 and 4.9-Pr2) exhibit major trypsin activity. It is supposed that selective pressure is so little to diverge specificities and create chymotrypsins from one of the ancestral Pr2 genes. A comparison of the overall rates of hydrolysis suggests that Pr2 are much less able to degrade cuticle, which estimate that trypsin-related enzymes could participate in more selective functions (St. Leger *et al.*, 1996b). An example of this is the enzymatic activity of *B. bassiana* that enables germinating hyphae to penetrate randomly the cuticle of the leaf surface of corn plants. The hyphae are able to move within xylem and preserve fungal virulence against the European corn borer, *Ostrinia nubilalis* (Hübner) (Lepidoptera: Crambidae) (Wagner and Lewis, 2000).

INSECT DEFENSE MECHANISMS

The insects exhibit some defense mechanisms against fungal infection. One of the most important defense mechanisms is the presence of the insect cuticle which acts as an infection barrier. In addition the presence of inhibitory compounds, such as phenols, quinones, and lipids in the cuticle contribute to insect resistance (Hajek and St. Leger, 1994).

Quinones and melanin which are oxidized phenolic products, together with phenols may inactivate fungal enzymes. Melanine cuticle could exert some resistance to penetration and also it may exhibit some resistance degradation caused by enzymes hindering diffusion of nutrients. Also hyphal apical walls become melanized and loose plasticity necessary for growth (St. Leger *et al.*, 1988).

Melanin is non toxic to fungal growth, but it may be combined with biodegradable components of the cuticle creating a physical barrier to enzymatic activity (St. Leger *et al.*, 1988).

However, insects possess other defense mechanisms. One of them is the cellular defense reaction that includes phagocytosis, nodule formation, and encapsulation (Cho *et al.*, 1999). The other mechanism is the humoral response that includes the phenoloxidase (PO) cascade system of melanization, antimicrobial peptides, reactive oxygen and nitrogen intermediates and some other molecules. As many humoral components are produced by hemocytes, it is clear that these cells are the first line of defense when foreign organisms enter the hemocoel (Hyllier, 2009).

The phenoloxidase (PO) cascade system catalyzes the oxidation of phenol to quinone, which is then polymerized non-enzymatically to melanin. The melanin encapsulates invading foreign organisms and immobilizes them. It has been evidenced in *Tenebrio molitor* Linnaeus (Coleoptera: Tenebrionidae) that an activated PO is involved in the cell clump/cell adhesion, which is a cellular defense mechanism that occurs in response to a variety of foreign substances (Lee *et al.*, 1999).

Dendrolimus pini (Linnaeus) (Lepidoptera: Lasiocampidae), *Galleria mellonella* (Linnaeus) (Lepidoptera: Pyralidae), and *Calliphora vicina* Jean-Baptiste Robineau-Desvoidy (Diptera: Calliphoridae), exhibited PO activity in the hemolymph when exposed to sporulating cultures of *Conidiobolus coronatus*. However in the cellular response during confrontation between the insect and the fungus, phagocytic activity of plasmocytes, and melanized cells aggregate (nodules) formation, were only found in *G. mellonella*. Lysozyme-like activity increased in the three species (Bogús *et al.*, 2007).

A protein which could be probably part of the early-staged encapsulation-relating proteins in a coleopteran insect was found. This diapauses protein 1-like protein maybe acts in the insect cellular defense reaction as the diapauses protein 1 is implicated in cuticle formation during molting and in the formation of adult proteins during metamorphosis. When foreign materials are introduced in the insect body, the insect reacts rapidly and forms capsules that include hemocytes and extracellular hemolymph components (Cho *et al.*, 1999).

GENETIC STRUCTURE

There are several reports indicating that genetic structure of entompathogenic fungal species is related to host insect (Bridge *et al.*, 1993). When fungal isolates were analyzed by molecular markers, it was observed that each individual exhibited a different and unique genotype (Bridge *et al.*, 1993; Fegan *et al.*, 1993), although many similarities were observed among them.

Fegan *et al.* 1993 found that similar RAPD profiles among *M. anisopliae* isolates may relate genotypes to specific geographical origins, and also may indicate host range.

In the same way the restriction patterns of ribosomal internal transcribed spacers (ITSs) of *Beauveria* and *Metarhizium* indicated that grouping of isolates was correlated to pathogenic specialization toward *Hoplochelus marginalis* Fairmaire (Coleoptera: Melolonthidae) and *Melolontha melolontha* Fabricius (Coleoptera: Scarabaeidae) (Neuvéglise *et al.*, 1997).

Bidochka *et al.* 2001 indicated that the population genetic structure of deuteromycetous fungi is associated to habitat selection and not host-insect selection, so the insect host related population structure is the result of the habitat influences in which the insect and fungus co-occur.

OTHER VIRULENCE FACTORS

The insecticidal compounds called the destruxins (DTX) are metabolic toxins secreted by the fungus *M. anisopliae* (Walhman and Davidson, 1993, Wang *et al.*, 2004). These toxins have proven to be toxic for many insect pests causing muscle paralysis. However given the variety of *M. anisopliae* strains, the production time of DTX is very different.

An isolate of *M. anisopliae* var *anisopliae*, that was pathogenic to the tobacco hornworm *M. sexta*, the desert locust *Schistocerca gregaria* (Forskal) (Orthoptera: Acrididae), and the vine weevil, *Otiorhynchus sulcatus* (Fabricius) (Coleoptera: Curculionidae), produced DTX after 3 days in culture (Kershaw *et al.*, 1999), while two different *M. anisopliae* strains produced DTX after 10-20 days on rice (Wang *et al.*, 2004).

It is possible that destruxins have a role in pathogenicity, however the variation in production time can cause toxin be unable to act before insect dies by fungal invasion in the hemocel. The use of these toxins may not be suitable for implementation as part of biological pest control.

Some other cytotoxic metabolites are produced by entompathogenic fungi. The cyclic hexadepsipeptide beauvericin is known as a secondary metabolite produced by *B. bassiana*, and recently has been recognized as an important mycotoxin. The cytotoxic effect of beauvericin on SF-9 cells (immortalized pupal ovarian cells of *S. frugiperda*) caused a marked decrease in cell viability at 1 µM concentration; and the effect became more evident at higher concentrations. Its application in biological control programs must be considered seriously as this mycotoxin also affects mammals (Calo *et al.*, 2003).

It might be possible that this metabolite is a contributing factor to larval mortality, but no beauvericin was detected in the hemolymph of *H. zea* larvae after injection of viable spores. It seems to be that beauvericin is not produce before 5 days of fungal growth. Bassianolide, that is another cyclic depsipeptide produced by *B. bassiana*, resulted toxic against *H. zea* (Champlin and Grula, 1979).

Characteristics of Some Important Entomopathogenic Fungi

Some important fungi currently used or some others with potential in biological control of insect pests are briefly described below.

B. bassiana

This fungus (Fig. 3.6) is one of the most studied organisms in biological control of insect pests. Its wide activity spectrum allows to control a variety of insect that include scarab beetles (Fig. 3.7), leaf feeding beetles, whiteflies, mealybugs, leafhoppers and plant hoppers, aphids, thrips, psyllids, diamond back moth, borers, grasshoppers, among many others. The use of this fungus has been approved to be applied in forestry, pastures, greenhouses, commercial landscapes and all food/feed commodities (EPA, 1999).

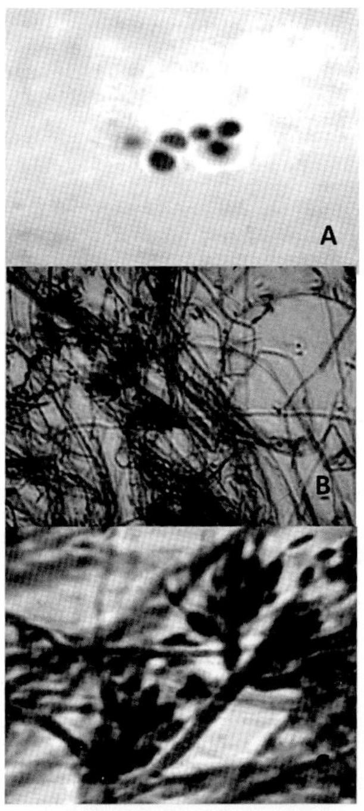

Fig. 3.6. *Beauveria bassiana* photographs at optical microscope. A) Conidia (100X), B) Mycelium with fruiting bodies (40X), C) Fruiting bodies (100X)

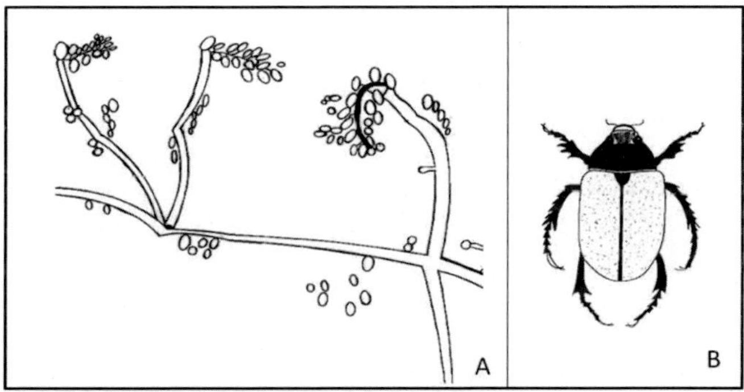

Fig. 3.7. Schematic drawing of fruiting body of *Beauveria bassiana* (A) and a susceptible beetle (B)

M. anisopliae

This fungus (Fig. 3.8) is widely distributed in nature and it is characterized for its wide host range activity. Insects attack by this fungus, are regularly covered by conidia which color vary from light to dark green. For this reason it is known as green muscardine. Conidia are uninucleated in simple conidiophores. Conidia are formed at the tips of branched structures called synema. It is a soil-dwelling fungus that causes fatal disease to root weevils, locust, and grasshoppers (Alves, 1998) (Fig. 3.9) among others.

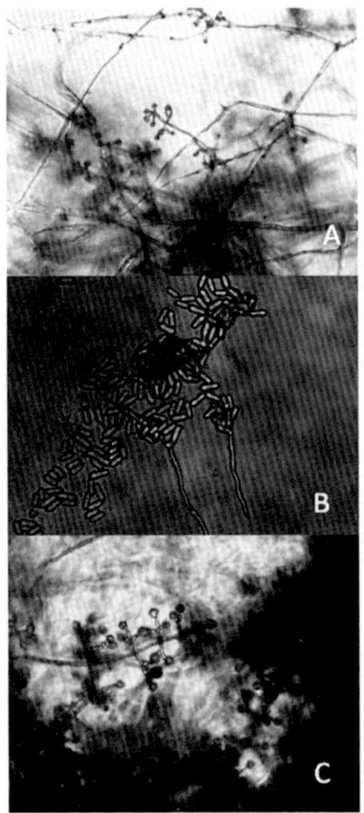

Fig. 3.8. *Metarhizium anisopliae* photographs at optical microscope. A) Mycelium with fruiting bodies (40X), B) Conidia (100X), C) Fruiting bodies (100X)

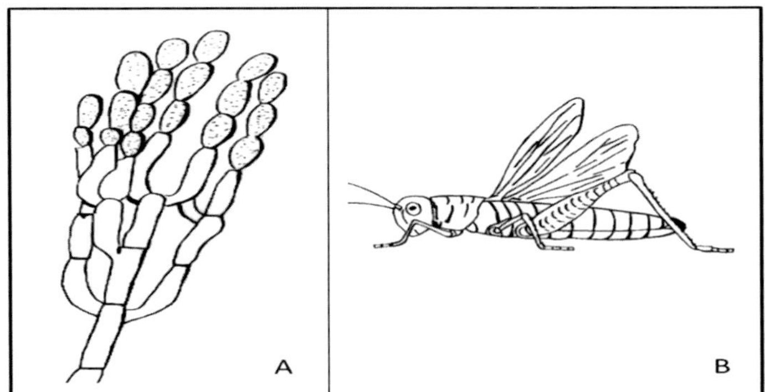

Fig. 3.9. Schematic drawing of fruiting body of *Metarhizium anisopliae* (A), susceptible locust (B)

P. fumosoroseus

Some species formerly classified in *Paecilomyces* sect. *Isiaroidea* has been reclassified based on phylogenetic analysis, however no generic re-assignments have been made for these hypocrealean anamorphs, and they are treated as *Paecilomyces* species (Sung *et al.*, 2007). These species (Fig. 3.10) have a much narrower host range, but they it control effectively the diamond back moth *Plutella xilostella*, the fall armyworm (*S. frugiperda*) (Altre and Vandenberg, 2001), *Diuraphis noxia*, (Vandenberg and Cantone, 2004), *Bemisia tabaci*, *Bemisia argentifolii*, and *Trialeurodes vaporariorum* (Souza Azevedo *et al.*, 2000).

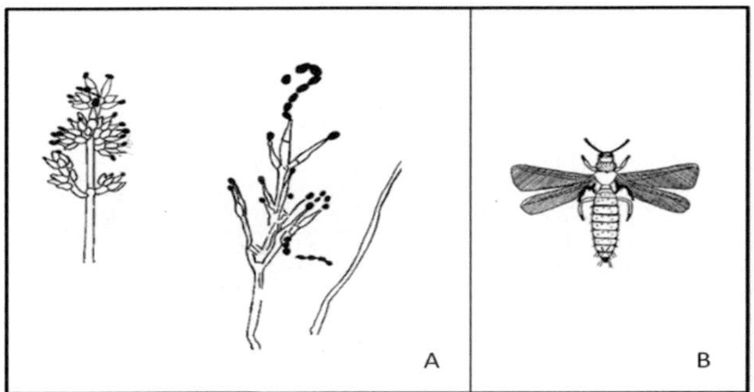

Fig. 3.10. Schematic drawing of fruiting body of *Paecilomyces fumosoroseus* (A) and a susceptible whitefly (B)

Lecanicillium lecanii (Verticillim)

The long-accepted classification of *Verticillium* species recognized two sections: *Verticillium* and *Prostata*, however, due to the current standards of systematic, most entomopathogens from this group are now placed in *Lecanicillium*, whose type species *L. lecanii* is also recognized to be a species complex which include at least three species (*L. lecanii*, *L. muscarium*, and *L. longisporum*) (ARSEF 2009). This fungus (Fig. 3.11) has a wide host range and it has been used as a useful

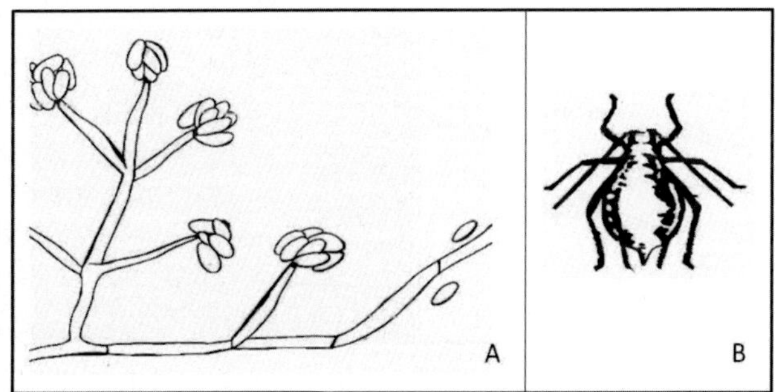

Fig. 3.11. Eschematic drawing of fruiting body of *Lecanicillium lecanii* (A), and a susceptible aphid (B)

model in the study of the evolution of pathogenic mechanisms. The insect pathogenic strains produce higher levels of subtilisin-like proteases (Bidochka *et al.*, 1999).

Nomuraea rileyi

This fungus (Fig. 3.12) occurs naturally worldwide in many economically important crops. It is though that the yeast-like stage can be associated to virulence. This pathogen attacks more than 32 insect species belonging to orders Lepidoptera, Coleoptera and Orthoptera. Around 90% of hosts are lepidopterans (Alves, 1998). *N. rileyi* produces greenish-colored conida (Sung *et al.*, 2007).

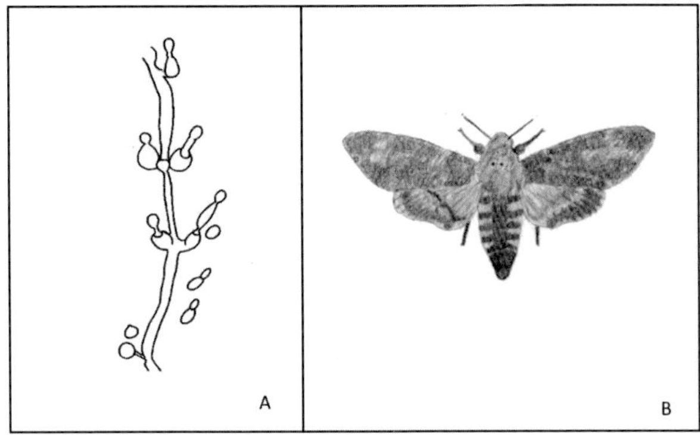

Fig. 3.12. Eschematic drawing of conidiophore of *Nomurea rileyi* (A) and a susceptible lepidopteran (B)

Fusarium sp.

This entomopathogen (Fig. 3.13) occurs naturally in populations of scales, whiteflies and mites in citrus (Alves *et al.*, 2002).

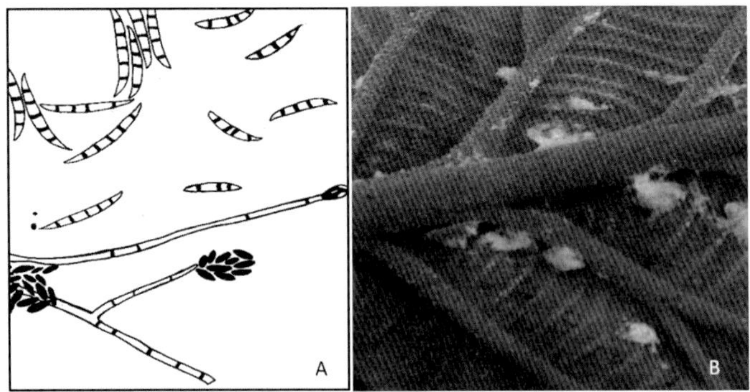

Fig. 3.13. Schematic drawing of conidia and conidiophores from different species of *Fusarium* (A), and a photograph showing scales in a leaf (B)

Tolypocladium sp.

This genus is characterized by producing single or whorled conidiogenous cells called phialides (Fig. 3.14). It belongs to the Clavicipitaceae clade A, although it can be found in other clades of Clavicipitaceae. It is an anamorphic form of the genus *Cordyceps* (Sung *et al.*, 2007). *T cylindrosporum* attacks dipterans, mainly *Aedes* sp., and blastopores are more infectious than conidia (Alves, 1998). This species that can grow axenically in simple media (Sung *et al.*, 2007).

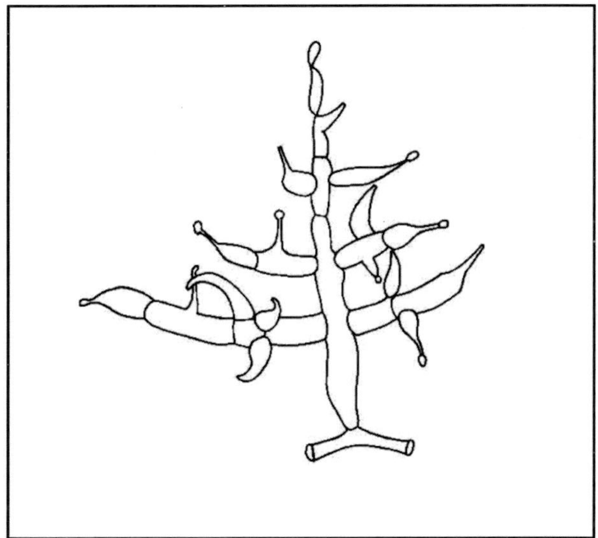

Fig. 3.14. Schematic drawing if conidiophores and conidia from *Tolypocladium*

REFERENCES

Altre, J.A. and Vandenberg, J.D. (2001). Penetration of cuticle and proliferation in hemolymph by *Paecilomyces fumosoroseus* isolates that differ in virulence against lepidopteran larvae. *Journal of Invertebrate Pathology*, **78(2)**: 81-86.

Alves, S.B. (1998). Fungos entompatogênicos. *In*: Alves, ÇS.B. (*Ed.*), Controle *microbiano de insetos*. Piracicaba, FEALQ, Brazil, pp 289-381.

Alves, S.B., Pereira, R.M., Lopes, R.B. and Tamai, M.A. (2002). Use of entomopathogenic fungi in Latin America. *In*: Upadhyay, R.K. (*Ed.*), Advances in Microbial Control of Insect Pests. Kluwer Academic/Plenum Publishers, New York, pp. 193-211.

Ansari, M.A., Vestergaard, S., Tirry, L. and Moens, M. (2004). Selection of a highly virulent isolate, *Metarhizium anisopliae* CLO53, for controlling *Hoplia philanthus*. *Journal of Invertebrate Pathology*, **85**: 89-96.

ARSEF (ARS Collection of entomopathogenic fungal cultures). (2009). Catalog of species. USDA-ARS Biological Integrated Pest Management Research. Robert W. Holley Center for Agriculture and Health. Ithaca, New York.

Barson, G. (1977). Laboratory evaluation of *Beauveria bassiana* as a pathogen of the larval stage of the large elm bark beetle, *Scolytus scolytus*. *Journal of Invertebrate Pathology*, **29**: 361-366.

Bidochka, M.J., Kamp, A.M., Lavender, T.M., Dekoning, J. and Amritha de Croos, J.N. (2001). Habitat association in two genetic groups of the insect-pathogenic fungus *Metarhizium anisopliae*: uncovering cryptic species?. *Applied and Environmental Microbiology*, **67(3)**: 1335-1342.

Bidochka, M.J. and Khachatourians, G.G. (1994). Protein hydrolisis in grasshopper cuticles by entomopathogenic fungal extracelular proteases. *Journal of Invertebrate Pathology,* **63**:7-13.

Bidochka, M.J., St. Leger, R.J., Stuart, A. and Gowanlock, K. (1999). Nuclear rDNA phylogeny in the fungal *Verticillium* and its relationship to insect and plant virulence, extracelular proteases and carbohydrases. *Society for General Microbiology*, **145**: 955-963.

Bogús, M.I., Kedra, E., Bania, J., Szczepanik, M., Czygier, M., Jablónski, P., Pasztaleniec, A., Samborski, J., Mazgajska, J. and Polanowski, A. (2007). Different defense strategies of *Dendrolimus pini, Galleria mellonella*, and *Calliphora vicina* against fungal infection. *Journal of Insect Physiology,* **53**: 909-922.

Boucias, D.G., Pendland, J.C. and Latge, J.P. (1988). Nonspecific factors involved in attachment of entomopathogenic deuteromycetes to host insect cuticle. *Applied and Environmental Microbiology,* **54(7)**: 1795-1805.

Bridge, P.D., Williams, M.A.J., Prior, C. and Paterson, R.R.M. (1993). Morphological, biochemical and molecular characteristics of *Metarhizium anisopliae* and *M. flavoride*. *Journal of General Microbiology*, **139**: 1163-1169.

Broom, J.R., Sikorowski, P.P. and Norment, B.R. (1976). A mechanism of pathogenicity of *Beauveria bassiana* on larvae of the imported fire ant, *Solenopsis richteri*. *Journal of Invertebrate Pathology*, **28**: 87-91.

Calo, L., Fornelli, F., Nenna, S., Tursi, A., Caiaffa, M.F. and Macchia, L. (2003). Beauvericin cytotoxicity to the invertebrate cell line SF-9. *Journal of Applied Genetics*, **44(4)**: 515-520.

Catalogue of Life. 2010. *Biodiversity occurrence data.* Annual Checklist.

Website:http://www.catalogueoflife.org *(Accessed through GBIF Data Portal, data.gbif.org, 2010-04-22).*

Champlin, F.R. and Grula, E.A. (1979). Noninvolvement of beauvericin in the entomopathogenicity of *Beauveria bassiana*. *Applied and Environmental Microbiology.* **37(6)**: 1122-1125.

Clark, T.B., Kellen, W.R., Fukuda, T. and Lindegren, J.E. (1968). Field and laboratory studies on the pathogenicity of the fungus *Beauveria bassiana* to three genera of mosquitoes. *Journal of Invertebrate Pathology*, **11**: 1-7.

Cheung, P.Y.K. and Grula, E.A. (1982). *In vivo* events associated with entomopathology of *Beauveria bassiana* for the corn earworm (*Heliothis zea*). *Journal of Invertebrate Patholology*, **39**: 303-313.

Cho, M.Y., Lee, H.S., Lee, K.M., Homma, K., Natori, S. and Lee, B.L. (1999). Molecular cloning and functional properties of two early-stage encapsulation-relating proteins from the coleopteran insect, *Tenebrio molitor* larvae. *European Journal of Biochemistry*, **262**: 737-744.

Down, R.E., Cuthbertson, A.G.S., Mathers, J.J. and Walters, F.A. (2009). Dissemination of the entomopathogenic fungi, *Lecanicillium longisporum* and

L. muscarium, by the predatory bug, *Orius laevigatus*, to provide concurrent control of *Myzus persicae*, *Frankliniella occidentalis* and *Bemisia tabaci*. *Biological Control,* **50**: 172-178.

Dubois, T., Li, Z., Jiafu, H. and Hajek, A.E. (2004). Efficacy of fiber bands impregnated with *Beauveria brongniartii* cultures against the Asian longhorned beetle, *Anoplophora glabripennis* (Coleoptera: Cerambycidae). *Biological Control*, **31(3)**: 320-328.

Enkerli, J., Widmer, F. and Keller, S. (2004). Long-term persistence of *Beauveria brongniartii* strains applied as biocontrol agents against european cockchafer larvae in Switzerland. *Biological Control,* **29**: 115-123.

EPA (US Environmental Protection Agency). 1999. *Beauveria bassiana* strain GHA (128924) Fact sheet.

Fang, W., Zhang, Y., Yang. X., Zheng, X., Duan, H., Li, Y. and Pei, Y. (2004). *Agrobacterium tumefaciens*-mediated transformation of *Beauveria bassiana* using an herbicide resistance gene as a selection marker. *Journal of Invertebrate Pathology*, **85**: 18-24.

Fargues, J., Vidal, C., Smits, N., Rougier, M., Boulard, T., Mermier, M., Nicot, P., Reich, P., Jeannequin, B., Ridray, G. and Lagier, J. (2003). Climatic factors on entomopathogenic hyphomycetes infection of *Trialeurodes vaporariorum* (Homoptera: Aleyrodidae) in mediterranean glasshouse tomato. *Biological Control*, **28**: 320-331.

Federici, B.A. (1999). A perspective on pathogens as biological control agents for insect pests. *In*: Bellows, T.S. and Fisher, T.W. (*Eds.*), *Handbook of Biological Control, Principles and applications of biological control*. Academic Press, New York, pp. 417-548.

Fegan, M., Manners, M., Maclean, D.J., Irwin, J.A.G., Samuels, K.D.Z., Holdom, D.G. and Li, D.P. (1993). Random amplified polymorphic DNA markers reveal a high degree of genetic diversity in the entomopathogenic fungus *Metarhizium anisopliae* var. *anisopliae. Journal of General Microbiology*, **139**: 2075-2081.

García-Gutiérrez, C., González-Maldonado, M.B., Medrano-Roldán, H. and Chaírez-Hernández, I. 2004. Evaluación de la cepa BBP1 de *Beauveria bassiana*, Mycotrol®, Meta-Sin®, y azinfos metílico contra *Cydia pomonella* L. (Lepidoptera: Tortricidae) en laboratorio y campo. *Folia Entomológica Mexicana*, **43(1)**: 1-7.

Gillespie, J.P., Bateman, R. and Charnley, A.K. (1998). Role of cuticle-degrading proteases in the virulence of *Metarhizium spp*. for desert locust, *Schistocerca gregaria. Journal of Invertebrate Pathology*, **71**: 128-137.

Guarro, J., Gené, J. and Stchigel, A.M. (1999). Developments in fungal taxonomy. *Clinical Microbiology Reviews*, **12(3)**: 454-500.

Gupta, S.C., Leathers, T.D., El-Sayed, G.N. and Ignoffo, C.M. (1994). Relationships among enzyme activities and virulence parameters in *Beauveria bassiana* infections of *Galleria mellonella* and *Trichoplusia ni*. *Journal of Invertebrate Pathology*, **64**: 13-17.

Hajek, A.E. and St. Leger, R.J. (1994). Interaction between fungal pathogens and insect hosts. *Annual Review of Entomology*, **39**: 293-322.

Hardham, A.R. (2001). Cell biology of fungal infection of plants. *In*: Howard, R.J. and Gow, N.A.R. (*Eds.*), *The Mycota: biology of the fungal cell*. Springer, Berlin Heidelberg New York, pp 91-123.

Hawksworth, D.L. (2002). *Straminipila*, a new kingdom name for 'Oomycete' fungi. *Mycological Research*, **106**: 1121-1122. Doi: 10.1017/S0953756202227033.

Hillyer, J.F. (2009). Transcription in mosquito hemocytes in response to pathogen exposure. *Journal of Biology. BioMed Central*, **8**: 51.

Holder, D.J., Kirkland, B.H., Lewis, M.W. and Keyhani, N.O. (2007). Surface characteristics of the entomopathogenic fungus *Beauveria (Cordyceps) bassiana*. *Society for General Microbiology*, **153**: 3448-3457.

Hung, S.-Y. and Boucias, D.G. (1992). Influence of *Beauveria bassiana* on the cellular defense response of the beet armyworm, *Spodoptera exigua*. *Journal of Invertebrate Pathology*, **60**: 152-158.

Jackson, M.A. and Jaronski, S.T. (2009). Production of microsclerotia of the fungal entomopathogen *Metarhizium anisopliae* and their potential for use as a biocontrol agent for soil–inhabiting insects. *Micological Research*, **113**: 842-850.

Joshi, L. and St. Leger, R.J. (1999). Cloning, expression, and substrate specificity of mecpa, a zinc carboxypeptidase that is secreted into infected tissues by the fungal entomopathogen *Metarhizium anisopliae*. *The Journal of Biological Chemistry,* **274(14)**: 9803-9811.

Kersahw, M.J., Moorhouse, E.R., Bateman, R, Reynolds, S.E. and Charnley, A.K. (1999). The role of destruxins in the pathogenicity of *Metarhizium anisopliae* for three species of insects. *Journal of Invertebrate Pathology*, **74**: 213-223.

Khachatourians, G.G. (1992). Virulence of five *Beauveria bassiana* strains, *Paecilomyces farinosus*, and *Verticillium lecanii* against the migratory grasshopper, *Melanoplus sanguinipes*. *Journal of Invertebrate Pathology*, **52**: 212-214.

Lee, H.S., Cho, M.Y., Lee, K.M., Kwon, T.H., Homma, K., Natori, S. and Lee, B.L. (1999). The pro-phenoloxidase of coleopteran insect, *Tenebrio molitor,* larvae was activated during cell clump/cell adhesion of insect celular defense reactions. *Federation of European Biochemical Societies Letters*, **444**: 255-259.

Lord, J.C. (2001a). Desiccant dusts synergize the effect of *Beauveria bassiana* (Hyphomycetes: Moniliales) on stored-grain beetles. *Journal of Economic Entomology*, **94(2)**: 367-372. (a).

Lord, J.C. (2001b). Response of the wasp *Cephalonomia tarsalis* (*Hymenoptera*: Bethylidae) to *Beauveria bassiana* (Hyphomycetes: Moniliales) as free conidia or infection in its host, the sawtoothed grain beetle, *Oryzaephilus surinamensis* (Coleoptera: Silvanidae). *Biological Control*, **21**: 300-304.

Luz, C., Rocha, L.F.N., Nery, G.V., Magalhaes, B.P. and Tigano, M.S. (2004). Activity of oil-formulated *Beauveria bassiana* against *Triatoma sordida* in peridomestic areas in Central Brazil. *Memòrias do Instituto Oswaldo Cruz*, **99(2)**: 201-218.

Neuvéglise, C. Brygoo, Y. and Riba, G. (1997). 28s rDNA group-I introns: a powerful tool for identifying strains of *Beauveria brongniartii*. *Molecular Ecology*, **6**: 376-381.

Parker, B.L. Skinner, M., Costa, S.D., Gouli, S., Reid, W. and Bouhssini, M.E. (2003). Entomopathogenic fungi of *Eurygaster integriceps* Puton (Hemiptera: Scutelleridae): collection and characterization for development. *Biological Control*, **27**: 260-272.

Screen, S.E. and St. Leger, R.J. (2000). Cloning, expression, and substrate specificity of a fungal chymotrypsin. *The Journal of Biological Chemistry*, **275(9)**: 6689-6694.

Shan, P.A. and Pell, J.K. (2003). Entomopathogenic fungi as biological control agents. *Applied Microbiology and Biotechnology*, **61**: 413-423.

Siebeneicher, S.R., Vinson, S.B. and Kenerley, C.M. (1992). Infection of the red imported fire ant by *Beauveria bassiana* through various routes of exposure. *Journal of Invertebrate Pathology*, **59**: 280-285.

Small, N.C.-L. and Bidochka, M.J. (2005). Up-regulation of Pr1, a subtilisin-like protease, during conidiation in the insect pathogen *Metarhizium anisopliae*. *Mycological Research*, **109(3)**: 307-313.

Souza Azevedo, A.C., Sosa-Gomez, D.R., Rodrigues Faria, M. and Pelegrinelli Fungaro, M.H. (2000). Effects of double stranded RNA on virulence of *Paecilomyces fumosoroseus* (Deuteromycotina: Hyphomycetes) against the silverleaf whitefly *Bemisia tabaci* strain B (Homoptera: Aleyrodidae). *Genetics and Molecular Biology*, **23(1)**: doi: 10.1590/S1415-47572000000100010.

St. Leger, R.J., Charnley, A.K. and Cooper, R.M. (1987). Characterization of cuticle-degrading proteases produced by the entomopathogenic *Metarhizium anisopliae*. *Archives of Biochemistry and Biophysics*, **253**: 221-232.

St. Leger, R.J., Charnley, A.K. and Cooper, R.M. (1986a). Cuticle-degrading enzymes of entomopathogenic fungi: mechanisms of interaction between pathogen enzymes and insect cuticle. *Journal of Invertebrate Pathology*, **47**: 295-302.

St. Leger, R.J., Charnley, A.K. and Cooper, R.M. (1986b). Cuticle-degrading enzymes of entomopathogenic fungi: synthesis in culture on cuticle. *Journal of Invertebrate Pathology*, **48**: 85-95.

St. Leger, R.J., Cooper, R.M. and Charnley, A.K. (1986c). Cuticle-degrading enzymes of entomopathogenic fungi: regulation of production of chitinolytic enzymes. *Journal of General Microbiology*, **132**: 1509-1517.

St. Leger, R.J., Cooper, R.M. and Charnley, A.K. (1988). The effect of melanization of *Manduca sexta* cuticle on growth and infection by *Metarhizium anisopliae*. *Journal of Invertebrate Pathology*, **52**: 459-470.

St. Leger, R.J., Joshi, L., Bidochka, M.J., Rizzo, N.W. and Roberts D.W. (1996a). Characterization and ultrastructural localization of chitinases from *Metarhizium anisopliae*, *M. flavoviride*, and *Beauveria bassiana* during fungal invasion of host (*Manduca sexta*) cuticle. *Applied and Enviromental Micobiology*, **62(3)**: 907-912.

St. Leger, R.J., Joshi, L., Bidochka, M.J., Rizzo, N.W. and Roberts, D.W. (1996b). Biochemical characterization and ultrastructural localization of two extracellular trypsins produced by *Metarhizium anisopliae* in infected insect cuticles. *Applied and Enviromental Micobiology*, **62(4)**: 1257-1264.

St. Leger, R.J., Joshi, L. and Roberts, D.W. (1997). Adaptation of proteases and carbohydrases of saprophytic, phytopathogenic and entomopathogenic fungi to the requirements of their ecological niches. *Microbiology*, **143**: 1983-1992.

St. Leger, R.J., Joshi, L. and Roberts, D.W. (1998). Ambient pH is a major determinant in the expression of cuticle-degrading enzymes and hydrophobin by *Metarhizium anisopliae*. *Applied and Enviromental Micobiology*, **64(2)**: 709-713.

St. Leger, R.J., May, B., Allee, L.L., Frank, D.C., Staples, R.C. and Roberts, D.W. (1992). Genetic differences in allozymes and in formation of infection structures among isolates of the entomopathogenic fungus *Metarhizium anisopliae*. *Journal of Invertebrate Pathology*, **60**: 89-101.

St. Leger, R.J., Nelson, J. O. and Screen, S.E. (1999). The entomopathogenic fungus *Metarhizium anisopliae* alters ambient pH, allowing extracellular protease production and activity. *Microbiology*, **145**: 2691-2699.

Sung, G-H., Hywel-Jones. N.L., Sung, J.-M., Luangsa-ard, J.J., Shreshtha, B. and Spatafora, J.W. (2007). Phylogenetic classification of *Cordyceps* and the clavicipitaceous fungi. *Studies in Mycology* **57**: 5-59. (Available at *http://www.studiesinmycology.org*).

Ugine, T.A., Wraight, S.P. and Sanderson, J.P. (2005). Acquisition of lethal doses of *Beauveria bassiana* conidia by western flower thrips, *Frankliniella occidentalis*, exposed to foliar spray residues of formulated and unformulated conidia. *Journal of Invertebrate Pathology*, **90**: 10-23.

Vandenberg, J.D. and Cantone, F.A. (2004). Effect of serial transfer of three strains of *Paecilomyces fumosoroseus* on growth *in vitro*, virulence, and host specificity. *Journal of Invertebrate Pathology*, **85(1)**: 40-45.

Vega, F.E., Posada, F., Aime, M.C., Pava- Ripoll, M., Infante, F. and Rehner, S.A. (2008). Entomopathogenic fungal endophytes. *Biological Control*. **46**: 72-82.

Wagner, B.L. and Lewis, L.C. (2000). Colonization of *Zea mays*, by the entomopathogenic fungus *Beauveria bassiana*. *Applied and Environmental Microbiology*, **66(8)**: 3468-3473.

Wahlman, M. and Davidson, B.S. (1993). New destruxins from the entomopathogenic fungus *Metarhizium anisopliae*. *Journal of Natural Products*, **56(4)**: 643-647.

Wang, C., Skrobek, A. and Butt, T.M. (2004). Investigations on the destruxin production of the entomopathognic fungus *Metarhizium anisopliae*. *Journal of Invertebrate Pathology*, **85**: 168-174.

{4}

Recent Research Trends in the Use of Predators for Biological Control

Jesusa Crisostomo Legaspi, Benjamin C. Legaspi, Jr. and Alvin M. Simmons

ABSTRACT

We discuss four recent trends in the study of predators in biological control: 1) Intraguild predation (IGP) – IGP is the "killing and eating of species that use similar resources and are thus potential competitors". IGP effects can result in a multitude of unexpected and counter-intuitive effects. Multiple species released where predation is the dominant interaction may result in both increased costs and decreased control. 2) Predator movement – The release of natural enemies assumes dispersal towards target pests. However, sometimes predators are unable or unwilling to disperse, as in the case of refuge crops with very favorable conditions. Techniques have been developed for monitoring and quantifying insect movement. Precise knowledge of the movements of predators and the effects of agronomic practices are useful for optimizing biological control; 3) Genetically modified insects – Scientists are rightfully proceeding with caution in the use of transgenic insects as biological control agents because some actions may be irreversible. Complexities in engineering more efficient natural enemies may serve as a useful mechanism to slow scientific progress while the necessary legal and regulatory controls are being formulated. 4) Climate change – Global warming can produce distribution shifts in plants, herbivores and predators. Species expansion may lead to escape from historical natural enemies. The coupling or uncoupling of control mechanisms may affect ranges and abundance. Many predict that global warming will result in an increase in the frequency and intensity of pest outbreaks. Recent scientific developments ensure that the study of predators in biological control will continue to thrive.

INTRODUCTION

The use of predators for biological pest control has a long history. An ancient cultured citrus ant, *Oecophylla smaragdina* F. (Hymenoptera: Formicidae), has been used in China for almost 1700 years to protect citrus fruits against damage by phytophagous pests. A nest would be secured on one tree where bamboo strips would serve as bridges connected to adjacent trees to facilitate ant movement across the orchard (Huang and Yang, 1987). Since its ancient origins, the use of biological control in general has undergone many changes due to advances in technology, as well as increasing need because of problems associated with traditional chemical pest control (reviewed by Gurr *et al.,* 2000). The use of biological control agents, parasitoids, predators and pathogens, is well-documented (*e.g.*, Helyer *et al.,* 2003; Hajek, 2004; Jervis, 2007; van Driesche *et al.,* 2008). Rather than attempting to duplicate existing excellent references, we will focus on some recent interesting research trends in biological control using predators. We will address the following four areas of current research:

1. Predation.
2. Analysis of predator movement.
3. Genetically modified insects, with emphasis on transgenic predators.
4. Possible effects of anthropogenic (man-made) climate change, especially global warming.

For each topic, we briefly present background information, and focus on implications in the study and use of predators for biological control. We realize the last two topics can be controversial because of social and political ramifications. However, we avoid political or ethical discussions and present a hopefully unbiased summary of the state of science.

INTRAGUILD PREDATION

Polis and Holt (1992) define intraguild predation (IGP) as the "killing and eating of species that use similar resources and are thus potential competitors". Here, the term "guild" refers to a group of organisms sharing a common food resource, and "predation" is used in a broad sense to include inter-specific interactions, such as parasitism (Polis *et al.,* 1989; Rosenheim *et al.,* 1995). The concept of IGP and recognition of its ubiquity is largely attributed to the works of Gary Polis and Robert Holt (Polis and Holt, 1992; Polis *et al.,* 1989). Within the context of biological control, the simplest case of IGP would be that of a three-species system consisting of an herbivorous pest, a parasitoid attacking the pest, and a predator feeding on both the herbivorous pest and the parasitoid (either adult parasitoids, or immature stages developing within a parasitized host) (Polis and Holt, 1992) (Fig. 4.1). In this system, the parasitoid is the intraguild prey (IG prey), because it is eaten by the predator (IG predator). In addition to predation, the IG predator competes with the IG prey for the herbivorous pest; therefore, the IG predator interacts with IG prey through both predation and competition. Early theoretical models of IGP were modifications of Lotka-Volterra equations with additional terms to express short-term effects of IGP (Holt and Polis, 1997).

Several general predictions were made based on models and IGP theory:

1. *IGP should be relatively rare in natural systems*: Stable 3-species equilibria are difficult to achieve. Very often, one of the predators is excluded. To achieve stable 3-species equilibrium, the IG prey should be superior at exploiting the shared resource. The IG prey should have the advantage of being a superior competitor for the shared resource, to offset the disadvantage of being eaten by the IG predator; whereas the IG predator should gain significantly from its consumption of the IG prey (Holt and Polis, 1997; Borer *et al.,* 2007). An IG predator superior at exploiting the common resource would result in exclusion of the IG prey (even without predation).
2. *IGP results in increases in the shared herbivore prey:* In a stable IG prey – resource system, addition of an IGP predator would result in an outbreak in the pest because of suppression of the IG prey. Conversely, removal of the top predator should ultimately result in depression of the shared resource which can be freely exploited by the IG prey (the superior competitor) (Polis and Holt, 1992). Analytical models reviewed by Rosenheim *et al.,* (1995) suggest that in cases of failure of biological control in IGP-based systems, predators are likely contributors. Vance-Chalcraft *et al.,* (2007) suggest that the occurrence and magnitude of pest suppression or release is dependent on the relative abilities of the predators to suppress the shared prey. Ordinary differential equations suggest that stage-specific predation can result in predator coexistence on a single prey. Under certain conditions, a predator may not persist without the presence of its competitor (de Roos *et al.,* 2008).

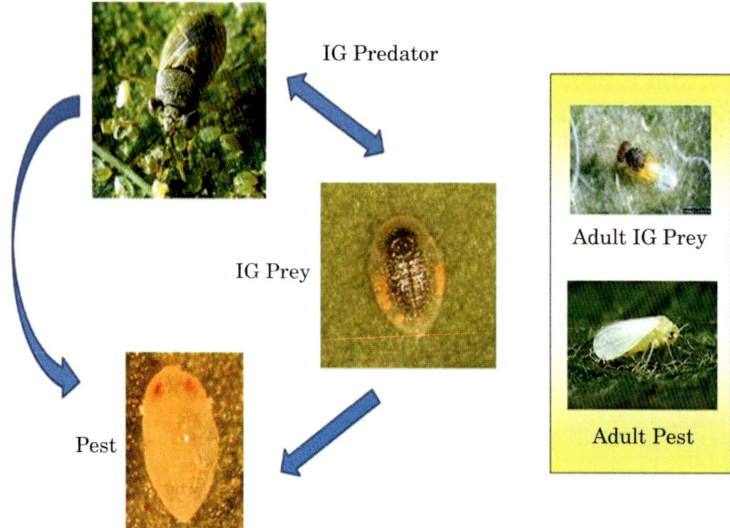

Fig. 4.1. *Intraguild predation*: IG Predator (*Geocoris punctipes*) (Photo by Jack Dykinga, USDA-ARS); IG Prey (*Encarsia* sp.) and shared resource pest (*Bemisia tabaci*). Adult stages of IG prey and pest are shown in inset (Adult *Encarsia* sp. Photo by David Cappaert, Michigan State University, East Lansing Michigan; adult *Bemisia* photo by Scott Bauer, USDA-ARS). The IG predator feeds on the IG prey as it develops within the host, as well as on the shared resource directly. The IG prey parasitizes the shared resource. Therefore, the IG predator engages the IG prey both as a predator and competitor for the shared resource pest. (Other photos by Ignacio Baez, USDA-ARS)

3. *Disruption of biological control is more likely in systems where predation is the dominant interaction*: The term "coincidental IGP" has been applied where the IG predator consumes the shared resource pest harboring developing parasitoids (Rosenheim and Harmon, 2006). "Omnivorous IGP" is used where the IG prey is consumed directly (*e.g.*, as adult parasitoids). Therefore, multispecies introductions should exclude strong IG predators so as to cause the strongest suppression of the target pest (Polis and Holt, 1992). An IG predator attacking an adult parasitoid, or another predator (IG prey) does not cause mortality in the resource pest. In contrast, IGP interactions between IG predators and immature parasitoids (or between two parasitoids) necessarily results in the death of the prey host, thereby strengthening biological control. High mortality of the IG prey has no effect on the pest, which can lead to a breakdown of biological control.

Empirical evidence collected over several years has been found to contradict some of these predictions.

1. *IGP is ubiquitous*: Although predicted to be unstable, IGP is commonly found throughout freshwater, marine and terrestrial food webs (Arim and Marquet, 2004; Polis and Holt, 1992; Polis *et al.,* 1989). The two critical factors determining the frequency and type of IGP within a food web are differences in relative body sizes and degrees of trophic specialization (Polis *et al.,* 1989). For example, in coccinellid beetles, larger individuals are more likely to attack smaller ones (Pell *et al.,* 2008; Moser and Obrycki, 2009). IGP tends to occur where generalist predators are larger than IG prey or where there is decreased abundance of non-guild prey where hungry predators expand their diets to include guild members.
2. *IGP does not always disrupt biological control*: IGP interactions have resulted in variable effects on biological control. In some situations, disruption in biological control is attributed to IGP, as predicted by theory. For example, IGP by generalist predators, especially *Nabis* spp. (Hemiptera: Nabidae) and *Zelus* spp. (Hemiptera: Reduviidae), on lacewing larvae, *Chrysoperla carnea* (Stephens) (Neuroptera: Chrysopidae), was likely a factor in the disruption of biological control by lacewing larvae, feeding on the cotton aphid, *Aphis gossypii* (Glover) (Hemiptera: Aphididae) (Rosenheim and Wilhoit, 1993). Similar disruption is suggested in spider mites, *Tetranychus* spp. (Acari: Tetranychidae), that undergo periodic outbreaks in California cotton, despite the presence of at least four predators [*Galandromus occidentalis* (Nesbitt), (Acari: Phytoseiidae), *Frankliniella occidentalis* (Pergande), (Thysanoptera: Thripidae), *Orius tristicolor* (White), (Hemiptera: Anthocoridae) and *Geocoris* spp. (Hemiptera: Lygaeidae)], each with demonstrated ability to suppress mite populations (Rosenheim *et al.*, 2005). However, other studies have suggested that IGP between a superior IG predator often is not detrimental to biological control of the target prey as predicted by theory, but instead may slightly enhance it (Royer *et al.,* 2008) and that experimental removal of IG predators does not always result in predicted decrease in the shared resource (Janssen *et al.,* 2006; Rosenheim and Harmon, 2006).

Many reasons have been proposed for the apparent discrepancies between theoretical predictions and empirical observations (Janssen *et al.,* 2006):

1. *Time scales* – Predictions based on equilibrium conditions assume time scales sufficient in duration for the system to reach

equilibrium. In contrast, experimental data and empirical observations are conducted over much shorter durations and may reflect transient conditions. Comparisons of systems with and without IGP may produce misleading results because observations are made when population cycles are out-of-phase in the short-term (Janssen *et al.*, 2006; Rosenheim and Harmon, 2006). However, data from short-term experiments may still prove useful in understanding equilibrium conditions if predator and prey densities used approximate those in natural equilibrium conditions (Vance-Chalcraft *et al.*, 2007).

2. *Spatial scales* – Models and theory assume homogenous and mixed populations, in contrast to actual populations comprised of patches of interacting species and a mosaic of shared resources. Experimental design may affect degree of homogeneity in sampled populations. Small arenas may have well-mixed populations, whereas large field sites may have patches of plants and insects (Janssen *et al.*, 2006). Increased spatial heterogeneity can result in suppression of a prey resource even in the presence of multiple IG predators by providing increased refuges for IG prey thereby reducing predation loss due to IG predators (Denno and Finke, 2006). Mathematical analysis of the effects of spatial heterogeneity on IGP systems is given by Bampfylde and Lewis (2007).
3. *Food web complexity* – Quite often, models assume simple 3-species interactions. Real-life guilds often have multiple asymmetric trophic interactions (Janssen *et al.*, 2006), including multiple resource species (Okuyama, 2009). Some empirical evidence suggests that increased complexity of food webs reduces the favorable effects of IGP in pest suppression (Janssen *et al.*, 2006). However, detailed studies in a natural ecosystem, *Spartina* cordgrass, and its arthropod community demonstrated that food-web connectance (fraction of feeding links) and not predator diversity *per se* resulted in decreased prey suppression by the predator complex (Denno and Finke, 2006).
4. *Behavioral responses* – IG prey can reduce predation pressure by avoiding IG predators. For example, the predatory mite *Neoseiulus cucumeris* (Oudemans) (Acari: Phytoseiidae) (IG prey) and heteropteran bug, *Orius laevigatus* (Fieber) (IG predator) are used as biological control agents against western flower thrips, *F. occidentalis*, where *Orius* also feeds on *N. cucumeris*. IG prey recognized IG predators from a distance based on their diet. IG prey avoided odors of IG predators that had consumed shared prey thrips, but did not avoid odors of IG predators that had fed on

other IG prey (Magalhães *et al.,* 2005). IG prey may also employ tactics to avoid IG predation. *Harmonia axyridis* (Pallas) (Coleoptera: Coccinellidae) larvae reflexively exude hemolymph ("reflex bleeding"), to avoid IGP by *Coccinella septempunctata brucki* Mulsant (Coleoptera: Coccinelidae) (Sato *et al.,* 2009). IG prey may exhibit escape behaviors from IG predators (Polis *et al.,* 1989).

One prediction that seems to be supported by empirical evidence is that "coincidental IGP" is less disruptive to biological control than "omnivorous IGP", *i.e.*, guilds where predation are the dominant relationships tend to disrupt biological control (Denno and Finke, 2006). Meta-analysis (statistical analysis across multiple datasets) revealed that inclusion of a coincidental IG predator significantly enhances biological control, at least in the short term; whereas inclusion of an omnivorous IG predator had little effect (Rosenheim and Harmon, 2006) – an increase in pest densities might have been more consistent with theory. Meta-analysis by Vance-Chalcraft *et al.,* (2007) showed that: 1) increased IGP resulted in decreased prey suppression; 2) IG prey was more effective at suppressing prey than the IG predator; and 3) effects of the IGP predator can result in release of the pest ("trophic cascades", Hairston *et al.,* 1960). Therefore, food webs dominated by predation are more likely to result in prey release and caution must be exercised in deploying multiple species of predators when strong predation interactions are likely, *i.e.* when the prospective control agents are likely to eat each other.

Critical to the eventual effect on the pest shared resource is IG predator's preference for consuming the IG prey rather than the herbivorous pest. In some cases, IG predators prefer parasitized hosts, as was reported in *Geocoris punctipes* (Say) (Hemiptera: Geocoridae), *Orius insidiosus* (Say) (Hemiptera: Anthocoridae) and *Hippodamia convergens* Guérin-Méneville (Coleoptera: Coccinellidae) feeding on *Bemisia tabaci* (Gennadius) (Hemiptera: Aleyrodidae) nymphs parasitized by *Eretmocerus* sp. nr. *emiratus* (Naranjo, 2007). IG predators may show no predation preference between parasitized or unparasitized whitefly nymphs as in the case of *Delphastus catalinae* (Horn) (=*D. pusillus*) (Coleoptera: Coccinellidae) feeding on *B. tabaci* and the parasitoid *Encarsia sophia* (Girault & Dodd) [= *E. transvena* (Timberlake)] (Hymenoptera: Aphelinidae) (Zang and Liu, 2007). The most common result is that avoidance of parasitized hosts increases with developmental stage of the parasitoid, as in of the aphidophagous hoverfly *Episyrphus balteatus* DeGeer (Diptera: Syrphidae) selecting prey *Acyrthosiphon pisum* Harris (Hemiptera: Aphididae), parasitized by *Aphidius ervi* Haliday (Hymenoptera: Aphidiidae) (Almohamad *et al.,* 2008).

Apart from effects on population densities, IGP may induce niche shifts in the IG prey to mitigate the effects of predation. Niche shifts may be induced to avoid predation from the IG predator through an evolutionary response, short-term behavioral avoidance or as a reflection of mortality (Polis *et al.,* 1989; Polis and Holt, 1992). Niche shifts are common and often result in the IG prey being displaced into less favorable times and habitats. For example, survival of the coccinellid predators *Coleomegilla maculata* (DeGeer) and *Coccinella septempunctata* (L.) feeding on the aphid, *A. pisum,* declined in the presence of *H. axyridis* (Moser and Obrycki, 2009). This incompatibility is a possible reason for the observed spatial and temporal separations among these species that have been observed in the field.

In summary, perhaps there is not as much discrepancy in IGP studies as it might appear. Meta-analyses as performed by Rosenheim and Harmon (2006) and Vance-Chalcraft *et al.,* (2007) suggest that fundamental underlying principles may hold regardless of the complexities with different IGP assemblages. Specific predatory relationships and their strengths will give different outcomes and each assemblage may be unique. Clearly, the study of IGP has important ramifications for applied biological control. Predictions by theory are not always matched by empirical observations for several reasons. Because each system is different, IGP effects can result in a multitude of unexpected and counter-intuitive effects. However, IGP does not always result in the disruption of biological control; it may also result in enhanced control. The outcomes in given assemblages appear to be specific not only to species, but also to temporal and spatial conditions. Perhaps the most important implications of IGP are in the area of multiple species releases for biological control. In cases where predation effects are strong, practitioners must be careful because the deployment of IG predators may result not only in increased costs, but actually in decreased control.

ANALYSIS OF PREDATOR MOVEMENT

In agricultural systems, the success of biological control agents may depend on the movement of pests and natural enemies, their relations to each other, and to the underlying cropping patterns. Natural enemies may be deployed using several methods (van Driesche and Bellows, 2001). Agents may be released by hand, into field cages or liberated freely, using mechanical systems from either the ground or air, or using "banker" plants (containing populations of reproducing natural enemies; Frank, 2009). Timing of release is often made to synchronize with vulnerable stage of the host, or when the natural enemy is mobile or otherwise prepared to attack the target pest. Endemic natural enemies may also be promoted through "refuge crops" (which sustain beneficial insect populations during periods of adversity). Whether natural enemies are introduced or endemic populations are maintained, they will be ineffective if they do not disperse towards pests. While it is reasonable to assume that natural enemies will eventually move towards pests (Corbett, 1998), this is not always true (McLachlan and Wratten, 2003; Jervis *et al.,* 2004). Therefore, it is imperative that natural enemy movement be studied using techniques such as marking and tracking (Lavandero *et al.,* 2004a, 2004b). Here, we summarize recent efforts to investigate movement of predators towards pests.

When reporting population data, ecologists typically describe temporal population dynamics as statistical averages from static sampling points or as spatial distributions at given points in time. Data on abundance and distribution may represent different aspects of the same populations, but such static representations fail to capture important dynamics. To record movement data, organisms usually need to be marked, except in such cases as release of an agent not endemic to the release site. Techniques for marking insects have been described comprehensively (*e.g.*, Southwood and Henderson, 2000; Hagler and Jackson, 2001). Briefly, these methods include the use of tags (labels), mutilation, paint, ink, dust, dyes, pollen, genetic markers (*e.g.*, mutations), rare elements (*e.g.* rubidium chloride, RbCl), radioactive-isotopes, and more recently, proteins (*e.g.*, ELISA - enzyme-linked immunosorbent assay) and genetically modified insects (Hagler and Jackson, 2001). RbCl is the most commonly used trace element for insects (Hagler and Jackson, 2001). Rubidium replaces potassium in plant tissue when watered or applied as foliar applications. Natural enemies feeding on plant nectar or pollen thus become labeled (Scarratt *et al.,* 2008). In cropping systems using C_3 (*e.g.*, cotton, wheat, soybeans) and C_4 (*e.g.*, corn, sorghum) photosynthetic pathways, naturally-occurring carbon isotopes are ingested by herbivores and have been used to track movements of natural enemies such as coccinellid beetles (Prasifka and Heinz, 2004).

Quantitative techniques have been developed to analyze the complexities of movement data (Turchin, 1998). Furthermore, current computer hardware and software have facilitated the development of spatially-explicit models that may be used to simulate insect movement through crop matrices. A spatio-temporal model of the boll weevil, *Anthonomus grandis* Boheman (Coleoptera: Curculionidae) and its exotic parasitoid, *Cactolaccus grandis* (Burks) (Hymenoptera: Pteromalidae) was developed using integrodifference equations assuming dispersal by random diffusion (Legaspi *et al.,* 1998). Simulations showed predicted effects of different release strategies for the parasitoid. Corbett (1996) developed a general model for spatial dynamics of natural enemies in diversified agroecosystems. The proposed framework can be used to construct more complicated models for specific systems. A parameterized model was developed to simulate movement of the checkered beetle, *Thanasimus dubius* (F.) (Coleoptera: Cleridae), attacking the southern pine beetle, *Dendroctonus frontalis* Zimmermann (Coleoptera: Scolytidae) (Cronin *et al.,* 2000). Predator movement was best simulated using random diffusion and a fast- and slow-moving form of the insect, whereas simple diffusion was sufficient to model prey movement. Skirvin (2004) developed a template model to investigate the movement of predators as affected by plant architecture and canopy connectedness. Simulations on a simplified plant canopy and randomly searching predators showed that increased plant connectedness results in a greater search, but that more complex plant architectures lowers searching efficiency. The approach may be useful in studying the impact of canopy structure and predator search for sparsely distributed prey within complex plant canopies (Skirvinm, 2004).

Refuge crops can be important in biological control because of their role in providing natural enemies with shelter and nutritional requirements through pollen, nectar, water and other food sources, especially during adverse conditions (Bugg and Pickett, 1998; Ferro and McNeil, 1998; van Driesche and Bellows, 2001; Jervis *et al.,* 2004; Landis *et al.,* 2000). Floral attractants such as color, size, shape, and odor may be used to lure natural enemies to refuge crops (Kevan and Baker, 1999). The application of supplemental food, such as pollen, makes predators such as *Orius laevigatus* Fieber (Hemiptera: Anthocoridae) more likely to remain on host plants (Skirvin *et al.,* 2007). However, the proximity of natural enemies lured by volatiles or sheltered in refuge crops does not guarantee subsequent dispersal to economic crops of interest (McLachlan and Wratten, 2003; Jervis *et al.,* 2004). In the case of parasitoids, studies using marked insects have demonstrated both the feasibility of using markers to track natural enemy movement, as well as the migration of natural enemies from refuges to target crops. The movement of the *Eretmocerus eremicus* and *Encarsia* spp. (Hymenoptera: Aphelinidae) from overwintering refuges into adjacent crops of cantaloupe and cotton was studied in the desert agricultural region of southeastern California through the use of RbCl markers (Pickett *et al.,* 2004). Marking studies demonstrated that the refuge strips supported early-season parasitoids which later migrated into adjacent crops (Pickett *et al.,* 2004). RbCl marking was also used in New Zealand to measure movement of the parasitoid *Dolichogenidea tasmanica* (Cameron) (Hymenoptera: Braconidae) from flowering buckwheat, to attack larvae of the lightbrown apple moth, *Epiphyas postvittana* Walker (Lepidoptera: Tortricidae) (Scarrat *et al.,* 2008). Knowledge of parasitoid movements from floral resources may be useful in designing appropriate strategies for the deployment of flowering plants in conservation biological control (Scarrat *et al.,* 2008). In Australia, the movement of *Diadegma semiclausum* (Hellén) (Hymenoptera: Ichneumonidae) which parasitizes the diamondback moth, *Plutella xylostella* (L.) (Lepidoptera: Plutellidae), from refuge flowering crops to the *Brassica* crop was triggered by plowing the refuge (Schellhorn *et al.,* 2008). Field tests were performed to evaluate sunflower, *Helianthus annuus* L. (Asterales: Asteraceae), as refuge crops for predators, mostly *Orius insidiosus*, against thrips in bell pepper, *Capsicum annuum* L. (Solanales: Solanaceae) (var. 'Camelot') (Legaspi and Baez 2008) (Fig. 4.2). Both pest and predators remained in the sunflower, thereby mitigating its effectiveness as a refuge crop. Therefore, careful manipulation of planting schedules or management of refuge crops may be useful in influencing short-term migration patterns of the whitefly and its natural enemies.

Fig. 4.2. A) Field trials comparing different varieties of sunflower for attractiveness to *Orius* and thrips populations B) Sunflower used as refuge crop with bell peppers C) *Orius insidiosus* adult feeding on thrip (Photos by Ignacio Baez and Jesusa C. Legaspi, USDA-ARS)

Studies on the movement of predators as biological control agents have produced mixed results. Combinations of protein-marking enzyme-linked immunosorbent assays (ELISA) and a predator gut content ELISA can simultaneously monitor predator movement and feeding activity (Hagler and Naranjo, 2004). Under field conditions in central

Arizona, the ELISA techniques were used to monitor movement and feeding of commercially-purchased *Hippodamia convergens*. After 15 d, the beetles had dispersed from the release sites. A predator gut content ELISA showed significant predation on *B. tabaci* relative to native beetles. Dispersal of the predatory mite *Phytoseiulus persimilis* Athias-Henriot (Acari: Phytoseiidae) to control the phytophagous mite *Tetranychus urticae* Koch in greenhouse roses was facilitated by the use of a mechanical disperser or inter-plant bridges consisting of plastic flagging tape (Casey and Parrella, 2005). In contrast, the predatory mite *Amblyseius swirskii* Athias-Henriot (Acari: Phytoseiidae) did not disperse far from release plants in potted greenhouse chrysanthemums, whether in the presence or absence of prey thrips, *Frankliniella occidentalis* (Buitenhuis *et al.,* 2009).

Cropping patterns can have a significant effect on movement of natural enemies through adjacent fields. The coccinellid predators *H. convergens* and *Scymnus loweii* Mulsant, marked by RbCl sprays, moved preferentially into cotton from grain sorghum, *Sorghum bicolor* (L.), possibly because of crop phenology and densities of aphid prey (Prasifka *et al.,* 2004). In a commercial crop of hay lucerne (alfalfa), *Medicago sativa* L., unharvested vegetation was used as a refuge crop for natural enemies of *Helicoverpa* spp. (Leipdoptera: Noctuidae) (Hossain *et al.,* 2002). Harvesting triggered movement of natural enemies such as coccinellids, syrphids, carabids and spiders, as well as the pest. In that system, refuge strips were effective for natural enemies when they were spaced <30m apart.

In summary, the release of natural enemies such as predators assumes dispersal towards target pests. However, in some cases, predators are unable to disperse, or unwilling to do so, as in the case of refuge crops with very favorable conditions. Precise knowledge of the movements of predators, and the effects of agronomic practices such as harvest and cropping patterns are useful for optimal biological control. On a cautionary note, Rand *et al.,* (2006) warn of the largely unknown, and possibly disruptive, effects caused by the movement of natural enemies into adjacent natural ecosystems.

GENETICALLY MODIFIED INSECTS, WITH EMPHASIS ON TRANSGENIC PREDATORS

The term "genetic engineering" encompasses a multitude of technologies all built on the premise that DNA is a resource that can be manipulated to achieve desired goals in pure and applied science and medicine (Nicholl, 2008). Transgenic organisms represent a subset of genetically modified organisms in that often the DNA inserted originated from another species. The term "paratransgenisis" refers to genetic modification of microbes in the insect gut or reproductive system (*e.g.*, Bextine *et al.,* 2005) because the insect itself is not modified. Genetic engineering became possible in the late 1960s because of two key discoveries: isolation of the enzyme DNA ligase in 1967 that allowed scientists to join DNA segments, and isolation of restriction enzymes in the 1970s that allowed splicing DNA at desired points. The first recombinant DNA bacterium was created at Stanford University in 1973 in the form of *Escherichia coli* expressing a *Salmonella* gene (Cohen *et al.,* 1973). Others have reviewed transgenic plants and their compatibility with biological control or their effects on non-target organisms (Legaspi *et al.,* 2004; Lundgren *et al.,* 2009; Lövei *et al.,* 2009; O'Callaghan *et al.,* 2005; Sanvido *et al.,* 2007). Here, we address the topic of transgenic insects with emphasis on predators as biological control agents.

Techniques for genetic engineering of insects have been reviewed extensively (*e.g.*, Atkinson *et al.,* 2001; Handler and O'Brochta, 1991; Handler, 2000; Handler and James, 2000; Hoy, 1996, 2003a). The first successful heritable germ-line transformations of insects occurred mostly in the vinegar fly, *Drosophila melanogaster* Meigen (Diptera: Drosophilidae) in 1982 (Rubin and Spradlin, 1983). Successful transformations for other insects were achieved for another 13 years (Handler, 2000) (Fig. 4.3).

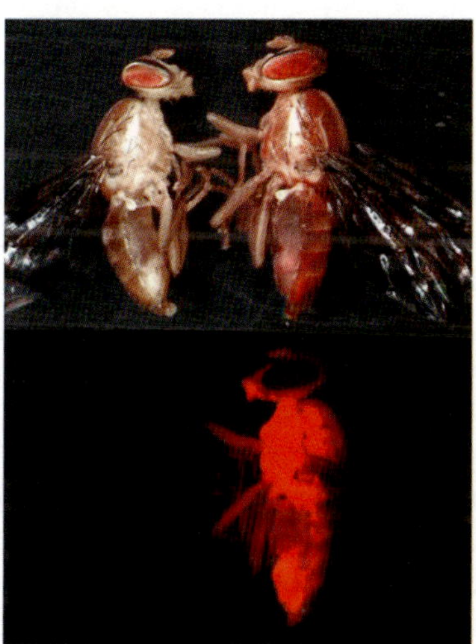

Fig. 4.3. Non-transgenic (left) and transgenic (right) *Anastrepha suspensa* (Loew) (Diptera: Tephritidae) transformed with a piggyBac-transposon vector marked with polyubiquitin regulated DsRed fluorescent protein (pBac{PUb-DsRed1} under brightfield (top) and Texas Red epifluorescent (bottom) optics (courtesy of A.M. Handler, USDA-ARS)

Transformation with recombinant DNA has two advantages relative to conventional breeding: 1) genetic improvement can be achieved rapidly; and 2) scientists are not constrained to traits or genes within a limited gene pool (Harrison and Bonning, 2000). Drawbacks may include the constraint that the traits be determined by single genes; inbreeding and loss of genetic variability, and instability of genetic material (Handler, 2001; Hoy, 2000a, 2000b). Significant transgenic insects include the following:

- Mosquitoes and other vectors of human and animal diseases so that they cannot transmit the targeted diseases (Ito *et al.,* 2002; Nirmala and James, 2003; Scott *et al.,* 2008).
- Strains for Sterile Insect Technique – improved strains of the New World screw-worm, *Cochliomyia hominivorax* (Coquerel) (Diptera: Calliphoridae) (Handler *et al.,* 2009) and tephritid fruit flies (Diptera: Tephritidae) (Scolari *et al.,* 2008; Zimowska *et al.,* 2009; Schetelig *et al.,* 2009); fluorescence marker gene in pink bollworm, *Pectinophora gossypiella* Saunders (Lepidoptera: Gelechiidae) (Miller *et al.,* 2007, 2008).
- Silk moths capable of producing new types of silk (Wen *et al.,* 2009).
- Insects capable of producing new drugs and vaccines (Tomita *et al.,* 2003; Wurm 2003).
- Enhanced natural enemies (Hoy, 2000a, 2000b, 2003a, 2003b).

The potential risks involved in the deployment of genetically-engineered insects need to be addressed carefully (Hoy, 1996, 2000a, 200b, 2003a, 2003b, PIFB, 2004). Potential risks will include instability in the modified gene or trait; possible horizontal transfer to other organisms (especially in case of pesticide or antibiotics resistance though these possibilities have been ameliorated to a large extent by new stabilized vectors; Handler *et al.,* 2004); host or prey specificity; unexpected performance in geographic area of deployment; unintended environmental effects; and difficulties in recovery from field sites (Hoy, 2003b). Risks with transgenic insects may be higher than those for plants because transgenic crops need to be agricultural inputs and maintenance, whereas insects may potentially establish independent of human intervention. However, new technologies have been suggested to ensure that transgenic natural enemies cannot persist indefinitely, thereby mitigating concerns about wide scale deployment (Meyer *et al.,* 2006).

From the initial work on *Drosophila*, genetic transformation is being pursued for diverse groups of insects (PIFB, 2004). Genetic engineering of insects as natural enemies was reviewed by Harrison and Bonning (2000). Potentially useful genes and genetic regulatory elements are being identified for pest insects (Hoy, 1996), although less effort is being expended on the identification of genes to improve the efficacy of beneficial insects (Hoy, 2000a). In theory at least, genetically improved biological control agents may consist of predators and parasitoids engineered for resistance to insecticides, environmental hardiness, increased longevity and fecundity, decreased dormancy periods, optimal sex ratios, improved host search capabilities, or other traits (Asokan, 2007; PIFB, 2004). Genetic improvement of natural enemies will be most promising under the following conditions: 1) the natural enemy is potentially effective, except for one limiting factor; 2) the limiting trait is controlled primarily by a single gene; 3) the gene can be obtained through selection, mutagenesis or cloning (Hoy, 1996). After release the strains should be fit and effective and preferably subject to reproductive isolation. The most promising trait for natural enemy improvement is therefore pesticide resistance (Hoy 1996). Prospects for other desired traits will be more difficult because the genetic basis for traits such as fecundity, sex ratio, longevity and diapauses are not as well known (Whitten and Hoy, 1999).

The most significant genetically modified predator is undoubtedly the western predatory mite, *Metaseiulus occidentalis* (Nesbitt) (Acari: Phytoseiidae). A gene delivery method called "maternal microinjection" was developed by Presnail and Hoy (1992) wherein genetic material was injected directly into gravid females (rather than eggs). The transgenic mite strain was released for field evaluation in 1996 after extensive safety reviews. As predicted, the transgenic mite did not persist in the field (Hoy, 2000a, 2000b). Maternal microinjection was also proved feasible in parasitoids, as in the case of *Cardiochiles diaphaniae* Marsh (Hymenoptera: Braconidae) (Presnail and Hoy, 1996). The work on transgenic *M. occidentalis* mite serves as a model for the transformation and deployment of transgenic natural enemies and may serve as a model for a "universal" DNA delivery system in arthropods (Hoy, 1996).

Scientists are rightfully proceeding with caution in the use of transgenic insects as biological control agents because some actions may be irreversible. The scientific complexities involved in engineering more efficient natural enemies may serve as a useful mechanism to slow scientific progress in this area while the necessary legal and regulatory controls are being formulated.

CLIMATIC CHANGE WITH EMPHASIS ON GLOBAL WARMING

In this discussion, we will accept the argument that anthropogenic climate change is a reality. We will discuss putative effects of climate change on insects in general, with an emphasis on global warming. We then focus on potential effects on predator-prey interactions, especially within the context of biological control. Global warming is the increase in the average temperature of the earth's surface and waters since the mid-20th century as a result of human activities, and its predicted continuing increase. According to the Intergovernmental Panel on Climate Change (IPCC), global surface temperature increased 0.74 ± 0.18 °C during the last century, mostly due to increasing concentrations of greenhouse gases resulting from fossil fuel burning and deforestation (IPCC 2007). Atmospheric CO_2 has increased by 31% over the past 200 years (Hance *et al.,* 2007). The reality of global warming has long been controversial and will continue to be debated by scientists, politicians and the general public for the foreseeable future (*e.g.*, Johansen, 2010). Here we are concerned only with research that addresses potential effects of global warming on predators as biological control agents.

The evidence that global warming affects biological organisms is quite convincing. In one study, 59% of 1598 species studied exhibited measurable changes to their phenologies or distributions over the past 20 to 140 years (Parmesan and Yohe, 2003). Species were included across diverse ecosystems (*e.g.*, temperate grasslands, marine intertidal zones, tropical cloud forests) from a wide variety of taxonomic groups (*e.g.*, birds, butterflies, alpine flowers, coral reefs). Thomas *et al.,* (2004) predicted that 15–37% of species in sample regions and taxa will be 'committed to extinction' based on climate-warming scenarios for 2050. Parmesan (2006) summarizes the following biotic effects of global warming:

- *Phenological changes* – Global warming can induce changes in the periodic life cycle events that are related to climate. The first appearance of butterflies within a given year is closely related to spring temperatures. Earlier first appearance dates have been documented in 26 of 35 species studied in the United Kingdom, and for all 17 species studied in Spain. In central California, 70% of 23 species studied have advanced date of first flight by an average of 24 days over the past 31 years.
- *Disruptions in synchrony across trophic levels* – The primary effect of climate change on many species is in effects on synchrony with prey or hosts and habitat resources. Timing is critical between the life cycle events of herbivores and their host plants, parasitoids or predators and their host insects, and insect pollinators and flowering plants. Few studies have addressed the possible effects of climate change on synchrony. However, 7 of 11 cases contained

evidence of increasing asynchrony resulting in decreased fitness (Visser and Both, 2005). Perhaps a counterexample is the parasitoid *Cotesia melitaearum* (Wilkinson) (Hymenoptera: Braconidae) which was brought in greater synchrony with its host *Melitaea cinxia* (L.) (Lepidoptera: Nymphalidae) because of warmer spring temperatures (van Nouhuys and Lei, 2004).

- *Shifts in species distribution ranges and local extinctions* – Warming trends are expected to show increasing species distributions towards the poles and towards higher elevations. Numerous studies have documented northward expansion of butterfly and moth ranges in Finland, Great Britain and across Europe. A study of Odonata in Great Britain revealed that 23 of 24 temperate species had expanded northern ranges by an average of 88 km. Populations of the butterflies *Euphydryas editha* (Boisduval) (Nyphalidae) and cool-adapted *Parnassius apollo* (L.) (Papilionidae) have shifted to higher elevations, with local extinctions at lower elevations (Parmesan, 1996). Predation rates by syrphid larvae decline rapidly with increasing latitude (Hodkinson, 1999). A large-scale study of 35 non-migratory European butterflies showed that 63% had shifted their ranges northward by 35 – 240 km since 1900 (Parmesan *et al.,* 1999).

The effects of global warming on the action of predators in biological control will be complex. Because predators occupy higher trophic levels, they are likely to be especially sensitive to changes in climate (Voigt *et al.,* 2003), because they depend on the lower levels to adapt to climatic changes (Hance *et al.,* 2007). Most reviews on the effects of global warming on natural enemies have focused on parasitoids (*e.g.*, Thomson *et al.,* 2010; Hance *et al.,* 2007; Stireman *et al.,* 2005), perhaps because the host-parasitoid synchrony is more coupled than that of the predator-prey and therefore leads to more profound or adverse consequences in the event of disruptions. Even within parasitoids, host-specific species are expected to suffer disproportionately relative to generalists when host populations fluctuate (Stireman *et al.,* 2005). Specialists are more vulnerable to missing windows of host vulnerability than generalists. Extension of this argument seems to support the contention that predators, especially the generalists, will be less impacted by global warming and other factors related to climate change because their biologies are not as strongly coupled with those of their prey.

Global warming, together with another component of climate change - increased levels of atmospheric CO_2 – can have potentially counterintuitive effects of predators which are mediated through plants and herbivores. Starting from the lowest trophic level, temperature increase can cause changes in plant phenology such as seed germination or breaking of dormancy (Thomson *et al.,* 2010). These changes can disrupt host plant – herbivore synchrony, to the detriment of the herbivore. Gypsy moth, *Lymantria dispar* L. (Lepidoptera: Lymantriidae) egg hatch before bud burst in oaks, *Quercus* spp, leads to starvation, while egg hatch too long afterwards leads to reduced foliage quality and reduced moth fecundity.

Increases in atmospheric CO_2 can also reduce the nutritional quality of plant material. To compensate, herbivores may increase predation rates leading to higher levels of plant damage. However, the reduced nutritional quality may also result in lower development rates, reduced fecundity and higher mortality. Reduced fitness in the herbivore trophic level results in declines in the fitness of natural enemy parasitoids and predators. Decreases in prey size do not always result in decreased success of natural enemies (Thomson *et al.,* 2010). The omnivorous bug, *Oechalia schellenbergii* Guérin-Méneville (Hemiptera: Pentatomidae) consumed more prey cotton bollworm, *Helicoverpa armigera* (Hübner) (Lepidoptera: Noctuidae) under elevated CO_2 levels. Host pea plants had reduced nitrogen levels, resulting in smaller herbivores, which were more easily subdued by the predator (Coll and Hughes, 2008). Thomson *et al.,* (2010) suggest that under many scenarios, increased CO_2 levels may actually increase the effectiveness of generalist predators. However, searching time to locate prey could be increased due to greater plant foliage that might have to be searched.

Increases in developmental time due to poorer quality plant nutrition may lengthen windows of herbivore susceptibility to predators. However, increases in developmental time of predators may also result in more severe outbreaks of herbivore pests. Increases in temperature can influence the production and release of volatile compounds (Thomson *et al.,* 2010). Parasitoids and predators use volatile plant cues to locate and evaluate hosts. Climatic changes, including higher temperatures, can cause slight but significant changes in the relative ratios of the volatiles emitted (Hance *et al.,* 2007). Generally, temperatures between 22 – 27°C are optimal for the emission of volatiles. Higher temperatures can result in increases in the number of insect generations by either increasing the developmental rate of herbivorous insects, or by prolonging the crop growing season. Subsequent effects on natural enemies of herbivores are unclear. Temperature differences can alter dynamics of predator-prey interactions. For example, reproduction of the pea aphid, *Acyrthosiphon pisum* exceeds predation rate by *Coccinella septempunctata* below 11°C, whereas the reverse is true at higher temperatures (Hance *et al.,* 2007).

Climatic changes will not affect all organisms interacting directly in the same way. A climate-controlled research facility, called the "Ecotron", is used to study effects of prolonged increases in CO_2 (220 ppm above ambient) and temperature (2°C above ambient) on ecosystems (Fig. 4.4). In a 4-year study, above-ground changes in flora and fauna were minimal, but major effects were found in soil micro-arthropods (*i.e.* Collembola) (Lawton *et al.,* 1993). Harmon *et al.,* (2009) induced heat shocks in field populations of *Acyrthosiphon pisum* and the predators *Coccinella septempunctata* and *Harmonia axyridis*. Heat shocks are short periods of high temperatures that can result in reduced fecundity in sensitive strains. The predators differed in behavioral responses to aphid abundance. As a result, population density of *C. septempunctata* strongly correlates with that of the aphid, whereas *H. axyridis* does not. As predicted, population decline of the aphid induced

Fig. 4.4. NERC – CPB Ecotron: The Natural Environment Research Council - Centre for Population Biology at Silwood Park in the United Kingdom uses a controlled environment facility called the "Ecotron" to study the effects of climate change on ecological communities. The facility consists of 16 separate chambers and was operational in July 1991. (Photos courtesy of T. Hefin Jones, University of Cardiff, Wales)

by heat shock was reduced additively by predation from *C. septempunctata*, whereas reduced predation from *H. axyridis* mitigated the decline. Clones with bacterial endosymbionts were also exposed to ambient of heat shock treatments to demonstrate potential for rapid evolution for heat shock tolerance. Such non-linear species-specific responses to climate result in complicated ecological and evolutionary responses to climate change.Global warming and other climate changes can also produce distribution shifts in the complex of plants, herbivores and predators. The outcomes of such distribution shifts will be dependent on the ability of the herbivores to track plant distribution shifts, and the ability of parasitoids and predators to track the shifts in host or prey species. Subsequent insect redistribution will be dependent on factors such as mobility and climatic tolerances. Expansion of herbivores may lead to escape from historical natural enemies, although they may encounter new association natural enemies in the newly colonized range. The uncoupling or coupling of inter-specific interactions may have significant effects of species ranges and abundance (Davis *et al.,* 1998).

The northward expansion of the butterfly *Aricia agestis* Denis & Schiffermüller (Lepidoptera: Lycaenidae) due to global warming resulted in lower parasitism rates (Menendez *et al.,* 2008). Citing the success of invasive ladybeetles, *Harmonia axyridis*, Molina-Montenegro *et al.,* (2009) suggest that invasive predators that expand distribution ranges as a result of global warming may be more efficient in food search than native counterparts. Although predictions and generalizations are difficult to make, many workers predict that global warming will result in an increase in the frequency and intensity of pest outbreaks (*e.g.*, Stireman *et al.,* 2005).

CONCLUSIONS

The use of predators as biological control agents has certainly come a long way since the ancient Chinese used ants to control citrus pests. The practice of using predators continues to evolve because of new and exciting trends in science and technology – either directly because of recent developments, as in transgenic predators and climate change; or as applications of new technology to old ideas, as in insect movement from release points or refuge crops; or in increased appreciation of the complexity of ecological systems, as in the study of intraguild predation webs. The trends we selected for review herein reflect our interests, although other recent developments could warrant attention, among them: habitat manipulation, bioclimatic and spatial modeling, multi-trophic analyses, and invasion theory. The relentless pressures of human population increase and demand for food, coupled with increasing demands for environmentally-friendly and sustainable agriculture will ensure that the theory and application of biological control will continue to produce critical and exciting science.

ACKNOWLEDGEMENTS

We are grateful for comments kindly provided by T. Hefin Jones (Cardiff University, United Kingdom) and Al M. Handler (USDA-ARS Center for Medical, Agricultural and Veterinary Entomology, Gainesville, Florida, USA). This article presents the results of research only. The use of trade, firm, or corporation names in this publication is for the information and convenience of the reader. Such use does not constitute an official endorsement or approval of the United States Department of Agriculture or the Agricultural Research Service of any product or service to the exclusion of others that may be suitable.

REFERENCES

Almohamad, R., Verheggen, F.J., Francis, F., Hance, T. and Haubruge, E. (2008). Discrimination of parasitized aphids by a hoverfly predator: effects on larval performance, foraging, and oviposition behavior. *Entomologia Experimentalis et Applicata*, **128**: 73-80.

Arim, M. and Margquet, P.A. (2004). Intraguild predation: a widespread interaction related to species biology. *Ecology Letters*, **7**: 557-564.

Asokan, R. (2007). Genetic engineering of insects. *Resonance*, **12**: 47-56.

Atkinson, P.W., Pinkerton, A.C. and O'Brochta, D.A. (2001). Genetic transformation systems in insects. *Annual Review of Entomology*, **46**: 317-346.

Bampfylde, C.J. and Lewis, M.A. (2007). Biological control through intraguild predation: case studies in pest control, invasive species and range expansion. *Bulletin of Mathematical Biology*, **69**: 1031-1066.

Bextine, B., Lampe, D., Lauzon, C., Jackson, B. and Miller, T.A. (2005). Establishment of a genetically marked insect-derived symbiont in multiple host plants. *Current Microbiology*, **50**: 1-7.

Borer, E.T., Briggs, C.J. and Holt, R.D. (2007). Predators, parasitoids, and pathogens: a cross-cutting examination of intraguild predation theory. *Ecology*, **88**: 2681-2688.

Bugg, R.L. and Pickett, C.H. (1998). Introduction: enhancing biological – habitat management to promote natural enemies of agricultural pests. *In*: Pickett, C.H. and Bugg, R.L. (*Eds.*), *Enhancing biological control: habitation management to promote natural enemies of agricultural pests.* Univ. California Press, Berkeley, California, pp. 1-23.

Buitenhuis, R., Shipp, L. and Scott-Dupree, C. (2010). Dispersal of *Amblyseius swirskii* Athias-Henriot (Acari: Phytoseiidae) on potted greenhouse chrysanthemum. *Biological Control,* **52**: 110-114.

Casey, C.A. and Parrella, M.P. (2005). Evaluation of a mechanical dispenser and interplant bridges on the dispersal and efficacy of the predator, *Phytoseiulus persimilis* (Acari) Phytoseiidae) in greenhouse cut roses. *Biological Control,* **32**: 130-136.

Cohen, S.N., Chang, A.C.Y., Boyer, H.W. and Helling, R.B. (1973). Construction of biologically functional bacterial plasmids *in vitro*. *Proceedings of the National Academy of Sciences of the United States of America,* **70**: 3240-3244.

Coll, M. and Hughes, L. (2008). Effects of elevated CO_2 on an insect omnivore: a test for nutritional effects mediated by host plants and prey. *Agriculture, Ecosystems and Environment,* **123**: 271-279.

Corbett, A. (1998). The importance of movement in the response of natural enemies to habitat manipulation. *In*: Pickett, C.H. and Bugg, R.L. (*Eds.*), *Enhancing biological control: habitat management to promote natural enemies of agricultural pests.* Univ. California Press, Berkeley, California, pp. 25-48.

Cronin, J.T., Reeve, J.D., Wilkens, R. and Turchin, P. (2000). The pattern and range of movement of a checkered beetle predator relative to its bark beetle prey. *Oikos*, **90**: 127-138.

Davis, A.J., Lawton, J.H., Shorrocks, B. and Jenkinson, L.S. (1998). Individualistic species responses invalidate simple physiological models of community dynamics under global environmental change. *Journal of Animal Ecology*, **67**: 600-612.

Denno, R.F. and Finke, D.I. (2006). Multiple predator interactions and food-web connectance: implications for biological control. *In*: Brodeur, J. and Boivin, G. (*Eds.*), *Trophic and guild interactions in biological control.* Springer, Dordrecht, The Netherlands, pp. 45-70.

De Roos, A.M., Schellekens, T., van Kooten, T. and Persson, L. (2008). Stage-specific predator species help each other to persist while competing for a single prey. *Proceedings of the National Academy of Sciences of the United States of America*, **105**: 13930-13935.

Ferro, D.N. and McNeil, J.N. (1998). Habitat enhancement and conservation of natural enemies of insects. *In*: Barbosa, P. (*Ed.*), *Conservation biological control.* Academic Press, New York, pp. 123-132.

Frank, S.D. (2010). Biological control of arthropod pests using banker plant systems: past progress and future directions. *Biological Control*, **52**: 8-16.

Gurr, G.M., Barlow, N.D., Memmott, J., Wratten, S.D. and Greathead, D.J. (2000). A history of methodological, theoretical and empirical approaches to biological control. *In*: Gurr, G. and Wratten, S. (*Eds.*), *Biological control: measures of success.* Kluwer Acad. Publ., The Netherlands, pp. 3-37.

Hagler, J.R. and Jackson, C.G. (2001). Methods for marking insects: current techniques and future prospects. *Annual Review of Entomology*, **46**: 511-543.

Hagler, J. and Naranjo, S. (2004). A multiple ELISA system for simultaneously monitoring intercrop movement and feeding activity of mass-released insect predators. *International Journal of Pest Management*, **50**: 199-207.

Hajek, A. (2004). *Natural enemies: an introduction to biological control.* Cambridge Univ. Press, United Kingdom.

Hairston N.G., Smith, F.E. and Slobodkin, L.B. (1960). Community structure, population control and competition. *American Naturalist*, **94**: 421-425.

Hance, T., van Baaren, J., Vernon, P. and Boivin, G. (2007). Impact of extreme temperatures on parasitoids in a climate change perspective. *Annual Review of Entomology*, **52**: 107-126.

Handler, A.M. (2001). A current perspective on insect gene transformation. *Insect Biochemistry and Molecular Biology*, **32**: 111-128.

Handler, A.M. and O'Brochta, D.A. (1991). Prospects for gene transformation in insects. *Annual Review of Entomology*, **36**: 159-183.

Handler, A.M. and James, A.A. (2000). *Insect transgensis: methods and applications*. CRC Press, Boca Raton, Florida.

Handler, A.M., Zimowska, G.J. and Horn, C. (2004). Post-integration stabilization of a transposon vector by terminal sequence deletion. *Nature Biotechnology*, **22**: 1150-1154.

Handler, A.M., Allen, M.L. and Skoda, S.R. (2009). Development and utilization of transgenic New World screwworm, *Cochliomyia hominivorax*. *Medical and Veterinary Entomology*, **23**: 98-105.

Harrison, R.L. and Bonning, B.C. (2000). Genetic engineering for biocontrol agents for insects. *In*: Rechcigl, J.E. and Rechcigl, N.A. (*Eds*.), *Biological and biotechnological control of insect pests*. CRC Press, Boca Raton, Florida, pp. 243-280.

Harmon, J.P., Moran, N.A. and Ives, A.R. (2009). Species response to environmental change: impacts of food web interactions and evolution. *Science*, **323**: 1347-1350.

Helyer, N., Brown, K. and Cattlin, N.D. (2003). *A color handbook of biological control in plant protection*. Timber Press, Portland, Oregon.

Hodkinson, I. (1999). Species response to global environmental change or why ecophysiological models are important: a reply to Davis *et al*., *Journal of Animal Ecology*, **68**: 1259-1262.

Holt, R.D. and Polis, G.A. (1997). A theoretical framework for intraguild predation. *American Naturalist*, **149**: 745-764.

Hossain, Z., Gurr, G.M., Wratten, S.D. and Raman, A. (2002). Habitat manipulation in lucerne *Medicago sativa*: arthropod population dynamics in harvested and 'refuge' crop strips. *Journal of Applied Ecology*, **39**: 445-454.

Hoy, M.A. (1996). Novel arthropod biological control agents. *In*: Persley, G.J. (*Ed*.), *Biotechnology and integrated pest management*. CAB International, Wallingford, United Kingdom, pp. 164-185.

Hoy, M.A. (2000a). Deploying transgenic arthropods in pest management programs: Risks and realities. *In*: Handler, A.M. and James, A.A. (*Eds*.), *Insect Transgenesis. Methods and Applications*. CRC Press, Boca Raton, Florida, pp. 335-367.

Hoy, M.A. (2000b). Transgenic arthropods for pest management programs: Risks and realities. *Experimental and Applied Acarology*, **24**: 463-495.

Hoy, M.A. (2003a). *Insect molecular genetics: an introduction to principles and applications*. Academic Press, San Diego, California.

Hoy, M.A. (2003b). Session on: transgenic insects for pest management programs: status and prospects. *Environmental Biosafety Research*, **1**: 1-4.

Huang, H.T. and Yang, P. (1987). Ancient cultured citrus ant used as biological control agent. *BioScience*, **37**: 665-671.

[IPCC] Intergovernmental Panel on Climate Change (2007). Summary for policymakers. *In*: Solomon, S., Qin, D., Manning, M., Chen, A., Marquis, M., Averyt, K.B., Tignor, M. and Miller, H.L. (*Eds*.), *Climate change 2007: the physical science basis. Contribution of working group I to the fourth assessment report of the Intergovernmental Panel on Climate Change*. Cambridge University Press, Cambridge, United Kingdom and New York, NY, USA.

Ito, J., Ghosh, A., Moreira, L.A., Wimmer, E.A. and Jacobs-Lorena, M. (2002). Transgenic mosquitoes impaired in transmission of a malaria parasite. *Nature*, **417**: 452-455.

Janssen, A., Montserrat, M., HilleResLambers, R., de Roos, A., Pallini, A. and Sabelis, M.W. (2006). Intraguild predation usually does not disrupt biological control. *In*: Brodeur, J. and Boivin, G. (*Eds.*), *Trophic and guild interactions in biological control.* Springer, Dordrecht, The Netherlands, pp. 21-44.

Jervis, M.A. (2007). *Insects as natural enemies: a practical perspective.* Springer, Dordrecht, The Netherlands.

Jervis, M.A., Lee, J.C. and Heimpel, G.E. (2004). Use of behavioural and life-history studies to understand the effects of habitat manipulation. Chap. 5 *In* Gurr, G.M., Wratten, S.D. and Altieri, M.A. (*Eds.*) *Ecological engineering for pest management*: *advances in habitat manipulation for arthropods.* Cornell University, Ithaca, New York. pp. 65-100.

Johansen, B.E. (2010). *Hard science and hot air*: *dissecting the global warming debate.* Potomac Books, Washington D.C.

Kevan, P.G. and Baker, H.G. (1999). Insects on flowers. *In*: Huffaker, C.B. and Gutierrez, A.P. (*Eds.*), *Ecological entomology.* Wiley & Sons, New York, pp. 553-584.

Kimura, K. (1997). The current status of transgenic research in insects. *In*: Nilsson, A. (*Ed.*). *Transgenic animals and food production: Proceedings from an International Workshop in Stockholm May 1997*, Royal Swedish Academy of Agriculture and Forestry, Stockholm.

Landis, D.A., Wratten, S.D. and Gurr, G.M. (2000). Habitat management to conserve natural enemies of arthropod pests in agriculture. *Annual Review of Entomology*, **45**: 175-201.

Lavandero, B., Wratten, S., Hagler, J. and Jervis, M. (2004a). The need for effective marking and tracking techniques for monitoring the movements of insect predators and parasitoids. *International Journal of Pest Management*, **50**: 147-151.

Lavandero, B., Wratten, S.D., Hagler, J. and Tylianakis, J. (2004b). Marking and tracking techniques for insect predators and parasitoids. *In*: Gurr, G.M., Wratten, S.D. and Altieri, M.A. (*Eds.*), *Ecological engineering for pest management*: *advances in habitat manipulation for arthropods.* Cornell University, Ithaca, New York, pp. 117-131.

Lawton, J.H., Naeem, S., Woodfin, R.M., Brown, V.K., Gange, A., Godfray, H.C.J., Heads, P.A., Lawler, S., Magda, D., Thomas, C.D., Thompson, L.J. and Young, S. (1993). The Ecotron: a controlled environmental facility for the investigation of population and ecosystem processes. *Philosophical Transactions*: *Biological Sciences*, **341**: 181-194.

Legaspi, J.C. and Baez, I. (2008). Intercropping sunflower varieties with bell pepper: effect on populations of *Orius insidiosus* (Hemiptera: Anthocoridae) and thrips. *Subtropical Plant Science*, **60**: 13-20.

Legaspi, Jr., B.C., Allen, J.C., Brewster, C.C., Morales-Ramos, J.A. and King, E.G. (1998). Areawide management of the cotton boll weevil: use of a spatiotemporal model in augmentative biological control. *Ecological Modelling*, **110**: 151-164.

Legaspi, J.C., Legaspi, Jr., B.C. and Sétamou, M. (2004). Insect resistant transgenic crops expressing plant lectins. *In*: Koul, O. and Dhaliwal, G.S. (*Eds.*), *Transgenic crop protection: concepts and strategies.* Science Publishers Inc. Plymouth, UK, pp. 85-116,

Lövei, G.L., Andow, D.A. and Arpaia, S. (2009). Transgenic insecticidal crops and natural enemies: a detailed review of laboratory studies. *Environmental Entomology*, **38**: 293-306.

Lundgren, J.G., Gassmann, A.J., Bernal, J., Duan, J.J. and Ruberson, J. (2009). Ecological compatibility of GM crops and biological control. *Crop Protection*, **28**: 1017-1030.

Magalhães, S., Tudorache, C., Montserrat, M., van Maanen, R., Sabelis, M.W. and Janssen, A. (2005). Diet of intraguild predators affects antipredator behavior in intraguild prey. *Behavioral Ecology*, **16**: 364-370.

McLachlan, A.R.G. and Wratten, S.D. (2003). Abundance and species richness of field-margin and pasture spiders (Araneae) in Canterbury, New Zealand. *New Zealand Journal of Zoology*, **30**: 57-667.

Meyer, J.L., Hoy, M.A. and Jeyaprakash, A. (2006). Insertion of a yeast metallothionein gene into the model insect *Drosophila melanogaster* (Diptera: Drosophilidae) to assess the potential for its use in genetic improvement programs with natural enemies. *Biological Control*, **36**: 129-138.

Moser, S.E. and Obrycki, J.J. (2009). Competition and intraguild predation among three species of Coccinellids (Coleoptera: Coccinellidae). *Annals of the Entomological Society of America*, **102**: 419-425.

Menendez, R., Gonzalez-Megias, A., Lewis, O.T., Shaw, M.R. and Thomas, C.D. (2008). Escape from natural enemies during climate-driven range expansion: a case study. *Ecological Entomology*, **33**: 413-421.

Miller, T.A., Lampe, D.J. and Lauzon, C.R. (2007). Transgenic and paratransgenic insects in crop protection. *In*: Ishaaya, I, Nauen, R. and Horowitz, A.R. (*Eds.*), *Insecticides design using advanced technologies*. Springer, Heidelberg, Germany, pp. 87-103.

Miller, T.A., Lauzon, C.R. and Lampe, D.J. (2008). Technological advances to enhance agricultural pest management. *In*: Askoy, S. (*Ed.*), *Transgenesis and the management of vector-borne disease*. Springer, New York, pp. 141-150.

Molina-Montenegro, M.A., Briones, R. and Cavieres, L.A. (2009). Does global warming induce segregation among alien and native beetle species in a mountain top? *Ecological Research*, **24**: 31-36.

Naranjo, S.E. (2007). Intraguild predation on *Eretmocerus* sp. nr. *emiratus*, a parasitoid of *Bemisia tabaci*, by three generalist predators with implications for estimating the level and impact of parasitism. *Biocontrol Science and Technology*, **17**: 605-622.

Nicholl, D.S.T. (2008). *An introduction to genetic engineering*. Ambridge Univ. Press, Cambridge, United Kingdom.

Nirmala, X. and James, A.A. (2003). Engineering plasmodium-refractory phenotypes in mosquitoes. *Trends in Parasitology*, **19**: 384-387.

O'Callaghan, M., Glare, T.R., Burgess, E.P.J. and Malone, L.A. (2005). Effects of plants genetically modified for insect resistance on nontarget organisms. *Annual Review of Entomology*, **50**: 271-292.

Okuyama, T. (2009). Intraguild predation in biological control: consideration of multiple resource species. *BioControl*, **54**: 3-7.

Parmesan, C. (2006). Ecological and evolutionary responses to recent climate change. *Annual Review of Ecology and Systematics*, **37**: 637-669.

Parmesan, C., Ryrholm, N., Stefanescu, C., Hill, J.K., Thomas, C.D., Descimon, H., Huntley, B., Kaila, L., Kullberg, J., Tammaru, T., Tennent, W.J., Thomas, J.A. and Warren, M. (1999). Poleward shifts in geographical ranges of butterfly species associated with regional warming. *Nature*, **399**: 578-583.

Parmesan, C. and Yohe, G. (2003). A globally coherent fingerprint of climate change impacts across natural systems. *Nature*, **421**: 37-42.

Pell, J.K., Baverstock, J., Roy, H.E., Ware, R.L. and Majerus, M.E.N. (2008). Intraguild predation involving *Harmonia axyridis*: a review of current knowledge and future perspectives. *BioControl*, **53**: 147-168.

Pickett, C.H., Roltsch, W. and Corbett, A. (2004). The role of a rubidium marked natural enemy refuge in the establishment and movement of *Bemisia* parasitoids. *International Journal of Pest Management*, **50**: 183-191.

[PIFB] Pew Initiative on Food and Biotechnology (2004). *Bugs in the system? Issues in the science and regulation of genetically modified insects.* Washington, DC.

Polis, G.A. and Holt, R.D. 1992. Intraguild predation: the dynamics of complex trophic interactions. *Trends in Ecology and Evolution*, **7**: 151-154.

Polis, G.A., Myers, C.A. and Holt, R.D. (1989). The ecology and evolution of intraguild predation: potential competitors that eat each other. *Annual Review of Ecology and Systematics*, **20**: 297-330.

Prasifka, J.R. and Heinz, K.M. (2004). The use of C3 and C4 plants to study natural enemy movement and ecology, and its application to pest management. *International Journal of Pest Management,* **50**: 177-181.

Prasifka, J.R., Heinz, K.M. and Sansone, C.G. (2004). Timing, magnitude, rates and putative causes of predator movement between cotton and grain sorghum field. *Environmental Entomology*, **33**: 282-290.

Presnail, J.K. and Hoy, M.A. (1992). Stable genetic transformation of a beneficial arthropod, *Metaseiulus occidentalis* (Acari: Phytoseiidae), by a microinjection technique. *Proceedings of the National Academy of Sciences of the United States of America*, **89**: 7732-7736.

Presnail, J.K. and Hoy, M.A. (1996). Maternal microinjection of the endoparasitoid *Cardiochiles diaphaniae* (Hymenoptera: Braconidae). *Annals of the Entomological Society of America*, **89(4)**: 576-580.

Rand, T.A., Tylianakis, J.M. and Tscharntke, T. (2006). Spillover edge effects: the dispersal of agriculturally subsidized insect natural enemies into adjacent natural habitats. *Ecology Letters*, **9**: 603-614.

Rosenheim, J.A. and Harmon, J.P. (2006). The influence of intraguild predation on the suppression of a shared prey population: an empirical reassessment. *In*: Brodeur, J. and Boivin, G. (*Eds.*), *Trophic and guild interactions in biological control.* Springer, Dordrecht, The Netherlands, pp. 1-20.

Rosenheim, J.A., Kaya, H.K., Ehler, L.E., Marois, J.J. and Jaffee, B.A. (1995). Intraguild predation among biological-control agents: theory and evidence. *Biological Control*, **5**: 303-335.

Rosenheim, J.A. and Wilhoit, L.R. (1993). Predators that eat other predators disrupt cotton aphid control. *California Agriculture*, **47**: 7-9.

Royer, T.A., Giles, K.L., Lebusa, M.M. and Payton, M.E. (2008). Preference and suitability of greenbug, *Schizaphis graminum* (Hemiptera: Aphididae) mummies parasitized by *Lysiphlebus testaceipes* (Hymenoptera: Aphidiidae) as food for *Coccinella septempunctata* and *Hippodamia convergens* (Coleoptera: Coccinellidae). *Biological Control*, **47**: 82-88.

Rubin, G.M. and Spradling, A.C. (1983). Vectors for P element-mediated gene transfer in *Drosophila. Nucleic Acids Research*, **11**: 6341-6351.

Sanvido, O., Romeis, J. and Bigler, F. (2007). Ecological impacts of genetically modified crops: ten years of field research and commercial cultivation. *Advances in Biochemical Engineering/Biotechnology*, **107**: 235-278.

Sato, S., Kushibuchi, K. and Ysauda, H. (2009). Effect of reflex bleeding of a predatory ladybird beetle, *Harmonia axyridis* (Pallas) (Coleoptera:

Coccinellidae), as a means of avoiding intraguild predation and its cost. *Applied Entomology and Zoology*, **44**: 203-206.

Scarratt, S.L., Wratten, S.D. and Shishehbor, P. (2008). Measuring parasitoid movement from floral resources in a vineyard. *Biological Control*, **46**: 107-113.

Schellhorn, N.A., Bellati, J., Paull, C.A. and Maratos, L. (2008). Parasitoid and moth movement from refuge to crop. *Basic and Applied Ecology*, **9**: 691-700.

Schetelig, M.F., Scolari, F., Handler, A.M., Gasperi, G. and Wimmer, E.A. (2009). New genetic tools for improving SIT in *Ceratitis capitata*: embryonic lethality and sperm marking. *Proceedings 7th International Symposium on Fruit Flies of Economic Importance.* Salvador, Brazil. pp. 299-305.

Scolari, F., Schetelig, M.F., Gabrieli, P., Siciliano, P., Gomulski, L.M., Karam, N., Wimmer, E.A., Malacrida, A.R. and Gasperi, G. (2008). Insect transgenesis applied to tephritid pest control. *Journal of Applied Entomology*, **132**: 820-831.

Scott, T.W., Harrington, L.C., Knols, B.G.J. and Takken, W. (2008). Applications of mosquito ecology for successful insect transgenesis-based disease prevention programs. *In*: Askoy, S. (*Ed.*), *Transgenesis and the management of vector-borne disease*. Springer, New York, pp. 151-168.

Skirvin, D.J. (2004). Virtual plant models of predatory mite movement in complex plant canopies. *Ecological Modelling*, **171**: 301-313.

Skirvin, D.J., Kravar-Garde, L., Reynolds, K., Jones, J., Mead, A. and Fenlon, J. (2007). Supplemental food affects thrips predation and movement of *Orius laevigatus* (Hemiptera: Anthocoridae) and *Neoseiulus cucumeris* (Acari: Phytoseiidae). *Bulletin of Entomological Research,* **97**: 309-315.

Southwood, T.R.E. and Henderson, P.A. (2000). *Ecological methods*. Blackwell Science, Oxford, England.

Stireman, J.O. III, Dyer, L.A., Janzen, D.H., Singer, M.S., Lill, J.T., Marquis, R.J., Ricklefs, R.E., Gentry, G.L., Hallwach, W., Coley, P.D., Barone, J.A., Greeney, H.F., Connahs, H., Barbosa, P., Morais, H.C. and Diniz, R. (2005). Climatic unpredictability and parasitism of caterpillars: Implications of global warming. *Proceedings of the National Academy of Sciencesof the United States of America*, **102**: 17384-17387.

Thomas, C.D., Cameron, A., Green, R.E., Bakkenes, M. Beaumont, L.J., Collingham, Y.C., Erasmus, B.F.N., de Siqueira, M.F., Grainger, A., Hannah, L., Hughes, L., Huntley, B., van Jaarsveld, A.S., Midgley, G.F., Miles, L., Ortega-Huerta, M.A., Peterson, A.T., Phillips, O.L. and Williams, S.E. (2004). Extinction risk from climate change. *Nature*, **427**: 145-148.

Thomson, L.J., Macfadyen, S. and Hoffman, A.A. (2010). Predicting the effects of climate change on natural enemies of agricultural pests. *Biological Control* (in press).

Tomita, M., Munetsuna, H., Sato, T., Adachi, T., Hino, R., Hayashi, M., Shimizu, K., Nakamura, N., Tamura, T. and Yoshizato, K. (2003). Transgenic silkworms produce recombinant human type III procollagen in cocoons. *Nature Biotechnology*, **21**: 52-56.

Turchin, P. (1998). *Quantitative analysis of movement*: *measuring and modeling population redistribution in animals and plants*. Sinauer Assoc., Sunderland, Massachusetts.

Van Driesche, R.G. and Bellows, T.S. (2001). *Biological control.* Kluwer Acad. Publ., Dordrecht, The Netherlands.

Van Driesche, R., Hoddle, M. and Center, T. (2008). *Control of pests and weeds by natural enemies*: *an introduction to biological control.* Wiley-Blackwell, Hoboken, New Jersey.

Van Nouhuys, S. and Lei, G. (2004). Parasitoid-host metapopulation dynamics: the causes and consequences of phonological asynchrony. *Journal of Animal Ecology*, **73**: 526-535.

Vance-Chalcraft, H., Rosenheim, J.A., Vonesh, J.R., Osenberg, C.W. and Sih, A. (2007). The influence of intraguild predation on prey suppression and release: a meta-analysis. *Ecology*, **88**: 2689-2696.

Visser, M.E. and Both, C. (2005). Shifts in phenology due to global climate change: the need for a yardstick. *Proceedings of the Royal Society B*, **272**: 2561-2569.

Voigt, W., Perner, J., Davis, A.J., Eggers, T., Schumacher, J., Bährmann, R., Fabian, B., Heinrich, W., Köhler, G., Lichter, D., Marstaller, R. and Sander, F.W. (2003). Trophic levels are differentially sensitive to climate. *Ecology*, **84**: 2444-2453.

Wen, H., Lan, X., Zhang, Y., Zhao, T., Wang, Y., Kajiura, Z. and Nakagaki, M. (2009). Transgenic silkworms (*Bombyx mori*) produce recombinant spider dragline silk in cocoons. *Molecular Biology Reports.* DOI 10.1007/s11033-009-9615-2.

Whitten, M.J. and Hoy, M.A. (1999). Genetic improvement and other genetic considerations for improving the efficacy and success rate of biological control. *In*: Bellows, T.S. and Fisher, T.W. (*Eds.*), *Handbook of Biological Control.* Academic Press, San Diego, California, pp. 271-296.

Wurm, F. (2003). Human therapeutic proteins from silkworms. *Nature Biotechnology*, **21**: 34-35.

Zang, L.S. and Liu, T.-X. (2007). Intraguild interactions between an oligophagous predator, *Delphastus catalinae* (Coleoptera: Coccinellidae), and a parasitoid, *Encarsia sophia* (Hymenoptera: Aphelinidae), of *Bemisia tabaci* (Homoptera: Aleyrodidae). *Biological Control*, **41**: 142-150.

Zimowska, G.J., Xavier, N. and Handler, A.M. (2009). The β2-tubulin gene from three tephritid fruit fly species and use of its promoter for sperm marking. *Insect Biochemistry and Molecular Biology*, **39**: 508-515.

{5}

Biological Control Agents for Lepidoptera Pests

Lilia H. Morales-Ramos, Hugo A. Luna-Olvera, Katiushka Arévalo-Niño, Benito Pereyra-Alférez, Sergio M. Salcedo-Martínez and Luis J. Galán-Wong

ABSTRACT

The global consensus to reduce inputs of chemical pesticides which are perceived as hazardous to the environment has provided opportunities for the development of novel, benign, sustainable crop protection strategies. The use of pathogenic microorganisms against insects offers a promising alternative in insect management programs specially those focused to reduce lepidopteran pest populations where a wealth of biological material can be exploited by humans. Lepidopteran larvae are among the most serious insect pests and cause a significant reduction of production in many crops every year, however, their control by biopesticides can be particularly effective when incorporated into integrated pest management programs. Examples of successful applications of such entomopathogens are found throughout the world, but widespread commercial production of innovative insecticides awaits further studies of the biology, ecology and pathogenicity of the agents. These microbial insecticides could decrease our dependence on chemical pesticides.

INTRODUCTION

Lepidoptera (moths and butterflies) is the second largest order in the class Insecta. This order has more than 180,000 species in 128 families and 47 superfamilies. Estimates of the number of species suggest that the order may have more species and is among the four largest, successful orders, along with the Hymenoptera, Diptera, and the Coleoptera. As pollinators of many plants, adult moths and butterflies are usually beneficial insects that feed on nectar using their siphoning proboscis. Their caterpillars, however, almost always have chewing mouth parts that are suitable for feeding on various parts of a plant. Most lepidopteran larvae are herbivores; some species eat foliage, some burrow into stems or roots, and some are leaf-miners (Heppner, 2008; Mallet and Way, 2002; Powell, 2003).

Lepidoptera, like all holometabola, undergo complete metamorphosis, going through a four-stage life cycle: egg, larva/ caterpillar, pupa/chrysalis and adult. Their life cycle can include inactive periods or diapauses in any of the pre-adult stages, which help them, overcome unsuitable environmental conditions (Powell, 2003).

A common classification of the order Lepidoptera involves their differentiation into butterflies and moths. A butterfly is an insect notable for its unusual life cycle, with a larval caterpillar stage, an inactive pupal stage and a spectacular metamorphosis into its familiar and colorful winged adult form. Most species fly during the day, so they regularly attract attention. Butterflies comprise the true butterflies (superfamily Papilionoidea), the skippers (superfamily Hesperioidea) and the moth-butterflies (superfamily Hedyloidea). Butterflies exhibit polymorphism, mimicry, and aposematism. Moreover, some migrate over long distances. Butterflies are important economically as agents of pollination. In addition, a few species are pests, because they can damage domestic crops and trees in their larval stage.

Since a moth is an insect closely related to butterflies, both belong to the order Lepidoptera. The number of species of moths has been estimated between 150,000 and 250,000, with thousands of species yet to be described. Moths, and particularly their caterpillars, are major agricultural pests in many parts of the world. Examples include corn borers and bollworms. Moths can be distinguished from butterflies in several ways, the principal is that most species are nocturnal, but there are crepuscular and diurnal species. The differences between butterflies and moths involve more than just their habits, but traditionally the names "Rhopalocera" (butterflies) and "Heterocera" (moths) have been used in taxonomy to denote the popular distinction based on habits (Heppner, 2008; Mallet and Way, 2002; Mallet and Willmott, 2003; Meyer, 2005; Powell, 2003).

LEPIDOPTERA TAXONOMY

Kingdom: Animalia • Phylum: Arthropoda • Class: Insecta • Subclass: Pterygota • Infraclass: Neoptera • Superorder: Endopterygota • Suborder: Glossata Infraorders:

- Dacnonypha
- Exoporia
- Lophocoronina
- Neopseustina
- *Heteroneura*: comprises over 99% of all butterflies and moths. This group is characterized by wing venation, which means they are dissimilar or nonhomoneurous in both pairs of wings (Heppner, 2008; Mallet and Willmott, 2003; Tree of Life Web Project, 2008) (Fig. 5.1).

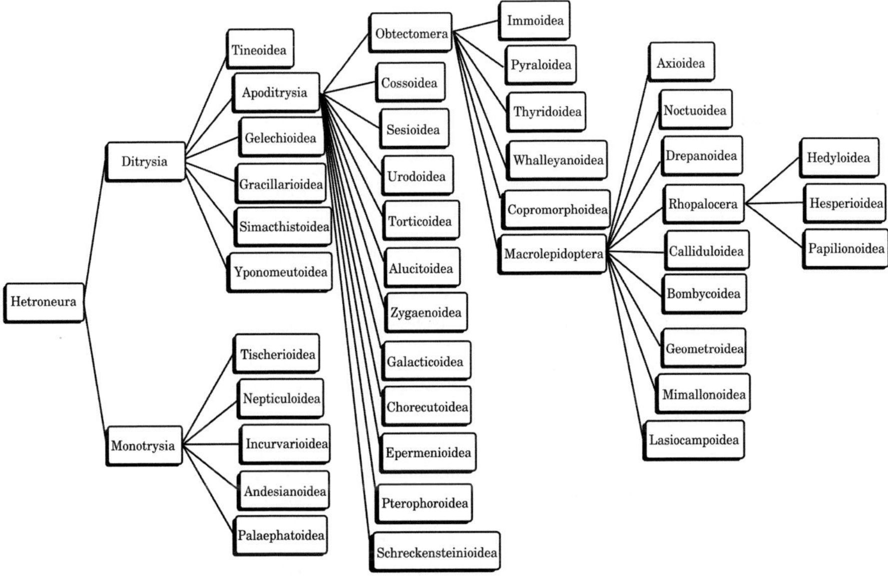

Fig. 5.1. Heteroneura classification

Division Ditrysia (Borner, 1925): contains both butterflies and moths. They are so named because the female has two distinct sexual openings: one for mating and the other for laying eggs. About 98% of described species of Lepidoptera belong to Ditrysia. The group can be divided into basal or *incertae sedis* "micro moths" and Apoditrysia which, includes mostly larger moths, as well as butterflies.

Division Monotrysia (Borner, 1939): This group contains only moths, and most of these (apart from the recently discovered family

Andesianidae) are small and are relatively understudied (compared to Ditrysia) in many regions of the world. The group is so named because the female has a single genital opening for mating and laying eggs, in contrast to the bulk of the Lepidoptera (Ditrysia), which have two female reproductive openings (Dugdale, 1974; Heppner, 2008).

Among the so-called "macrolepidoptera" are about 60% of all Lepidoptera species. Included here are the superfamilies Noctuoidea, Bombycoidea, Lasiocampidae, Mimallonoidea, Geometroidea and the Rhopalocera, which includes Papilionoidea (butterflies), Hesperioidea (skippers) and Hedyloidea (butterfly moths). The Macrolepidoptera include most of the economically important lepidopteran pests.

ECONOMIC IMPORTANCE OF LEPIDOPTERA

Some moths are farmed; the most notable of these is the silkworm, the larva of the domesticated moth *Bombyx mori* Linnaeus (Bombycidae, Bombycoidea). It is farmed for the silk with which it builds its cocoon. Since 2002, the silk industry has been producing over 130 million kilograms of raw silk each year, representing about 250 million U.S. dollars (FAO, 2008). Not all silk is produced by *Bombyx mori.* There are several species of Saturniidae (Bombycoidea) that are also farmed for their silk, such as the Ailanthus moth, *Samia cynthia* Drury, the Chinese Oak Silkmoth, *Antheraea pernyi* Guénerin-Méneville, the Assam Silkmoth, *Antheraea assamensis* Helfer and the Japanese Silk Moth, *Antheraea yamamai* Guénerin-Méneville. Another example is the mopane worm, the caterpillar of *Gonimbrasia belina* Westwood, from the family Saturniidae, which is a significant food resource in southern Africa (Cherry, 1993).

Although many Lepidoptera are valued for their beauty and a few are useful in commerce, the larvae of these insects are probably more destructive to agricultural crops and forest trees than any other group of insects. Examples of Lepidoptera considered as pests are found in the family Tineidae (Tineoidea), where several moths are commonly regarded as pests because their larvae eat fabric, such as clothes and blankets made from natural proteinaceous fibers, such as wool or silk. Pests include the webbing clothes moth *Tineola bisselliella* Hummel and the case-making clothes moth *Tinea pellionella* Linnaeus. In Nymphalidae (brushfooted butterflies, Papilionoidea), the largest butterfly family with 13 subfamilies of butterflies and moths with front legs reduced in size, where the fritillaries, admirals, emperors, and tortoiseshells are included, *Danaus chrysippus* Linnaeus larva is an important pest of the roostertree (*Calotropis procera* (Aiton) W.T. Aiton,

(Asclepiadaceae) in southern Iran. The adults are shiny, orange in color with black and white spots on their wings. And the larva of *Brassolis sophorae* Stichel, which feeds on the leaves of the palm species: *Acrocomia aculeate* Jacquin, *Archontophoenix alexandrae* Muell, *Arecastrum romanzoffianum* Chamisso does some harm to the host. Pyralidae (snout moths, Pyraloidea), the second largest family of Lepidoptera includes the European corn borer, *Ostrinia nubilalis* Hübner; the Indian meal moth, *Plodia interpunctella* Hübner; the sugarcane borer, *Diatraea saccharalis* Fabricius; the white grub, *Laniifera cyclades* Druce; zebra worm, *Olycella nephelepasa* Dyar and the greater wax moth *Galleria mellonella* Linnaeus. A highly problematic pest of *Toona australis* Muell (Meliaceae) is the Pyralidae moth *Hypsipyla robusta* Moore, which hampers attempts to establish Red Cedar (Maryland).

Noctuoidea

Noctuoidea has the largest number of species described for any Lepidopteran superfamily. It comprises eighth families, including the Arctiidae, Lymantriidae and the Noctuidae or noctuid ("night owl" or "owlet") moths.

The larvae of Noctuoidea, called cutworms or army worms, are worrisome pests of seedlings in nurseries; an example is the black cutworm *Agrotis ipsilon* Hufnagel. During its immature stages, it lives hidden in soil. The caterpillars emerge from the soil and feed on their host during the night. Typically, a young plant is neatly cut off right above the soil. Partial but complete defoliation or bending of the seedling is also common. Many other species, including the fall army worm *Spodoptera frugiperda* Smith, an important pest of corn, rice, peanut, cotton, soybean, alfalfa and forage grasses and the cabbage looper *Trichoplusia ni* Hübner, a pest of crucifers, are leaf feeders and stem borers larvae.

Arctiidae (tiger moths, Noctuoidea), are distinctive as adults, being white with black, red, yellow or orange markings. Many larvae are covered with long hairs (woolly bears). The family includes the fall web worm, *Hyphantria cunea* Drury an important pest of pecan and cherry trees.

In Lymantriidae (tussock moths, Noctuoidea), adults do not feed and their larvae are characterized by the presence of hair tufts along the body. Pests in this family include the caterpillar of the gypsy moth, *Lymantria dispar* Linnaeus, which causes severe damage to forests in the northeast United States, where it is an invasive species, whereas in tropical and subtropical climates, the brown tail moth, *Euproctis chrysorrhoea* Linnaeus, together with the diamondback moth, *Plutella xylostella* Linnaeus, are important pests, and the last is considered as the most problematic pest of brassicaceous crops. Other major Lymantriidae pests are *Lymantria ninayi* Bethune-Baker, *L. novaguinensis* Bethune-Baker, *Calliteara queenslandica* Strand, *Dasychira wandammena* Bethune-Baker, which are defoliators of *Pinus* spp. and *Eucalyptus* spp.

Gelechioidea

This is one of the largest families of micro-lepidoptera (Gelechiidae). These larvae feed on plants or plant products. Pests include the Angoumois grain moth, *Sitotroga cerealella* Olivier and the pink bollworm *Pectinophora gossypiella* Saunders.

Geometroidea

This is another large superfamily of Lepidoptera, whose larvae are often called inchworms or spanworms. It includes the winter moth, *Operophtera brumata* Linnaeus and the fall cankerworm, *Alsophila pometaria* Harris.

Some looper caterpillars (Geometridae), such as the Millionaire Moth, *Milionia isodoxa* Prout and other species of the genus *Milionia*, are defoliators of Hoop Pine. Others affect *Pinus* spp. like *Alcis papuensis* Warren and *Paradromulia nigrocellata* Warren while others attack Terminalias (*Terminalia* Combretaceae) and Kamarere (*Eucalyptus* Myrtaceae) such as *Hyposidera talcata* Guenée. Another potential host tree species of earth measuring worms is *Acacia mangium* Willd.

Tortricidae (leafrollers, tortrix moths, Tortricoidea) is the fourth largest moth family of Lepidoptera; it contains many pest species, including the codling moth, *Cydia pomonella* Linnaeus which causes extensive damage in temperate climates to fruit farms; the omnivorous leafroller, *Platynota stultana* Walsingham and the oriental fruit moth, *Grapholita molesta* Busck, whose larvae feed inside stems, leaves and fruits.

Sesiidae (clearwing moths, Sesioidea), are diurnally active adults that mimic wasps; many of them are pests of fruit and vegetable crops, including the peachtree borer, *Synanthedon exitiosa* Say and squash vine borer, *Melittia cucurbitae* Harris.

Bombycoidea

Among the eleven families of Bombycoidea, Lasiocampidae (lappet moths, Bombycoidea) have a global distribution; their larvae feed on the leaves of trees and some spin large webs or tents on the foliage. Important pests include the eastern tent caterpillar, *Malacosoma americana* Fabricius, and the forest tent caterpillar, *M. disstria* Hübner.

Saturniidae (giant silk moths, Bombycoidea), are large colorful moths, whose larvae feed on a wide range of trees and shrubs. Well known species include the cecropia moth, *Hyalophora cecropia* Linnaeus and the luna moth, *Actias luna* Linnaeus.

Sphingidae (hawk moths, Sphingoidea), are medium to large adults with long proboscis for collecting nectar, their larvae are frequently called hornworms. Pests include the tobacco hornworm, *Manduca sexta* Linnaeus and the tomato hornworm, *M. quinquemaculata* Haworth.

Pieridae (whites and sulfurs, Papilionoidea) adults are predominantly white or yellow with black markings. The imported cabbageworm, *Pieris rapae* Linnaeus is a pest throughout the world. A number of Pieridae butterflies like *Eurema blanda* Boisduval and other species of this genus defoliate *Albizia* spp. and other leguminous tree crops. Even though the host can be completely and repeatedly defoliated during severe outbreaks, the tree usually survives the attack (Golestaneh *et al.*, 2009; Schneider, 1999; Stauffer *et al.*, 1993).

Among the most important Lepidoptera pests to world agriculture are (in relative order of importance): *P. xylostella* L., *Pieris rapae* L., *Trichoplusia ni* Hübner, *Spodoptera frugiperda* Smith, *Estigmene acrea* Drury, *S. exigua* Hübner, *Helicoverpa zea* Boddie, *Evergestis rimosalis* Guenée, *Hellula rogatalis* Hulst and *Heliothis virescens* Fabricius in vegetable and field crops, and to world sylviculture mainly the gypsy moth *Lymantria dispar* L. and spruce budworm *Choristoneura fumiferana* Clemens are major forest pests (Cordero *et al.*, 2006; Hutchison *et al.*, 2006; Hamilton *et al.*, 2009).

INTEGRATED PEST MANAGEMENT (IPM)

Every year, significant portions of the world's commercially important agricultural crops, including foods, textile fibers and various domestic plants are lost due to pest infestation, resulting in losses of millions of dollars. Various strategies have been used in attempting to control such pests. One strategy is the use of broad spectrum pesticides, chemical pesticides with a broad range of applications. However, there is a disadvantage when using such chemical pesticides; they may destroy non-target organisms such as beneficial insects and parasites of destructive pests. Additionally, these chemical pesticides are frequently toxic to animals and humans, while targeted pests frequently develop resistance when repeatedly exposed to such substances. Another strategy has involved the use of Integrated Pest Management (IPM), which combines different management strategies and practices to grow healthy crops and minimize the use of pesticides. FAO promotes IPM as the preferred approach to crop protection and regards it as a pillar of both sustainable intensification of crop production and pesticide risk reduction. IPM, which helps minimize pesticide use, should encourage and promote research on, and the development of, alternatives posing fewer risks, such as: biological control agents and techniques, non-chemical pesticides and pesticides that are, as far as possible or desirable, target-specific, that degrade into innocuous constituent parts or metabolites after use and that are of low risk to humans and the environment (FAO, 2008). Among the non-chemical pesticides are the biopesticides, which are certain types of cells or compounds derived from such natural materials as animals, plants, bacteria, and certain minerals used to kill pests. A major category of such cells or compounds make use of naturally occurring pathogens to control insect, fungal and weed infestations of crops (EPA, 2009).

While the overall market for pesticides is showing a decline, the biopesticides market is growing rapidly, increasing from $672 million in 2005 to over $1 billion in 2010, which indicates an Average Annual Growth Rate (AAGR) of 9.9%. Biopesticides currently have 2.5% of the overall pesticides market, but their share of the market will have increased to over 4.2% by 2010. Use of conventional pesticides seems to be declining for a number of reasons; one of them is the growing consumer awareness of the detrimental effect of pesticides on the global environment. Many regulatory agencies have placed restrictions on the maximum residue level (MRL), that restricts or even negates the use of certain kinds of synthetic pesticides. The biopesticides are seeing increased usage because they are environmentally friendly; they are much safer than conventional pesticides, which are hazardous and dangerous chemicals, and are frequently scattered in large quantities. Biopesticides are very effective in smaller quantities and they decompose at a much faster rate than synthetics, minimizing their impact in the environment. Finally, biopesticides can supplement the conventional pesticides used in Integrated Pest Management (IPM) programs to minimize insecticide applications whenever possible, and if chemical control is necessary, include the use of reduced risk or IPM-compatible products. Several insecticides including spinosad, indoxacarb, novaluron, emamectin benzoate, and methoxyfenozide, which are classified as reduced-risk or IPM-compatible products, are efficacious against the lepidopteran pests. Other IPM-compatible insecticides, such as *Bacillus thuringiensis* subsp. *kurstaki*, acetamiprid, and azadirachtin have demonstrated efficacy against lepidopteran pests (Cordero *et al.*, 2006).

Biopesticides include microbial pesticides, plant-incorporated protectants, and other agents, which include natural predators, entomopathogenic nematodes, and parasitoids. Those most widely used are different subspecies of the bacterium, *Bacillus thuringiensis* Berliner (Bt), which constitute approximately 80% of the microbial used for insect control. Of the remaining 20%, a few baculoviruses have been successful as classical biological control agents, or have achieved moderate success as viral insecticides in several developing countries. Fungi, protozoa, and parasitic nematodes have been much less successful. The reason why pathogens are not used more widely as biological control agents or microbial insecticides in industrialized countries is due primarily to insufficient methods for cost-effective mass production. This becomes apparent by considering the different ways pathogens can be used for insect control, and by understanding the expectations used to evaluate their performance (Butt *et al.*, 1999; Lacey and Goettel, 1995; Menn and Hall, 1999; McSpadden, 2006; Thakore, 2006; Uri, 1999).

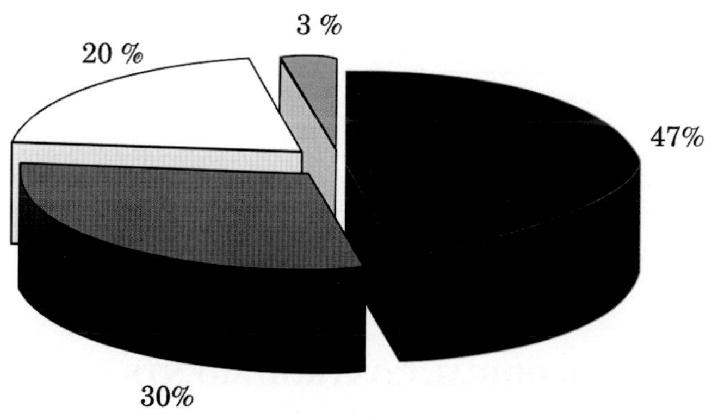

Fig. 5.2. Pesticides global market
Source: CPL Business Consultants, 2006

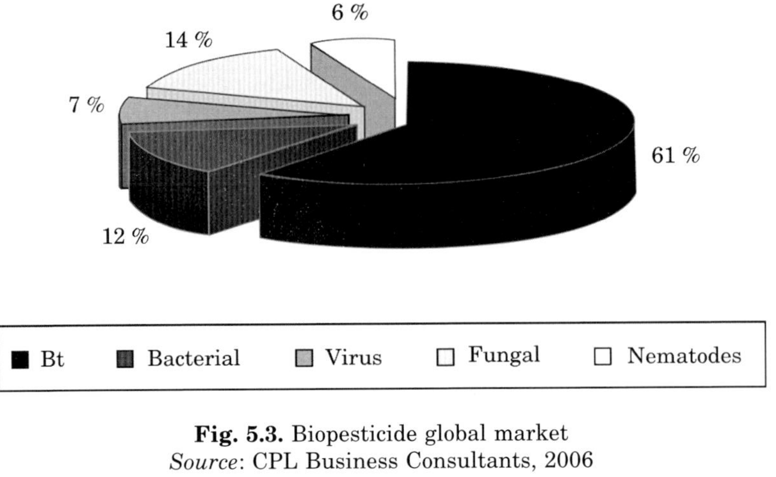

Fig. 5.3. Biopesticide global market
Source: CPL Business Consultants, 2006

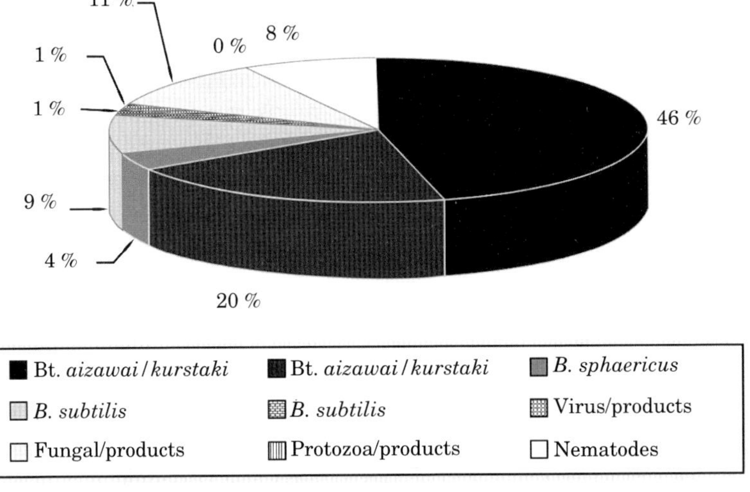

Fig. 5.4. Biopesticide market by product type
Source: McSpadden, 2006

BACTERIA AS BIOLOGICAL CONTROL AGENTS

The most widely used microbial pesticides are derived from the bacterium Bt Berliner. The δ-endotoxins produced by this bacterial agent are used to control a wide range of leaf-eating caterpillars and beetles, as well as mosquitoes. The soil microbe Bt is a Gram-positive, spore-forming bacterium characterized by parasporal crystalline protein inclusions. These inclusions often appear microscopically as distinctively-shaped crystals. The proteins can be highly toxic to pests and specific in their toxic activity. Preparations of the spores and crystals of Bt subsp. *kurstaki* have been used for many years as commercial environmentally acceptable insecticides for lepidopteran pests, other subspecies that are toxic to Lepidoptera include *berliner, alesti, aizawai and tolworthi* (Beegle 1978; Couch, 1980; Miller *et al.*, 1983; Gaertner, 1990; Krieg *et al.*, 1983; Li *et al.*, 2001).

These proteinaceous parasporal crystals, also referred to as insecticidal crystal proteins, Bt inclusions, crystalline inclusions, inclusion bodies and Bt toxins, compose a large collection of insecticidal proteins produced by Bt which are toxic upon ingestion by a susceptible insect host. During the decade of 80's, research on the structure and function of Bt toxins had covered all of the major toxin categories, and although these toxins differed in specific structure and function, general similarities in these characteristics were assumed. Based on the accumulated knowledge of Bt toxins, a generalized mode of action for its toxins has been created and includes: ingestion, in which it has been shown that the crystal proteins are toxic to the insect only after ingestion; solubilization in the insect midgut, in which the alkaline pH and proteolytic enzymes solubilize the proteins, thereby allowing the release of components which are toxic to the insect; binding to the midgut cells, in which formation of a pore in the insect cells begins the disruption of cellular homeostasis, causing the cease of insect feeding and eventually, the insects death. For this reason, Bt has proven to be an effective and environmentally safe insecticide in dealing with various insect pests (English and Slatin, 1992).

Most delta-endotoxins are protoxins that are proteolytically converted into smaller toxic (truncated) polypeptides in the target insect midgut (Höfte and Whiteley, 1989, Whiteley and Schnepf, 1986). The delta-endotoxins are encoded by *cry* (crystal protein) genes. The Cry proteins had been divided into six classes and several subclasses based on structural similarities and pesticide specificity. The major classes were Lepidoptera-specific (CryI); Lepidoptera- and Diptera-specific (CryII); Coleoptera-specific (CryIII); Diptera-specific (CryIV); Coleoptera- and Lepidoptera-specific (CryV); and Nematode-specific (CryVI) (Feitelson *et al.*, 1992). Based on the degree of sequence similarity, the proteins were further classified into subfamilies; more closely related proteins within each family were assigned divisional letters. Thus proteins with less than 45% amino acid identity differ in primary rank (Cry1, Cry2, Cry3, etc.). Proteins with 45-78% sequence homology differ in their secondary rank (Cry1A, Cry1B, etc.) whereas proteins with 78-95% sequence homology differ in their tertiary rank

(Cry1Aa, Cry1Ab, etc.). Finally, proteins with greater than 95% sequence homology are given a new quaternary rank (Cry1Aa1, Cry1Aa2, etc.). A full list of all the Bt toxins discovered so far is available at http://www.biols.susx.ac.uk/Home/Neil_Crickmore/Bt/Index.html (Crickmore *et al.*, 1998; Garczynski and Siegel, 2007; Arora *et al.*, 2007).

Most Bt-based bioinsecticide products are produced using naturally occurring strains of Bt, and utilize only a small fraction of the known Cry proteins. In the United States and Canada, derivatives of Bt strain HDl subsp. *kurstaki* have become the major pesticide used to control the gypsy moth, *Lymantria dispar* L. and the spruce budworm, *Choristoneura fumiferan* Clemens, respectively, other forest pests controlled by Bt include the nun moth, *Lymantria monacha* Linneaus, the pine processionary moth, *Thaumetopoea pityocampa* Schiff and the European pine shoot moth, *Rhyacionia buoliana* F. Funghi. In agriculture, Bt products have been used successfully in the vegetable, cotton, and specialty crop (fruits, nuts) markets, almost exclusively for the control of foliar-feeding leptdopteran pests. Table 5.1 provides a

Table 5.1. Some registered Bt-based product for Lepidoptera pests

Bt Strain	***Name***	***Pest controlled***	***Producer***
aizawai	Agree	Diamondback moth	Thermo Triology or Ciba-Geigy
aizawai	Desing	Lepidopterous larvae	Thermo Triology
aizawai	Florbac	Lepidopterous larvae	Abbott
aizawai	Xen Tari	Diamond back moth	Abbott
kurstaki	Condor	Lepidopterous larvae	Ecogen
kurstaki	Able	Lepidopterous larvae	Thermo Triology
kurstaki	Bactospeine	Lepidopterous larvae	Abbott
kurstaki	Biobit	Lepidopterous larvae	Abbott, Dupont
kurstaki	Constar	Lepidopterous larvae	Thermo Triology
kurstaki	CRYMAX	Lepidopterous larvae	Ecogen
kurstaki	Cutlass	Lepidopterous larvae	Ecogen
kurstaki	Cutlass	Diamond back moth	Ecogen
kurstaki	Dipel	Lepidopterous larvae	Abbott
kurstaki	Foil	Diamond back moth/coleoptera	Ecogen
kurstaki	Foray	Lepidopterous larvae	Abbott
kurstaki	Futura	Lepidopterous larvae	Abbott
kurstaki	Javelin	Lepidopterous larvae	Sandoz or Thermo Triology
kurstaki	Lepinox	Lepidopterous larvae	Ecogen
kurstaki	Raven	Lepidopterous larva/coleopteran	Ecogen
kurstaki	Steward	Lepidopterous larvae	Thermo Triology
kurstaki	Thuricide	Lepidopterous larvae	Thermo Triology
kurstaki	Vault	Lepidopterous larvae	Thermo Triology
Pseudomonas	MATTCH	Lepidopterous larvae	Mycogen
Pseudomonas	MVP	Lepidopterous larvae	Mycogen

Baum *et al.* (1999).

listing of better-known Bt-based product for lepidopteran pests (Baum *et al.*, 1999).

Certain Bt toxin genes have been isolated and sequenced, and recombinant DNA-based Bt products have been produced and approved for use. In addition, with the use of genetic engineering techniques, new approaches for delivering Bt endotoxins to agricultural environments are under development. They also include the use of plants genetically engineered with endotoxin genes for insect resistance, and the use of stabilized intact microbial cells as Bt endotoxin delivery vehicles, The efficacy and environmental safety of the two major Bt transgenic crops, Bt cotton and Bt maize, has led to their widespread adoption in the U.S. These crops produce Cry proteins such as Protein Cry1Ac (in cotton) and Protein Cry1Ab and others (in corn) to control lepidopteran pests such as the cotton budworm, *Heliothis virescens* and the European corn borer, *Ostrinia nubilalis* respectively (Cordero *et al.*, 2006, Gaertner and Kim, 1988; Lacey *et al.*, 2007; Federici, 2007).

VIRUS AS BIOLOGICAL CONTROL AGENTS

Viruses that cause natural epizootic diseases within insect populations are among the entomopathogens which have been developed as biological pesticides. A variety of viruses are known to be valuable biological control agents for insects. Baculoviruses are a large group of viruses which are infectious only in arthropods. The majority of baculoviruses used as biological control agents belong to two genera, *Polyhedrovirus* and *Granulovirus*. The representatives of these genera are known as *nucleopolyhedroviruses* (*NPV*) and *granuloviruses* (*GV*), respectively. More recently, it has been suggested that the nomenclature should be more closely allied to taxonomic groupings and should include four genera: *Alphabaculovirus* (lepidopteran-specific *NPV*), *Betabaculovirus* (lepidopteran-specific granuloviruses), *Gammabaculovirus* (hymenopteran-specific *NPV*) and *Deltabaculovirus* (dipteran-specific *NPV*) (Cory and Evans, 2007), although this scheme has yet to be adopted.

The nuclear polyhedrosis viruses are insect-specific DNA viruses, in which enveloped rod-shaped nucleocapsids, containing a circular double-stranded DNA genome ranging from 50 to 100 Mda in size, are occluded within a protein matrix known as a polyhedron. *NPV*s may be singly-embedded (*SNPV*) or multiply-embedded (*MNPV*), depending on the number of nucleocapsids per envelope. In nature, infection is initiated when an insect ingests food contaminated with baculovirus particles, typically in the form of occlusion bodies (OB). When they are consumed by susceptible insects, the protein crystal of the occlusion bodies is dissolved in the alkaline environment of the insect midgut, releasing individual virus particles which invade epithelial cells lining the midgut. Within the cell, the baculovirus migrates to the nucleus, where replication takes place. Generally, two forms of baculovirus, occluded and extracellular virus (ECV), are produced during viral replication. Some baculoviruses infect virtually every tissue in the host insect so that at the end of the infection process, the entire insect is liquefied, releasing extremely large numbers of occlusion bodies, which are then responsible for spreading the infection to other insects. The most common and effective type of insect viruses are the baculoviruses, which as a group, are known to infect over 600 insect species worldwide. Most baculoviruses infect caterpillars, which are the immature form of moths and butterflies. Insect viruses are potent population regulators of many caterpillar pests. All are highly specific in their host range, usually limited to a single type of insect. Host ranges of some baculoviruses include *Orgyia antique* Linnaeus, *Euproctis chrysorrhoea* Linnaeus, *Pieris rapae* Linnaeus, *Cydia pomonella* Linnaeus, *Helicoverpa zea* Boddie, *Anticarsia gemmatalis* Hübner, *Mamestra brassicae* Linnaeus, *Spodoptera* spp, *Anagrapha falcipera* Kirby and *Autographa californica* Speyer (Cory and Evans, 2007).

Some viruses have been intensively studied as biological pest control agents. Although they have been sold commercially for this purpose, the widespread use of such products has been limited by their poor persistence, slow speed of the kill, and high cost per unit area treated. Viruses have not exhibited unfavorable effects on plants, mammals, birds, fish, or even on non-target insects. This is especially desirable when beneficial insects are included in an overall IPM program, or when an ecologically sensitive area is being treated (Cory and Evans, 2007; D'Amico, 1976; Granados and Federici, 1986). Insect viruses are obligate disease-causing organisms that can only reproduce within a host insect. They can provide safe, effective and sustainable control of a variety of insect pests, although they are most effective as part of a diverse integrated pest management program. Some viruses are produced as commercial products, most notably for fruit pests, but many others are naturally occurring and can initiate outbreaks without additional inputs. Table 5.2 shows some registered virus-based products for lepidoptera pests.

Insect viruses are unable to infect mammals, including humans, which makes them very safe to handle. Most insect viruses are relatively specific, so the risk of non-target effects on beneficial insects is very low. Many viruses occur naturally and may already be present in the

Table 5.2. Some registered virus-based product for Lepidoptera pest

Virus used	*Name*	*Pest controlled*	*Producer*
Af NPV	CLV LC	Celery lopper *Anagrapha falcipera* Kirby	Certis USA
Ag NPV	VPN	*Anticarsia gemmatalis*	Agricola El Sol, Brazil
Ao GV	Capex	*Andoxophyses orana* Hübner	Andermatt Biocontrol, Switzerland
Cp GV	Virosoft CP4	Codling moth *Cydia pomonella*	BioTEPP, Inc
Cp GV	Madex	Codling moth *Cydia pomonella*	Andermatt Biocontrol, Switzerland
Cp GV	Granupom	Codling moth *Cydia pomonella*	AgrEvo, Europe
Cp GV	Carpovirusine	Codling moth *Cydia pomonella*	NPP, France
GV	CYD-X	Codling moth	Certis USA
Hz NPV	Gemstar LC, Biotrol, Elcar	Tobacco budworm *Helicoverpa zea* Boddie and Cotton bollworm *Heliothis virescens* Fabricius	Certis USA
Ld NPV	Gypchek	Gypsy moth *Lymantria dispar*	Thermo Trilogy
Ld NPV	Dispavirus	Gypsy moth *Lymantria dispar*	Canadian Forest Service
Mb *NPV*	Mamestrin	*Mamestra brassiccae* Linnaeus Cabbage moth, American bollworm, diamond back moth, potato tuber moth, and grape Berry moth	NPP, France
Ns NPV	Neochek-S	*Neodiprion serifer* Geoffr	Canadian Forest Service
Ns NPV	Sentifervirus	*Neodiprion serifer*	CanadianForest Service
Ns NPV	Lecontivirus	*Neodiprion serifer*	Canadian Forest Service
Op NPV	TM Biocontrol	Douglas fir tussock moth *Orgyia psuedotsugata* McDunnough	Thermo Trilogy
Se NPV	Spod-XLC	Beet armyworm *Spodoptera exigua* Hübner	Certis USA
Se NPV	Spodex	Beet armyworm *Spodoptera exigua*	Thermo Trilogy
Sl NPV	Spodopterin	*Spodoptera littoralis* Boisduval	NPP, France

Baum *et al.*, 1999; D'Amico, 1976; McNeil, 2009.

environment. Even in cases where they are applied, successful infections can perpetuate the disease outbreak making repeat applications within a season unnecessary. But they exhibit some disadvantages such as most insect viruses take several days to kill their host insect, during which the pest is still causing damage. Insect death is also dose dependent, and very high doses are often necessary for adequate control. As insects age, they can become less susceptible to virus infection, so viruses are usually only effective against early larval life stages. Although viruses can persist in the environment for months or years, exposed virus particles, like those on the surface of plants, are quickly inactivated by direct sunlight or high temperatures, which can limit their persistence within a given season. Also, some agricultural practices can reduce persistence between seasons, such as tillage, which buries virus particles in the soil. To sum up, insect viruses can provide safe, effective and sustainable control of a variety of insect pests, although they are most effective as part of a diverse integrated pest management program (McNeil, 2009).

FUNGI AS BIOLOGICAL CONTROL AGENTS

Most fungi used for the control of insect pests are found in the divisions:

- Zygomycota, class Entomophthorales (*Conidiobolus, Entomophaga, Entomophthora, Erynia, Neozygites, Pandora, Zoophthora*).
- Deuteromycota, class Hyphomycetes (*Hirsutella, Metarhizium, Paecilomyces, Verticillium, Beauveria*).
- Ascomycota (*Cordyceps, Hypocrella, Torrubiella*).
- Oomycota (*Lagenidium*), which were previously classified within the Fungi (Shah and Pell, 2003).

Many of the genera of entomopathogenic fungi currently used for the control of insect pests belong to the class hyphomycetes. Some species have been developed as commercial products because of their ability to be mass-produced. Most fungi in this class are usually found in the soil and can cause natural outbreaks when environmental conditions are favorable. They can infect a wide range of insect hosts. Specific fungal strains in commercial products target thrips, whiteflies, aphids, caterpillars, weevils, grasshoppers, ants, Colorado potato beetle, diamondback moth, gypsy moth and mealybugs.

There is another commonly encountered class of fungi called the Entomophthorales. These fungi can cause natural outbreaks in the populations of their insect hosts, however, due to the difficulty of being mass-produced, they are not considered for commercial production. They tend to be much more host specific; one well-known species only infects aphids. Despite the difficulties in producing them commercially, they can still have a large impact on the pest populations they infect. There are currently no commercially-based products available for organic vegetable production (Shah and Pell, 2003; Wraight *et al.*, 2007).

Fungal resistance structures such as mycelia, spores, or others are able to infect insects. Their virulence, insect host range, infection levels, germination rates and optimum temperature can vary among species and isolates. Members of the hyphomycetes are generally considered to be opportunistic pathogens infecting many species, considerable use of *B. bassiana* for control of European corn borer, *Ostrinia nubilalis* is reported in China and *Nomurea rileyi* is being used in large-scale field trials as a control for the cabbage looper and the velvet bean caterpillar.

Host death is commonly associated with toxin production overwhelming host defense responses. In contrast, other groups of fungi are thought to have evolved into higher parasitic forms. For example, infection and host death by entomophthorales tends to occur due to tissue colonization with little or no use of toxins, like *Entomophtora maimaiga,* which is distributed over much of the area infested by *L. dispar* in north-eastern USA (Shah and Pell, 2003).

Insects can become infected when they come into contact with spores on the plant surfaces, in the soil, in the air as airborne particles, or on the bodies of dead insects. Conidia should be in contact with the cuticle of a suitable host, and infect it by penetrating through the insect cuticle, often at joints or creases, where this protective covering is thinner. Once the fungus has passed through the cuticle of the external insect skeleton, enzyme activation reactions are initiated both by the host and the fungal parasite, followed by the invasion of the insect body and its circulatory system (hemolymph). The structures and processes for the invasion of insect tissues include the formation of germ tubes, infectious hyphae and penetration pegs.

Usually after the insect has died, the fungus grows out through the outer covering (exoskeleton) of the insect, preferably at thinner areas like joints or creases, and begins to produce spores. Insects killed by fungi often have a "fuzzy" appearance, caused by the growth of mycelia and conidia out of the exoskeleton.

Fungi are good biological control agents because they generally do not affect people or other mammals, so this feature makes them extremely safe to use. It is relatively easy to produce spores of an insect-parasitic fungus belonging to the class Hyphomycetes. So most commercial fungal products are spore-based formulations, and are sold at competitive prices (Table 5.3). Successful infections can be spread to other hosts, leading to high rates of persistence in a growing season. Fungal applications exhibit some disadvantages, such as the killing time, which is relatively long (~1 week for most fungi). The high

Table 5.3. Some registered fungi-based product for Lepidoptera pest

Fungus used	***Name***	***Pest controlled***	***Producer***
Beauveria bassiana	Conidia	Coffe berry borer	Live System Technology, Colombia
Beauveria bassiana	Ostrinil	Corn borer	Natural Plant Protection France
Beauveria bassiana	Corn Guard	European corn borer	Mycotech, USA
Beauveria bassiana	Botani Gard	Cotton pest	Mycotech, USA
Beauveria bassiana	Proecol	Army worm	Probioagro, Venezuela
Beauveria bassiana	Mycotrol	Crops	Mycotech, USA
Beauveria bassiana	Naturalis-L	Cotton pest	Troy Bioscience, USA
Metarhizium anisopliae	BIO 1020	Vine weevil	Licenced to Taensa, USA
Metarhizium anisopliae	Metaquino	Spittle bug	Brazil
Metarhizium anisopliae	Cobican	Sugar cane	Probioagro, Venezuela
Metarhizium anisopliae	Bioblast	Field	EcoScience
Paecilomyces fumosoroseus	PFR-97	Whitefly	ECO-tek, USA
Paecilomyces fumosoroseus	Pae-Sin	Whitefly	Agrobionsa, Mexico
Verticillium lecanii	Mycotal	Whitefly and thrips	Koppert,The Netherlands
Verticillium lecanii	Vertalec	Aphids	Koppert, The Netherlands

Baum *et al.*, 1999; Charnley and Collins, 2007·

concentrations of spores frequently needed to get an adequate control of pests in a crop, which makes difficult to reduce production costs. Their broad range of hosts can sometimes be a problem, especially if beneficial insects (*i.e.* predators, parasitoids, and pollinators) are present in the field; mortality in non-target insects can adversely affect the success of the overall biological control program. The environmental factors can also play an important role in the success of fungal products, relative high humidity is necessary for an effective control and a prolonged exposure to sunlight can also inactivate spores, reducing persistence in the crop (Butt and Copping, 2000; Wraight *et al.*, 2007; McNeil, 2009). Most entomopathogenic fungi are best used when total eradication of a pest is not required, but instead insect populations are controlled below an economic threshold, with some crop damage being acceptable. In addition entomopathogenic fungi play an essential role in IPM if they can be used in conjunction with other strategies for sustainable lepidopteran pest control (Shah and Pell, 2003).

CONCLUSIONS

Lepidopterous larvae are the most serious insect pests and represent a significant limitation for production of many crops so a research-based

Integrated Pest Management (IPM) program for Lepidopteran pests is necessary to determine when pest control is needed, and the best insecticide from the standpoint of safety management, reduced environmental risk, specificity (to conserve natural enemies) and economy. Lepidoptera IPM must be founded on a thorough understanding of the ecology of pest and beneficial species and their interaction with the crop. The IPM will provide a range of tactics which must be integrated by the producer to achieve economic and environmental sustainability.

Biopesticides can be particularly effective when incorporated into integrated pest and disease management programs and when used in conjunction with conventional pesticides. Biopesticide products typically have no pre-harvest intervals and very short reentry intervals, which permits a crop to be harvested immediately after the biopesticide is sprayed. This is particularly important in export crops that are shipped globally and are subject to international maximum residue levels.

REFERENCES

Arora, N., Agrawal, N., Yerramilli, V. and Bhatnagar, R.K. (2007). Biology and applications of *Bacillus thuringiensis* in integrated pest management. *In*: Ciancio, A. and Mukerji, K.G. (*Eds.*), *General Concepts in Integrated Pest and Disease Management*. Springer, Dordretch, The Netherlands, pp. 227-239.

Baum, J.A., Johnson, T.B. and Carlton, B.C. (1999). *Bacillus thuringiensis*: Natural and Recombinant Bioinsecticide Products. *In*: Hall, F.R. and Menn, J.J. (*Eds.*), *Methods in Biotechnology, Biopesticides: Uses and Delivery*. Humana Press Inc., Totowa, N.J., pp. 189-210.

Beegle, C. (1978). Use of entomogenous bacteria in agroecosystems. *Developments in Industrial Microbiology*, **20**: 97-104.

Butt, T.M., Harris, J.G. and Powell, K.A. (1999). Microbial Biopesticides. *In*: Menn, J.J. and Hall, F.R. (*Eds.*), *Methods in Biotechnology*: *Biopesticides Use and Delivery*. Humana Press Inc., Totowa, NJ, pp. 23-44.

Butt, T.M. and Copping, L. (2000). Fungal biological control agents. *Pesticide Outlook*, **11**: 186-191.

Cherry, R. (1993). Sericulture. *Bulletin of the Entomological Society of America*, **35**: 83-84.

Cordero, R.J., Kuhar, T.P. and Speese, J. (2006). Field efficacy of insecticides for control of lepidopteran pests on collards in Virginia. Plant Management Network. Available: http://www.plantmanagementnetwork.org/pub/php/research/2006/collard/. Visited on 02 October 2009.

Cory, J.S. and Evans, H.F. (2007). Viruses *In*: Lacey, L.A. and Kaya, H.K. (*Eds.*), *Field Manual of Techniques in Invertebrate Pathology*: *Application and Evaluation of Pathogens for Control of Insects and Other Invertebrate Pests*. Springer, Dordrecht, The Netherlands, pp. 149-174.

Couch, T.L. (1980). Mosquito pathogenicity of *Bacillus thuringiensis* var. *israelensis*. *Developments in Industrial Microbiology*, **22**: 61-76.

Charnley, A.K. and Collins, S.A. (2007). Entomopathogenic fungi and their role in pest control. *In*: Kubicek, C.P. and Druzhinina, I.S. (*Eds.*), *Environmental and Microbial Relationships The Mycota IV*, Springer-Verlag, Berlin-Heidelberg, pp. 159-187.

Crickmore, N., Zeigler, D.R., Feitelson, J., Schnepf, E., Van Rie, J., Lereclus, D., Baum, J. and Dean, D.H. (1998). Revision of the nomenclature for the *Bacillus thuringiensis* pesticidal crystal proteins. *Microbiology and Molecular Biology Reviews*, **62**: 807-813.

D'Amico, V. (1976). Baculoviruses, Biological Control: A Guide to Natural Enemies in North America. Available: http://www.nysaes.cornell.edu/ent/biocontrol/pathogens/baculoviruses. Visited on 02 October 2008.

Dugdale, J.S. (1974). Female genital classification in the classification of Lepidoptera. *New Zealand Journal of Entomology*, **1(2)**: 127-146.

English, L. and Slatin, S.L. (1992). Mode of action of delta endotoxins of *Bacillus thuringiensis*. A comparison with other bacterial toxins. *Insect Biochemistry and Molecular Biology*, **22**: 1-7.

EPA (Environmental Protection Agency) (2009). What are Biopesticides? Available: http://www.epa.gov/pesticides/biopesticides/whatarebiopesticides.htm. Visited on 7 November 2009.

FAO (Food and Agriculture Organization). (2008). Statistical yearbook provides a selection of indicators on food and agriculture by country: FAOSTAT. Available: http://faostat.fao.org. Visited on 02 October 2008.

Federici, B.A. (2007). Bacteria as biological control agents for insects: economics, engineering, and environmental safety *In*: Vurro, M. and Gressel, J. (*Eds.*), *Novel Biotechnologies for Biocontrol Agent Enhancement and Management*. Springer, Dordrecht, The Netherlands, pp. 25-51.

Feitelson, J.S., Payne, J. and Kirn, L. (1992). *Bacillus thuringiensis*: insects and beyond. *Bio / Technology*, **10**: 271-275.

Gaertner, F.H. and Kim, L. (1988). Current applied recombinant DNA projects. *Trends in Biotechnology*, **6(4)**: S4-S7.

Gaertner, F.H. (1990). Cellular delivery systems for insecticidal proteins: living and non-living microorganisms. *In*: Wilkins, R.M. (*Ed.*), *Controlled Delivery of Crop Protection Agents*. Taylor & Francis, New York, NY, pp. 245-255.

Garczynski, S.F. and Siegel, J.P. (2007). Bacteria. *In*: Lacey, L.A. and Kaya, H.K. (*Eds.*), *Field Manual of Techniques in Invertebrate Pathology: Application and Evaluation of Pathogens for Control of Insects and Other Invertebrate Pests*. Springer, Dordrecht, The Netherlands, pp. 175-197.

Golestaneh, S.R., Askary, H., Farar, N. and Dousti, A. (2009). The life cycle of *Danaus chrysippus* Linnaeus (Lepidoptera: Nymphalidae) on *Calotropis procera* in Bushehr-Iran. *Munis Entomology & Zoology*, **4(2)**: 451-456.

Granados, R.R. and Federici, B.A. (1986). *The Biology of Baculoviruses, Practical Application for Insect Control*. CRC Press Inc., Boca Raton, FL.

Hamilton, A.J., Waters, E.K., Kim, H.J., Pak, W.S. and Furlong, M.J. (2009). Validation of fixed sample size plans for monitoring lepidopteran pests of *Brassica oleracea* crops in North Korea. *Journal of Economic Entomology*, **102(3)**: 1336-43.

Heppner, J.B. (2008). Butterflies and moths *In*: Capinera, J.L. (*Ed.*), *Encyclopedia of Entomology*. Fourth Volume. Springer, Dordretch, The Netherlands, pp. 626-672.

Höfte, H. and Whiteley, H.R. (1989). Insecticidal crystal proteins of *Bacillus thuringiensis*. *Microbiological Reviews*, **53(2)**: 242-255.

Hutchison, W.D., Burkness, E.C., Carrillo, M.A., Hurley, T.M. and Pahl, G. (2006). Fresh-market cabbage: Increasing economic returns while reducing risk. Public. 08230. Univ. of Minnesota Extension Service, St. Paul, MN.

Krieg, A., Huger, A.M., Langenbruch, G.A. and Schnetter, W. (1983). *Bacillus thuringiensis var tenebrionis*: a new pathotype effective against larvae of coleopteran. *Journal of Applied Entomology*, **96**: 500-508.

Lacey, L.A. and Goettel, M.S. (1995). Current developments in microbial control of insect pests and prospects for the early 21st century. *Entomophaga*, **40(I)**: 3-27.

Lacey, L.A., Arthurs, S.P., Knight, A.L. and Huber, J. (2007). Microbial control of lepidopteran pests of apple orchards. *In*: Lacey, L.A. and Kaya, H.K. (*Eds*.), *Field Manual of Techniques in Invertebrate Pathology*: *Application and Evaluation of Pathogens for Control of Insects and Other Invertebrate Pests*. Springer, Dordrecht, The Netherlands, pp. 527-546.

Li, G., Zhang, X. and Wang, L. (2001). The use of *Bacillus thuringiensis* on forest integrated pest management. *Journal of Forestry Research*, **12(1)**: 51-54.

Mallet, J. and Way, S. (2002). The Lepidoptera Taxome Project Draft Proposals and Information. Centre for Ecology and Evolution, University College London. Available: http:// www.ucl.ac.uk/taxome. Visited on 02 October 2009.

Mallet, J. and Willmott, K. (2003). Taxonomy: renaissance or Tower of Babel?. *Trends in Ecology and Evolution*, **18(2)**: 57-59.

McNeil, J. (2009). Viruses as biological control agents of insect pests. Penn State University. Available: http://www.extension.org/article/18927. Visited 03 October 2009.

McSpadden, B.B. (2006). Biological Control: Current and Future Market Demands. Department of Plant Pathology. The Ohio State University Available: www.google.com.mx/search?hl=es&q=BMG_CSREESpres%5B1%5D.pdf. Visited on 02 October 2009.

Meyer, J.R. (2005). Lepidoptera: Butterflies/Moths. Department of Entomology, NC State University. Available: http://www.cals.ncsu.edu/course/ent425/ compendium/butter~1.html. Visited on 02 October 2009.

Menn, J.J. and Hall, F.R. Biopesticides (1999). Present status and future prospects. *In*: Menn, J.J. and Hall, F.R. (*Eds*.), *Methods in Biotechnology*: *Biopesticides Use and Delivery*. Humana Press, Totowa, NJ, pp. 1-10.

Miller, L.K., Lingg, A.J. and Bulla Jr., L.A. (1983). Bacterial, viral, and fungal insecticides. *Science*, **219**: 715-721.

Powell, J. (2003). Lepidoptera. *In*: Resh, V.H. and Cardé, R.T. (*Eds*.), *Encyclopedia of Insects*. Academic Press, New York, NY, pp. 631-664.

Schneider, M.F. (1999). Key to the Forest Insect Pests of Papua New Guinea. Papua New Guinea. Available: http://www.fzi.uni-freiburg.de/InsectPestKey-long%20version/info.htm. Visited on 02 October 2009.

Shah, P.A. and Pell, J.K. (2003). Entomopathogenic fungi as biological control agents. *Applied Microbiology and Biotechnology*, **61**: 413-423.

Stauffer, F., Clavilo, J.A. and Bevilacqua, M. (1993). Ataque de *Brassolis sophorae* (L., 1758) (Lepidoptera: Nymphalidae: Brassolinae) a las palmas del Parque del Este “Rómulo Betancourt”. *Boletín de Entomología Venezolana*, **8(1)**: 95-103.

Thakore, Y. (2006). The New Biopesticide Market: Report Highlights. BCC Research. Report ID: CHM029B, Available: http://www.bccresearch.com/report/CHM029B.html. Visited on 02 October 2009.

Tree of Life Web Project (2008). Ditrysia. *In*: The Tree of Life Web Project. http://tolweb.org/ Version 01 May 2008 (temporary). Available: http://tolweb.org/Ditrysia/11868/2008.05.01. Visited on 02 October 2009.

Uri, N.D. (1999). Pesticide Policy Influences. *In*: Hall, F.R. and Menn, J.J. (*Eds.*), *Methods in Biotechnology*: *Biopesticides*: *Use and Delivery*. Humana Press Inc., Totowa, NJ, pp. 55-73.

Whiteley, H.R. and Schnepf, H.E. (1986). The molecular biology of parasporal crystal body formation in *Bacillus thuringiensis*. *Annual Review of Microbiology*, **40**: 549-576.

Wraight, S.P., Douglas, G. and Goettel, M.S. (2007). Fungi. *In*: Lacey, L.A. and Kaya, H.K. (*Eds.*), *Field Manual of Techniques in Invertebrate Pathology*: *Application and Evaluation of Pathogens for Control of Insects and Other Invertebrate Pests*. Springer, Dordrecht, The Netherlands, pp. 223-248.

{6}

Biological Control of Coleopteran Pests

ZHANG JIBIN AND YU ZINIU

ABSTRACT

Biological control of coleopteran pests includes means such as the use of pathogens (bacteria, fungi), parasitoids and predators. Insecticides based on bacteria are the most popular biological pesticides against coleopteran pests. Representative agents of this category are Bacillus thuringiensis (Bt) and Bacillus popilliae. There are already more than 60,000 B. thuringiensis strains and at least 30 classes of active cry genes with activity against coleopteran pests. Several companies currently produce Bt products to control coleopterans. These companies produce Mone™, M-trak™, Trident™, Foi™, Ditera™ in the United States and Novodor in Switzerland. Currently, the United States has successfully developed a B. popilliae-based product called Doom™, which provided effective control against beetles in the United States, Canada, New Zealand and other countries. Fungal pesticides mainly include formulations based on Beauveria brongniartii and Metarhizium anisopliae. Many countries, such as the United States, Australia, Brazil and Japan, have produced fungal-based insecticides with high pathogenicity toward grubs. Seventeen products based on B. bassiana have been registered so far, and three more registered products based on M. anisopliae have been produced, BioPath™, Bio Blast™ and Biogreen™ in the United States. More recently, research has been focused on entomopathogenic nematodes and their application to control grubs. They are divided in two branches of pathogenic nematodes, Steinernematidae and Heterorhabditidae, both consisting of a number of species that exert good grub control. The two natural enemies of coleoptera pests, such as parasitoids and predators, will become useful elements of a strategic approach for coleopteran pest control.

INTRODUCTION

Coleopterans, also called beetles, are represented by 350,000 species worldwide and constitute the largest order of insects. Most of them are important pests in agriculture, forestry, fruit trees and horticulture, while only a few species are considered to be beneficial.

Biological Characteristics

Coleopteran females lay eggs either on the surface or in soil, species of Eumolpidae lay ootheca-encased eggs on plants; species of *Hydrophilus* lay eggs (that are often packaged in a cocoon) in the water. Although most eggs are laid in soil or plants, some individuals from the family Cerambycidae spawn in tree trunks and carve out a space for their eggs in the bark of the tree by biting with their maxillas. Some weevil beetles spawn by digging a hole on plants with their beaks. Larvae usually have three instars, for example, Carabidae larvae and Scarabaeidae larvae. For most species of leaf beetles however, there are four instars. Many coleopteran adults exhibit a phenomenon of mimicking death when they are scared, they rapidly draw their feet back and stay still. For example, when playing dead, some weevil beetles resemble guano.

Coleopterans live in diverse habitats. Their community structure is related intimately with the environment, and small individuals can live in places where other organisms cannot. For example, some species can exist in the leaves, bark crevices, soil gaps and dead wood piles, among other tight places. Other species (such as those from *Paederus* sp. (Coleoptera: Staphylinidae) live in humid areas, including weedy areas and farmlands. All of these characteristics have greatly improved their viability, which causes great difficulties for their control by humans.

Feeding Habits

A wide variety of feeding habits are present in larvae and adults. These include saprozoic (Histeridae), coprophagy (Scarabaeidae), necrophagy (Silphidae), phytophagous (Elateridae), predatory (Tenebrionidae) and parasitic (*Leptinus* sp.) (Coleoptera: Leptinidae). Many phytophagous insects are important pests of crops and forestry, for example, larvae of Elateridae and Scarabaeidae live in the soil and can damage seeds, roots and seedlings; some species, such as the larvae of Cerambycidae and Buprestidae, bore stems or boles, threatening cash crops like timber, fruit trees and sugar cane. Some species, such as the leaf beetles and a variety of adult beetles from several species, feed on leaves, while others, such as most Bruchidae pests, are important because they live in stored grains and eat the seeds of leguminous plants. There are also many predatory beetles that are the natural enemies of pests. Examples of these are most types of ladybugs, which prey on aphids, whiteflies, scale insects and other pests. Carabid beetles and Cicindelidae prey on various types of small-sized insects, especially the larvae of Lepidoptera. Blister beetle larvae parasitize locust eggs and honey-combs.

Harm to Agriculture and Forestry

Coleoptera, which includes 178 families, and 360,000 species, can be differentiated into four suborders according to the evolutionary process. The first suborder is Adephaga, which is carnivorous, the second suborder is Archostemata, which lives under bark and the third suborder is Polyphaga, which is by far the largest suborder, containing 85% of known species, many of which are phytophagous. The fourth suborder is Myxophaga, which are small or minute and associated with hygropetric habitats, drift material, or interstitial habitats among sand grains.

The two most abundant families in Adephaga are the Carabidae and Cicindelidae. The Carabidae live on the soil surface, ground or ant nests and plants; they feed on insects, spiders, or mollusks. Most adults have phototaxis and the hip glands of adult can release defensive substances such as formic acid or *p*-benzoquinone. Some of them are phytophagous, damaging cereals, strawberries, and potatoes.

Polyphaga consist mainly of the families Scarabaeidae, Elateridae, Coccinellidae, Bruchidae, Tenebrionidae, Chrysomelidae, and Cerambycidae. The major damage caused by members of these families is as follows: some Scarabaeidae and Chrysomelidae are critical pests in agriculture and forestry; some Elateridae live under the soil and are important pests for crops; some Buprestidae, and Cerambycidae eat wood and cause the loss of precious timber. In addition, some Bruchidae endanger the growth of leguminous plants. Longhorn beetles are the most dangerous pest in this suborder, they are phytophagous and most of them damage woody plants, including pine, cypress, willow, elm, citrus, apples, peaches and tea. Some coleopterans damage crops such as cotton, wheat, corn, sorghum, sugar cane and hemp, while others damage wood, construction, housing and furniture. They are important pests in forestry fields, crop cultivation and construction timber. In 1957, people found that the longhorn beetles destroyed lead sheets of telephone cables, resulting in the exposure of internal wires. They are also harmful to timber and dry goods. Books stored in bookstore were invaded by a variety of longhorn beetle larvae that then dug a tunnel through five books. Therefore, the damage done by longhorn beetles is very severe, especially for fruit trees and forests, and they are an extremely serious threat to human possessions.

In the 1990s, it was reported that the species *Echinocnemus squameus* (Billberg) (Coleoptera: Anthribidae), from the suborder Rhynchophora (now superfamily Curculionoidea), seriously damaged rapeseed, wheat and cotton crops, and is one of the major pests in paddy fields (Tan, 1992). The adults ruin leaves and the larvae ruin roots, causing serious hazards. Some individuals of Anthribidae are poisonous to the rice weevil *Sitophilus granarius* (Linnaeus) (Coleoptera: Curculionidae) and are present in warehouses. *Sympiezomias velatus* Chevrolet (Coleoptera: Curculionidae) is another agricultural pest that damages a wide variety of crops ranging from cotton, hemp and cereal to Swiss chard, sugar beet, melon, corn, peanut, soybean, sunflower, sorghum, tobacco and seedlings of fruit trees. The adults feed on the cotyledons, shoots, and leaves of cotton and other newly sprouted seedlings. These insects often cluster together while they eat the leaves and bite the tops off of stems; this clustering results in a circle or semicircle of notches on one leaf.

Overall on Biological Control of Coleopteran Pests

Biological control of coleopteran pests is a method that uses an organism to deal with pests. It includes means such as the use of predators, parasitoids, and pathogens (bacteria, fungi, viruses) that can reduce coleopteran pest populations. Its greatest advantage is that it does not pollute the environment.

Insecticides based on bacteria are the most popular biological pesticides against coleopteran pests. Representative agents of this category are *Bacillus thuringiensis* (Bt) and *Bacillus popilliae*. There are already more than 60,000 *B. thuringiensis* strains and at least 30 classes of active *cry* genes with activity against coleopteran pests. Several companies currently produce Bt products to control coleopterans. These companies include Mone™, M-trak™, Trident™, Foil™, Ditera™ of the United States and Novodor of Switzerland. Currently, the United States has successfully developed a *B. popilliae* product called Doom™, which was applied in the United States, Canada, New Zealand and other countries, where it was shown to provide effective control against beetles.

Beauveria bassiana is the most popular fungal pesticide, followed by *Metarhizium anisopliae*. Many countries, such as the United States,

Australia, Brazil and Japan, have produced fungal-based insecticides with high pathogenicity toward grubs. These bioinsecticides mainly include formulations based on *B. brongniartii* and *M. anisopliae*, while certain strains of *B. bassiana* have shown a high mortality rate against grubs. Seventeen products based on *B. bassiana* has been registered so far, and three registered products based on *M. anisopliae* have been produced, BioPath™, BioBlast™ and Biogreen™ in the United States. In the past twenty years, *B. bassiana* and *M. anisopliae* have been widely used to control grubs in many countries.

More recently, research has been focused on entomopathogenic nematodes and their application to control grubs. They are divided in two branches of pathogenic nematodes, *Steinernematidae* and *Heterorhabditidae,* both consisting of a number of species that exert good grub control.

Other biological control agents such as viruses, microsporidia and natural enemies have also been widely reported to control coleopterans.

Application of Pathogenic Microorganisms to Prevent Coleopteran Pests

Bacillus thuringiensis

Bacillus thuringiensis preparations make up the largest output of microbial pesticides in the world, and are widely applied to control pests in agriculture, forestry, storage and medicine (Yu, 1990; Zhang *et al.*, 1998b; Sun *et al.,* 2003). Eighty years ago, research was focused on finding strains toxic to Lepidoptera and Diptera pests. In 1983, Krieg *et al.,* discovered the first *B. thuringiensis* strain active against Coleoptera pests. After the 1980s; Krieg *et al.*, (1987); Herrnstadt *et al.*, (1986) successively isolated two strains corresponding to *B. thuringiensis* subsp. *tenebrionis* and *B. thuringiensis* subsp. *san diego* with toxic activity against Coleoptera while being inactive against Lepidoptera and Diptera. *B. thuringiensis* var. *san diego* has shown to be toxic to more than 20 species of coleopterans. Since then, new strains have been continuously discovered, and much research on these strains has been carried out. One of the most harmful Coleoptera pests is *Leptinotarsa decemlineata* Say (Coleoptera: Chrysomelidae), which causes damage to potato and tomato crops in several regions of America and Europe. This insect has developed resistance against a variety of chemical pesticides. The long-term use of chemical pesticides not only pollutes the environment, but also reduces the populations and numbers of natural enemies, resulting in an increased population of *L. decemlineata*.

There are also a number of Coleoptera pests that cause serious damages to forest and reserve materials, such as grain, food and forage. Therefore, it is very important to conduct research on *B. thuringiensis* preparations to control Coleoptera pests.

Crystal Protein against Coleoptera Pests

The current nomenclature distinguishes 174 holotype sequences that are grouped into 55 *cry* and 2 *cyt* families. A total of 47 toxins were tested against 39 species of Coleoptera in 190 bioassays. The toxins Cry1B, Cry1I, Cry3A, Cry3B, Cry3C, Cry7A, Cry8A, Cry8B, Cry8C, Cry8D, Cry8E, Cry8F, Cry8G, Cry9D, Cry14A, Cry18A, Cry22A, Cry22B, Cry23A, Cry34A, Cry34B, Cry35A, Cry35B, Cry36A, Cry37A, Cry43A, Cry43B, Cry55A, Cyt1A and Cyt2C were found to be active against coleopterans (Table 6.1). About 80% of the bioassays pertained to four Cry families: Cry3 (32%; 26.7% for Cry3Aa alone), Cry8 (24%), Cry1 (18.2%; 9.1% for Cry1B alone), and Cry34/35 (8%, alone or in combination). The broadest range of toxins was tested against *Diabrotica* sp. (Coleoptera: Chrysomelidae) (23 toxins) and *L. decemlineata* (16 toxins). The only toxin that was tested against a broad range of test species was Cry3Aa (23 species, from which 60% were susceptible) (van Frankenhuyzen, 2009). Several important insecticidal crystal proteins against coleopterans were described as follows (Liu *et al.*, 1998).

(1) Cry3A insecticidal crystal protein

B. thuringiensis subsp. *tenebrionis* strain BI256-82 has a single insecticidal crystal protein gene, *cry3A*, which encodes a 73.1 kDa protein. Under the effect of spore protease, the protein is degraded, forming a parasporal crystal of 67 kDa. When Coleoptera pests feed on parasporal crystals, the crystals are solubilized in the midgut, resulting in active and toxic insecticidal peptides of 55 kDa (Krieg *et al.*, 1983).

(2) Cry3B insecticidal crystal protein

B. thuringiensis subsp. *tolworthi*, H9 strain EG2838, and *B. thuringiensis* subsp. *kumamotoensis*, H18 strain EG4961, possess the insecticidal crystal protein genes *cry3B* and *cry3B2*, respectively, which encode 74.2 kDa and 74.4 kDa proteins. A mixture of the spores and crystals of these strains have exhibited toxicity against *L. decemlineata* larvae. The spore crystal mixture of EG4961 has demonstrated specific toxic activity against *Diabrotica undecimpunctata howardi* Barber (Coleoptera: Chrysomelidae) larvae (Donovan *et al.*, 1992).

Table 6.1. Insecticidal crystal proteins and genes from *Bacillus thuringiensis* against Coleoptera

Crystal protein	***MWkDa***	***cry gene***	***GenBank accession number***	***Bt strain***	***Target pests****	***Reference(s)***
Cry1Ba	139.5	*cry1Ba*	CAA29898.1	Bt subsp. *chinensis*	*Chrysomela scripta, Leptinotarsa decemlineata, Anthonomus grandis grandis*	Brizzard and Whiteley, 1988
Cry1Ia	81.2	*cry1Ia*	CAA44633	Bt subsp. *kurstaki*	*Leptinotarsa decemlineata, Anthonomus grandis grandis*	Tailor *et al.*, 1992
Cry3Aa	73.1	*cry3Aa*	AAA22541	Bt subsp.*teneb-rionis* strain BI256-82	*Leptinotarsa decemlineata, Callosobruchus maculatus Diabrotica balteata, Chrysomela scripta*	Sekar *et al.*, 1987
Cry3Ba	74.2	*cry3Ba*	CAA34983	Bt *tolworthi* 43F	*Leptinotarsa decemlineata*	Sick *et al.*, 1990
Cry3Bb (Cry3B2)	74.4	*cry3Bb* (*cry3B2*)	AAA22334	Bt EG4961	*Diabrotica undecimpunctata howardi, Leptinotarsa decemlineata*	Donovan *et al.*, 1992
Cry3Ca	129.4	*cry3Ca*	CAA42469	Bt *kurstaki* BtI109P	*Leptinotarsa decemlineata*	Lambert *et al.*, 1992
Cry8Aa	131.0	*cry8Aa*	AAA21117	Bt *kumamot-oensis*	*Leptinotarsa decemlineata Cotinis* spp.	Narva and Fu, 1992
Cry8Ba	133.5	*cry8Ba*	AAA21118	Bt *kumamot-oensis*	*Cotinis* spp., *Cyclocephala borealis, Cyclocephala pasadenae, Popillia japônica*	Narva and Fu, 1993
Cry8Bb	136.5	*cry8Bb*	CAD57542	Bt	*Diabrotica undecimpunctat, howardia, Diabrotica virgifera, Leptinotarsa decemlineata*	Abad *et al.*, 2002
Cry8Bc	137.2	*cry8Bc*	CAD57543	Bt	*Diabrotica undecimpunctata howardi, Diabrotica virgifera, Leptinotarsa decemlineata*	Abad *et al.*, 2002
Cry8Ca	130.4	*cry8Ca*	AAA21119	Bt *japonensis* Buibui	*Diabrotica undecimpunctata howardi*, *Anomala* spp., *Holotrichia oblita, Popillia japônica*	Sato *et al.*, 1995

Table 6.1. (*Contd...*)

Table 6.1. (*Contd...*)

Crystal protein	*MWkDa*	*cry gene*	*GenBank accession number*	*Bt strain*	*Target pests**	*Reference(s)*
Cry8Da	130.0	*cry8Da*	BAC07226	Bt *galleriae*	*Anomala cuprea, Anomala orientalis,*	Asano *et al*., 2003
Cry8Db	130.0	*cry8Db*	BAF93483	Bt BBT2-5	*Popillia japonica*	Yamaguchi *et al*., 2008
Cry8Ea	131.6	*cry8Ea*	AAQ73470	Bt 185	*Anomala corpulenta, Holotrichia oblita, Anomala cuprea*	Shu *et al*., 2009b
Cry8Fa	133.1	*cry8Fa*	AAT48690	Bt 185	*Holotrichia oblita, Holotrichia parallela*	Shu *et al*., 2009[a]
Cry8Ga	131.2	*cry8Ga*	AAT46073	Bt HBF-18	*Anomala corpulenta, Holotrichia oblita*	Shu *et al*., 2009b
Cry9Da	130.0	*cry9Da*	BAA19948	Bt *japonensis* N141	*Anomala cuprea*	Asano 1996
Cry22Aa	80.0	*cry22Aa*	CAD43579	Bt	*Anthonomus grandis grandis*	Isaac *et al*., 2001
Cry23Aa	29.0	*cry23Aa*	AAF76375	Bt	*Tribolium castaneum, Popillia japonica*	Donovan *et al*., 2000
Cry34/35	14.0 and 44.0	*cry34/35*	AAG41671	Bt PS149B1	*Diabrotica undecimpunctata howardi, Diabrotica virgifera*	Moellenbeck *et al*., 2001
Cyt1Aa	27.0	*cyt1A*	X03182	Bt *israelensis*	*Chrysomela scripta*	Waalwijk *et al*., 1985
Cyt2Ca	26.0	*cyt2C*	AAK50455	Bt	*Diabrotica undecimpunctat, howardia, Diabrotica virgifera, Leptinotarsa decemlineata, Tribolium castaneum, Popillia japonica*	Rupar *et al*., 2000

**Chrysomela scripta* Fabricius (Coleoptera: Chrysomelidae), *Anthonomus grandis grandis* Boheman (Coleoptera: Curculionidae), *Callosobruchus maculatus* (Fabricius) (Coleoptera: Bruchidae), *Diabrotica balteata* LeConte (Coleoptera: Chrysomelidae), *Cotinis* spp (Coleoptera: Scarabaeidae), *Cyclocephala borealis* Arrow (Coleoptera: Scarabaeidae), *Cyclocephala pasadenae* Cassey (Coleoptera: Scarabaeidae), *Popillia japonica* Newman (Coleoptera: Scarabaeidae), *Diabrotica virgifera virgifera* LeConte (Coleoptera: Chrysomelidae), *Anomala spp* (Coleoptera: Scarabaeidae), *Holotrichia oblita* Faldermann (Coleoptera: Scarabaeidae), *Anomala cuprea* (Hope) (Coleoptera: Scarabaeidae), *Anomala orientalis* Waterhouse (Coleoptera: Scarabaeidae), *Anomala corpulenta* Motschulsky (Coleoptera: Scarabaeidae), *Tribolium castaneum* (Herbst) (Coleoptera: Tenebrionidae).
(Donovan *et al*., 1992).

(3) Cry3C insecticidal crystal protein

The parasporal crystal of *B. thuringiensis* subsp. *galleriae*, strain BTS137J, is diamond-shaped, with a molecular weight of 129.4 kDa. The active fragment of 72 kDa has shown toxic activity against *L. decemlineata.*

(4) Cry3D insecticidal crystal protein

The Cry3D protein of *B. thuringiensis* subsp. *kurstaki*, H3ab strain BTI109P, encoded by its corresponding gene, has higher toxicity against *L. decemlineata* than other Cry3 proteins. Its LD_{50} value is 0.7 mg/ml; this LD_{50} value was obtained by measuring the trypsin hydrolysate activity of Cry3D by the leaf dipping bioassay (Lambert *et al.*, 1992).

(5) Cry1B insecticidal crystal protein

Cry1B crystal protein was found in *B. thuringiensis* subsp. *thuringiensis*, which is lethal to certain lepidopteran larvae, such as *Artogeia rapae* (Linnaeus) (Lepidoptera: Pieridae) and *Manduca sexta* (Linnaeus) (Lepidoptera: Sphingidae) (Brizzard *et al.*, 1991). Bradley *et al.*, (1995) discovered that the Cry1B crystal protein of *Bacillus thuringiensis* subsp. *thuringiensis* (strain BtS2, also called strain CT-43 in China, which is called *B. thuringiensis* subsp. *chinensis*), has dual specificity against coleopteran and lepidopteran larvae. The Cry1B crystal protein (139.5 KDa) has no toxicity against Coleoptera larvae unless it is degraded into a 65 kDa peptide.

Application

Several commercial bioinsecticides, such as M-one™, M-trak™, Trident™ and Foil™, have been produced in the United States of America. These bioinsecticides, based on the *san diego* and *tenebrionsis* subspecies, are designed for the control of leaf-feeding flea beetle species including the Colorado potato leaf beetle (*L. decemlineata*), cottonwood leaf beetle (*C. scripta*), Elm leaf beetle *Pyrrhalta luteola* Müller (Coleoptera: Chrysomelidae) and willow leaf beetle *Plagiodera versicolora* Laicharting (Coleoptera: Chrysomelidae), among others (Table 6.2).

Among the *B. thuringiensis* strains isolated in China, the C-001~006 and ES-017 strains exhibit different toxicity levels against *Tenebrio molitor* (Linnaeus) (Coleoptera: Tenebrionidae) and *A. corpulenta* larvae. A new strain of *B. thuringiensis*, strain HBF-1, isolated from Hebei province, China, has high insecticidal activity against *A. exoleta*

Table 6.2. *Bacillus thuringiensis*-based commercial products for the prevention of coleopteran pests

Company	***Product***
Mycogen	M-One
Sandoz	Trident
Abbott	Ditera
Novo Biokontrol	Novodor
Ecogen	Foil

Faldermann (Coleoptera: Scarabaeidae) and *A. corpulenta* larvae, and the mortality on first and second instars larvae of the cinnamon beetle treated with this strain was 100%. This isolate is the first discovered and reported to have specific insecticidal activity against beetle larvae in China.

The δ-endotoxin and β-exotoxin mixtures of *B. thuringiensis* subsp. *chinesensis* were toxic to *Aprionag germari* (Hope) (Coleoptera: Cerambycidae), *Batocera horsfieldi* (Hope) (Coleoptera: Cerambycidae) and *Anoplophora chinensis* Förster (Coleoptera: Cerambycidae) larvae. Its efficacy is better than that of *B. thuringiensis* preparations containing only δ-endotoxin, and its prevention and control capacity on *Mimastra cyanura* (Hope) (Coleopera: Chrysomelidae) is higher than that of chemical pesticides.

B. thuringiensis strain EG4961 has some toxic activity against the eleven stars leaf beetle of cucumber *D. undecimpunctata*, and the Colorado potato beetle *L. decemlineata* larvae and adults, and has a significant toxicity for *Apophylia thalassina* Faldermann (Coleoptera: Chrysomelidae), *P. versicolora* and Coccinellidae adults and larvae in Mexico.

The standard Coleoptera screening bioassay has indicated that *Phaedon brassicae* Baly (Coleoptera: Crhysomelidae), *P. versicolora* are highly sensitive to *B. thuringiensis* subsp. *tenebrionis* (H8ab), making them suitable as standards toxicity tests (Zhang *et al.*, 1998b).

In 1987, Herrnstadt *et al.*, cloned the first *cry3Aa* gene that was active against Coleoptera pests. After this, genetically engineering microbial agents and genetically modified potatoes with *cry*3Aa were put into market in 1991 and 1995, respectively, leading to a significant amount of potato beetle control.

In 1993, Perlak *et al.,* artificially synthesized the *cry*4 gene and introduced it into potato plants. Results indicated that insecticidal crystal protein was highly expressed in potato, and controlled the 2-3 instar potato beetles by up to 90%. Results from field tests from different regions showed that genetically modified potato plants are more effective than chemical pesticides in controlling beetles.

In 1999, it was found that *B. thuringiensis* subsp. *morrisoni* YM-03 (H8a8b), which are highly toxic against Coleoptera larvae, contained the *cry*3A gene. An engineered bacterium (Biot205) was produced by introduction of the *cry*3A gene. The recombinant strain caused high mortality (86.21%) to the elm blue leaf beetle *Pyrrhalta aenescens* (Fairmaire) (Coleoptera: Chrysomelidae).

Bacillus popilliae

Bacillus popilliae is an effective method of preventing Japanese beetle (*P. japonica*) larvae; it was found a little later than *B. thuringiensis*. In the 1930s, the strain was first isolated by Dutky from Japanese grubs that were infected by milky disease. *B. popilliae* is a pathogen of the golden tortoise beetle larvae. It is a highly efficient and environmentally friendly microbial insecticide promoted in the United States since 1939.

B. popilliae can infect fifty species of golden tortoise beetle larvae, and the infected larvae can move in a large area before death, increasing the infection probability of other beetle larvae. *B. popilliae* can maintain viability for several years in the soil and drought conditions. The advantages of using commercial preparations of *B. popilliae* are its narrow host range (they are effective only against Japanese beetles), and their complete safety for humans and other vertebrates.

Mechanism of Action

In natural conditions, *B. popilliae* infects insects through the digestive tract. After the golden tortoise beetle larvae feed on grassroots or other food contained the *B. popilliae* spores, the disease process develops in four stages: (1) Spore sprouting period. Spores become aclastic and the insect blood lymph becomes transparent. After one to three weeks, the hemolymph has a small amount of bacterial cells. (2) Period of vegetative propagation. During a period of 3 to 5 days, the hemolymph becomes full of bacteria and appears slightly turbid. (3) Spore formation period. Bacteria form spores for 5-10 days. (4) Spore maturation period. The hemolymph is filled with spores during a period of 14 to 21 days, leading to milky disease and appearing as an emulsion. Each larva contains 20 to 50 million spores at this stage. After this, the larvae die. Because the disease-causing parts during this process appear milky white, it is called milky disease and the bacteria are called emulsion bacillus.

Application

The first biological pesticide against the Japanese beetle based on *B. popilliae* came into the market in the United States in 1940. *B. popilliae* preparation was registered in the United States in 1950. In recent years, the insecticidal protein Cry18 gene of *B. popilliae* was cloned and transformed into other bacteria for coleopteran control (Zhang *et al.*, 1997). The United States had successfully developed a commercial preparation of *B. popilliae* called Doom™, which has been used successfully in large areas in the United States, Canada, New Zealand, and other countries infested with the Japanese beetle. For example, the United States used *B. popilliae* preparation to control coleoptera in more than ten million acres of farmland in 13 states and in the District of Columbia. In Maryland, USA, after application of the bacteria, the numbers of the grub in grass were reduced from 20-60 to 1-3 per square foot. Converted to square meter measurements, this would be a reduction of 11-33 from 215-645. This level was maintained for 9 years.

In China, *B. popilliae* subsp. *pengleai* (found in Provinces Henan, Hebei, Shandong and Shanxi) is a pathogen with a high infection rate for many cockchafer larvae, especially darkly dung beetles, copper lines dung beetles and four lines cockchafer. It can also infect big black dung beetles, monguli dung beetles and black velvet dung beetles. However, the infection rate of these species is low. Each acre of farmland uses 200 g powder containing 1 billion spores per gram. The highest grub reduction rate was 84.4%, and the lowest was 49.3%, with an average of 68.6%. It has also been shown that the optimum temperature for the occurrence of emulsion disease is 28-30°C in field conditions. When the temperature is below 16°C, there is no disease (Yu, 2000).

FUNGAL APPLICATION

There are more than 50 different fungal products registered, with nine able to control coleopterans (Yu, 2000) (Table 6.3).

In the last 20 years, many countries have carried out extensive research to examine the use of insect pathogenic fungi. The effect of fungi in controlling pests has improved greatly, but even more research is being conducted to develop and use of insect pathogenic fungi. Fungal insecticides mainly include *B. bassiana* and *M. anisopliae*.

***Beauveria* sp.**

There are two species of the genus *Beauveria*, *B. bassiana* and *B. tenella*. *B. bassiana* has shown virulence against insects from orders

Table 6.3. Fungal-based commercial products against Coleoptera

Fungi	*Product name™*	*Target pests*	*Manufacturer or country*
Beauveria bassiana	Biotrol FBB	Scale insects	America Abbott
Beauveria bassiana	Boverin	*Leptinotarsa decemlineata*	The former Soviet Union
Beauveria bassiana	Boverol	*Leptinotarsa decemlineata*	Czechoslovakia
Beauveria bassiana	Boverosil	*Leptinotarsa decemlineata*	Czechoslovakia
Beauveria bassiana	Naturalist	*Anthonomus grandis*	America Mycotech
Metarhizium anisopliae	Back-off-1	Scale insect, *Trialeurodes vaporariorum* (Westwood) (Hemiptera: Aleyrodidae)	America
Metarhizium anisopliae	BioPat	Coleoptera pests	America
Metarhizium anisopliae	BioBlas	Coleoptera pests	America
Metarhizium anisopliae	Biogreen	Coleoptera pests	Australia

Lepidoptera, Coleoptera, Hemiptera, and mites, while *B. tenella* only attacks some coleopteran pests, such as beetle larvae (Pu and Li, 1996).

Application

Beauveria belongs to the category of fungal agents, which is the most widely used category. After more than 20 years of extensive experiments and research, its target range has been greatly expanded. To the best of our knowledge, it has been used to kill borers, fruit tree pests and soil insects, among others (Gan *et al.*, 2007).

In the USA, attempts were made as early as 1962 by Nutrilite Products to obtain registration for *B. bassiana* as Biotrol™, but these attempts were unsuccessful. A wettable formulation of *B. bassiana* conidia, based on isolate ARSEF 252, was developed by Abbott Laboratories and used in pilot tests against the Colorado potato beetle during the 1980s.

Results from field experiments using the Abbott's wettable formulation against the Colorado potato beetle were reported, although these results differed from year to year. Although the larval populations of the first generation were larger with *B. bassiana* treatment, oviposition by the first and the second generations of adults were lower than that found for adults from the fenvalerate treatment group.

In the former Soviet Union, commercial preparations of *B. bassiana* (Boverin™) were applied to more than 10,000 ha of farmland, mainly against the Colorado potato beetle.

In Belgium, under greenhouse conditions, an 84% control of introduced larvae of the black vine weevil was obtained after an application of 2×10^8 conidia of *B. brongniartii* per liter of peat.

Metarhizium spp.

The most common control agent is *Metarhizium anisopliae* var. *anisopliae* (Metsch.) Sorokin. The soil-inhabiting larvae of scarab beetles are typical hosts of *Metarhizium* spp. and coevolution has led to some isolates being specific to one or two genera of scarab.

Application

In Australia, BioGreen™ and BioCane™ are the two registered mycoinsecticides for coleoptera pests. BioGreen™ is a granular product consisting of broken rice particles on which spores of *M. flavoviride* (Gams & Roszypal) grow. It has been used to control redheaded pasture cockchafer, *Adoryphorus couloni* (Burmeister) (Coleoptera: Scarabaeidae), in turf and pasture in Victoria and Tasmania. Another rice granule formulation of *M. anisopliae* var. *anisopliae* isolate FI-1045 is highly virulent against the greyback canegrub, *Dermolepida albohirtum* (Waterhouse) (Coleoptera: Scarabaeidae), which is the worst single pest in Australia, causing losses over $5 M per year. Extensive field trials over several seasons have shown that BioCane™ applied correctly at 33 kg/ha or 6.6×10^{13} conidia/ha leads to 50% to 60% control of that season's grubs.

In America, BioPat™ and Bioblast™ are two mycoinsecticides registered for the control of coleoptera. In China, spray spore suspensions of *M. anisopliae* control nearly 50% of the rice water weevil population 7-9 days after application. When *M. anisopliae* was used to control the North China black Melolonthidae larvae by the Shanxi Institute of Zoology, larval mortality was increased as spore concentration increased. At spore concentrations of 2.1×10^4, 1.9×10^5, 2.2×10^6, 2.0×10^7 and 1.8×10^8 cfu/ml, the mortality of larvae was 20%, 65%, 90%, 100% and 100%, respectively.

NEMATODES APPLICATION

Entomopathogenic nematodes search for their host insects based on the chemical stimulation of the host itself or of other substances. The smell of the excrement of the host or the carbon dioxide generated by its respiration may lure the nematodes in the infection period (Gaugler, 1981).

In China, a parasitic nematode was used to control *Actinidia chinensis*, which parasitizes *Ephedra equisetina* (Bunge). The larval mortality reached 94.4% to 100% in 4-6 days after application. In the forest, each wormhole contained 10,000 infective nematodes, leading to greater than 90% control. *Steinernema* was used to control *A. germari*, which occurs on the poplar. Each hole was injected with 6,000 *Steinernema* and the mortality was about 82.1% (Liu, 1993). Banana plants were treated with a *S. carpocapsae* preparation to control the black weevil, and larval mortality was between 73% and 90%, the mortality of pupa was between 68% and 92%, and the mortality of adults was between 25% and 80%.

Entomopathogenic nematodes are a new biological control factor with greater variation. More than 2,000 strains have been reported and experts dedicated to the transgenesis and molecular biology of nematodes hope to produce heat-resistant, anti-drought nematode strains in order to expand their production and application. Therefore, these will play a greater role in bio-ecological agricultural pest control systems.

PARASITOIDS APPLICATION

Parasitic enemies have been the most common in biological control of insect pests. Coleopteran larvae from subfamily Galerucinae are parasitized by the hymenopteran families Eulophidae, Braconidae and Encyrtidae, as well as by the dipteran Tachinidae. Adult Galerucinae are parasitized by braconids in the subfamily Euphorinae and tachinids in the tribe Blondeliini (Jolivet *et al.*, 1994).

The application of coleoptera pest parasitoids is popularly used in the field as follows: (1) In most areas of Europe and China, *Lariophagus distinguendus* (Förster) (Hymenoptera: Pteromalidae) is an important parasitoid of some warehouse coleoptera pests such as *Sitophilus zeamais* Motschulsky (Coleoptera: Curculionidae) and *S. oryzae* (Linnaeus) (Coleoptera: (Curculionidae). (2) Artificial breeding and release of *Scleroderma guani* (Hymenoptera: Bethylidae) has an important role in the control of small longicorns; large longicorn species such as mulberry, macular star sawyer, and clouds of natural flowers are parasitized by *Dastarcus helophoroides* (Fairmaire) (Coleoptera: Bothrideridae), which is an important parasitoid of beetles.

PREDATORS APPLICATION

The main predators of coleoptera are birds, ants, mites, beetles, maggots, scorpions, spiders and clams. A number of vertebrate predators, predominantly birds, have been shown to prey on Coleoptera pest such as *Diabrotica* spp. in North America. The red-winged blackbird *Agelaius phoeniceus* L. (Passeriformes: Icteridae) frequently preys on *D. longicornis* (Say) (Coleoptera: Chrysomelidae) in Canada (Bollinger and Caslick, 1985). The sticky seed heads of *Setaria verticillata* (Poaceae) were found to trap up to 30 *D. v. virgifera* beetles per head in maize fields in Hungary (Toepfer and Kuhlmann, 2004).

Ants are important predators of arthropods in many habitats and geographical regions, helping to control insect pests such as *Diabrotica* sp. (Paulson and Akre, 1992). In field experiments in Costa Rica, *Solenopsis geminata* (Fabricius) and *Pheidole* sp. (Hymenoptera: Formicidae) removed 80% of exposed *Diabrotica* eggs within 3 days (Risch, 1981). As common predators in agricultural settings worldwide, ants are quick to exploit coleopteran prey.

Carabids are the third group of *Diabrotica* predators in croplands and are abundant in temperate climates (Lovei and Sunderland, 1996; Lundgren and Riedell, 2008). More than 24 species of carabids have been shown to feed on *Diabrotica* in the laboratory or in the field. Predatory carabids are able to reduce populations of *D. undecimpunctata howardi* by as much as 50% in small vegetable plots.

One of the more effective predators to control *Dendroctonus valens* LeConte (Coleoptera: Curculionidae) in spruce is *Rhizophagus grandis* Gyllenhall (Coleoptera: Rhizophagidae). *D. valens* was introduced into China by log importation from the U.S. into China in the 1980s. It damages mainly *Pinus tabulaeformis* and *P. armandi,* and is listed among the six major pests in China. The application of *R. grandis* to control *D. valens* has been effective in Belgium, Britain, France and China (Yang, 2004).

CONCLUSIONS

Researches and applications of biological control of coleopteran pests have made a great progress in recent decades. Major commercial products are based on *Bacillus thuringinensis*, *Bacillus popilliae, Metarhizium anisopliae* and *Beauveria bassiana*. Entomopathogenic nematodes are a new biological control factor for coleopteran control and show good efficacy on coleopteran pests, which bore stems or boles in agriculture and forestry. The natural enemies of coleoptera pests, including parasitoids and predators will be useful elements of a strategic approach to the control of coleopteran pests. Overall, there is good evidence that accelerated exploration of biological control options may provide the advances in coleopteran pest management.

REFERENCES

Abad, A.R., Duck, N.B., Feng, X., Flannagan, R.D., Kahn, T.W. and Sims, L.E. (2002). Genes encoding novel proteins with pesticidal activity against coleopterans. US Patent: WO 0234774-A.

Asano, S., Yamashita, C., Iizuka, T., Takeuchi, K., Yamanaka, S., Cerf, D. and Yamamoto, T. (2003). A strain of *Bacillus thuringiensis* subsp. *galleriae* containing a novel *cry8* gene highly toxic to *Anomala cuprea* (Coleoptera: Scarabaeidae). *Biological Control*, **28**: 191-196.

Bollinger, E.K. and Caslick, J.W. (1985). Red-winged Blackbird (*Agelaius phoeniceus*) predation on northern corn rootworm beetles (*Diabrotica longicornis*) in field corn (*Zea mays*). *Journal of Applied Ecology*, **22**: 39-48.

Bradley, H., Harkey, M.A., Kim, M.K., Biever, K.D. and Bauer, L.S. (1995). The insecticidal CryIB crystal protein of *Bacillus thuringiensis* subsp. *thuringiensis* has dual specificity to coleopteran and lepidopteran larvae. *Journal of Invertebrate Pathology*, **65(2)**: 162-73.

Brizzard, B.L, Schnepf, H.E. and Kronstad, J.W. (1991). Expression of the *cry*IB crystal protein gene of *Bacillus thuringiensis*. *Molecular and General Genetics*, **231(1)**: 59-64.

Brizzard, B.L. and Whiteley, H.R. (1988). Nucleotide sequence of an additional crystal protein gene cloned from *Bacillus thuringiensis* subsp. *Thuringiensis Nucleic Acids Research*, **16(6)**: 2723-2724.

Donovan, W.P., Rupar, M.J., Slaney, A.C, Malvar, T., Gawron-Burke, M.C. and Johnson, T.B. (1992). Characterization of two genes encoding *Bacillus thuringiensis* insecticidal crystal proteins toxic to Coleoptera species. *Applied and Environmental Microbiology*, **58**: 3921-3927.

Donovan, W.P., Donovan, J.C. and Slaney, A.C. (2000). *Bacillus thuringiensis cryET33* and *cryET34* compositions and uses thereof. US Patent: US 6063756.

Gan, Y.Y., Qiang, J.H. and Jia, S.H. (2007). Technology on preventing *Monochamus alternatus* larvae with *Beauveria bassiana*. *China Forestry Science and Technology*, **21(1)**: 69-71.

Gaugler, R. (1981). Biological control potential of neoaplectanid nematodes. *Journal of Nematology*, **13**: 241-249.

Herrnstadt, C., Gilroy, T.E., Sobieski, D.A., Bennett, B.D. and Gaertner, F.H. (1987). Nucleotide sequence and deduced amino acid sequence of a coleopteran-active delta-endotoxin gene from *Bacillus thuringiensis* subsp. *san diego*. *Gene*, **57(1)**: 37-46.

Herrnstadt, C., Soares, G.G., Wilcox, E.R. and Edwards, D.L. (1986). A new strain of *Bacillus thuringiensis* with activity against coleopteran insects. *Biotechnology*, **4**: 305-308.

Isaac, B.C., Krieger, E.K., Mettus-Light, A.M., Sivasupramaniam, S. and Moshiri, F. (2001). Polypeptide composionns toxic to anthonomus insects, and methods of use. US Patent: WO 0187940-A.

Jolivet, P.H., Cox, M.L. and Petitpierre, E. (1994). Novel aspects of the biology of the Chrysomelidae. Kluwer Academics Publishers, Doderecht, The Netherlands.

Krieg, A., Huger, A.M. and Schnetter, W. (1987). *Bacillus thuringiensis* subsp. *tenebrionis* Stamm BI 256-82. *Journal of Applied Entomology*, **104**: 417-424.

Krieg, A., Huger, A.M., Langenbruch, G.A. and Schnetter. W. (1983). *Bacillus thuringiensis* var. *tenebrionis*: ein neuer, gegenüber Larven von Coleopteren wirksamer Pathotyp. *Zeitschrift für Angewandte Entomologie*, **96**: 500-508.

Lambert, B., Hofte, H., Annys, K., Jansens, S., Soetaert, P. and Peferoen, M. (1992). Novel *Bacillus thuringiensis* insecticidal crystal protein with a silent activity against coleopteran larvae. *Applied and Environmental Microbiology*, **58(8)**: 2536-2542.

Liu, M., Sun, M. and Yu, Z.N. (1998). The research on *Bacillus thuringiensis* preventing against coleoptera pests. *Chinese Journal of Biological Control*, **14(l)**: 38-42.

Liu, N.X. (1993). Entomopathogenic nematodes in Chinese fields and experimental research. *Insect Natural Enemies*, **15(2)**: 96-100.

Lovei, G.L. and Sunderland, K.D. (1996). Ecology and behavior of ground beetles (Carabidae). *Annual Review of Entomology*, **41**: 231-256.

Lundgren, J.G. and Riedell, W.E. (2008). Soybean nitrogen relations and root characteristics after *Cerotoma trifurcata* (Coleoptera: Chrysomelidae) larval feeding injury. *Journal of Entomological Science*, **43**: 107-116.

Moellenbeck, D.J., Peters, M.L., Bing, J.W., Rouse, J.R., Higgins, L.S., Sims, L., Nevshemal, T., Marshall, L., Ellis, R.T., Bystrak, P.G., Lang, B.A., Stewart, J.L., Kouba, K., Sondag, V., Gustafson, V., Nour, K., Xu, D., Swenson, J., Zhang, J., Czapla, T., Schwab, G., Jayne, S., Stockhoff, B.A., Narva, K., Schnepf, H.E., Stelman, S.J., Poutre, C., Koziel, M. and Duck, N. (2001). Insecticidal proteins from *Bacillus thuringiensis* protect corn from corn rootworms. *Nature Biotechnology*, **19(7)**: 668-672.

Paulson, G.S. and Akre, R.D. (1992). Evaluating the effectiveness of ants as biological control agents of pear psylla. *Journal of Economic Entomology*, **85**: 70-73.

Perlak, F.J., Stone, T.B., Muskopf, Y.M., Petersen, L.J., Parker, G.B., McPherson, S.A., Wyman, J., Love, S., Reed, G. and Biever, D. (1993). Genetically improved potatoes: protection from damage by Colorado potato beetles. *Plant Molecular Biology*, **22(2)**: 313-21.

Pu, Z.L. and Li, Z.Z. (1996). Insect Mycology. Anhui Science and Technology Press, Hefei, China.

Risch, S. (1981). Ants as important predators of rootworm eggs in the Neotropics. *Journal of Economic Entomology*, **74**: 88-90.

Rupar, M.J., Donovan, W.P., Tan, Y. and Slaney, A.C. (2000). *Bacilus thuringiensis* CryET29 compositions toxic to coleopteran insects and *Ctenocephalides* spp. US Patent: US 6093695-B.

Sato, R., Takeuchi, K., Ogiwara, K., Minami, M., Kaji, Y., Suzuki, N., Hori, H., Asano, S., Ohba, M. and Iwahana, H. (1995). Cloning, heterologous expression, and localization of a novel crystal protein gene from *Bacillus thuringiensis* serovar *japonensis* strain buibui toxic to scarabaeid insects. *Current Microbiology*, **28(1)**: 15-19.

Sekar, V., Thompson, D.V., Maroney, M.J., Bookland, R.G. and Adang, M.J. (1987). Molecular cloning and characterization of the insecticidal crystal protein gene of *Bacillus thuringiensis* var. *tenebrionis*. *Proceedings of the National Academy of Sciences in the United States of America*, **84(20)**: 7036-7040.

Shu, C., Yan, G., Wang, R., Zhang, J., Feng, S., Huang, D. and Song, F. (2009a). Characterization of a novel *cry8* gene specific to Melolonthidae pests:

Holotrichia oblita and *Holotrichia parallela*. *Applied Microbiology and Biotechnology*, **84(4)**: 701-707.

Shu, C., Yu, H., Wang, R., Fen, S., Su, X., Huang, D., Zhang, J. and Song, F. (2009b). Characterization of two novel *cry8* genes from *Bacillus thuringiensis* strain BT185. *Current Microbiology*, **58(4)**: 389-392.

Sick, A., Gaertner, F. and Wong, A. (1990). Nucleotide sequence of a coleopteran-active toxin gene from a new isolate of *Bacillus thuringiensis* subsp. *tolworthi*. *Nucleic Acids Research*, **18(5)**: 1305.

Sun, M., Zhang, L. and Yu, Z.N. (2003). Molecular biology of *Bacillus thuringiensis*. *In*: Upadhyay, R.K. (*Ed.*), *Advances in Microbial Control of Insect Pest*. Kluwer Academic/Plenum Publishers, New York, USA.

Tailor, R., Tippett, J., Gibb, G., Pells, S., Pike, D., Jordan, L. and Ely, S. (1992). Identification and characterization of a novel *Bacillus thuringiensis* delta-endotoxin entomocidal to coleopteran and lepidopteran larvae. *Molecular Microbiology*, **6**: 1211-1217.

Tan, Y.C. (1992). The prevention and control of pests and pathogens in hybrid rice: Chinese Agricultural Science and Technology Press, Beijing, China. pp. 79-85.

Toepfer, S. and Kuhlmann, U. (2004). Survey for natural enemies of the invasive alien Chrysomelid, *Diabrotica virgifera virgifera*, in Central Europe. *BioControl*, **49**: 385-395.

van Frankenhuyzen, K. (2009). Insecticidal activity of *Bacillus thuringiensis* crystal proteins. *Journal of Invertebrate Pathology*, **101**: 1-16.

Waalwijk, C., Dullemans, A.M., van Workum, M.E. and Visser, B. (1985). Molecular cloning and the nucleotide sequence of the Mr 28000 crystal protein gene of *Bacillus thuringiensis* subsp. *israelensis*. *Nucleic Acids Researchs*, **13(22)**: 8207-8217.

Yamaguchi, T., Sahara, K., Bando, H. and Asano, S. (2008). Discovery of a novel *Bacillus thuringiensis* Cry8D protein and the unique toxicity of the Cry8D-class proteins against scarab beetles. *Journal of Invertebrate Pathology*, **99(3)**: 257-262.

Yang, Z.Q. (2004). Advance in bio-control researches of the important forest insect pests with natural enemies in China. *Chinese Journal of Biological Control*, **20(4)**: 221-227.

Yu, Z.N. (1990). *Bacillus thuringiensis*: Science Press, Beijing, China.

Yu, Z.N. (2000). *Microbial pesticide and its industrialization. Science Press*, Beijing, China.

Zhang, J., Hodgman, T.C., Krieger, L., Schnetter, W. and Schairer, H.U. (1997). Cloning and analysis of the first *cry* gene from *Bacillus popilliae*. *Journal of Bacteriology*, **179(13)**: 4336-4341.

Zhang, J.B., Chen, G.Y., Ming, G.Z., Yuan, F.Y., Xu, B.Z. and Zhang, S.Q. (1998a). Efficacy of B.t.i floating briquette against *Culex quinquefasciatus* and *Aedes albopictus* larvae in small breeding sites in urban area, *Chinese Journal of Vector Biology and Control*, **9(3)**: 192-194.

Zhang, H.Y., Yu, Z.N. and Deng, W.X. (1998b). The screening of typical coleoptera pests measured biologically using *Bacillus thuringiensis* preparation. *Hubei Agricultural Science*, **(2)**: 40.

{7}

Biotechnology and Derived Products

Patricia Tamez-Guerra and Robert W. Behle

ABSTRACT

Microorganisms able to infect and kill insect pests, metabolites from plants and microorganisms, and transgenic crops are biotechnologically-derived products that are being promoted for use to control insect pests in lue of chemical insecticides. Products based on these technologies effectively control important pests of health, food and agriculture systems. Among products classified as bioinsecticides, entomopathogenic microorganisms, organism metabolites and transgenic seeds are considered as biotechnology derived agents for insect pest control. There are many bioinsecticides that are efficacious and currently available as commercial products in the agroindustry market. This review describes background of the general categories of bioinsecticides and lists current bioinsecticide products available for use to control insect pests.

INTRODUCTION

The increasing world population precludes the need to seek technologies to achieve high yields in agriculture, by controlling important pests of health and food products. Although chemical insecticides have contributed greatly to growth of the agricultural industry, problems

---

Commercial Endorsement Disclaimer
The use of trade, firm, or corporation names in this web site is for the information and convenience of the reader. Such use does not constitute an official endorsement or approval by the United States Department of Agriculture or the Agricultural Research Service of any product or service to the exclusion of others that may be suitable.

associated with applications of chemical insecticides have helped promote increased research, development, and use of bio-based technologies. Awareness of the negative implications of chemical insecticides on the ecosystems, human and animal health, and development of resistance by target insects, have played a role to encourage development of bioinsecticide products since late 1970's.

Additionally, companies focusing on the insecticide market have realized that the demand for organically produced food is increasing and profitable. Consequently, efforts to development safe, efficient and easy-to-use bioinsecticides have increased (Gips, 1986). Techniques in biotechnology have been used to improved products and, along with integrated pest management programs, may extend the working timeframe of bioinsecticides compared with chemicals, particularly by delaying the development on insecticide resistance by target pests (Menn and Hall, 1999). When comparing costs of research, development, and registration, marketing costs for bioinsecticides are about 3% that of chemical insecticides.

Reasons for this advantage of bioinsecticides include:

1. Reduced safety risk during production compared with chemical insecticides (processes do not involve hazard reagents or chemical reactions).
2. Evaluations of safety for human health and ecosystems are less rigorous compared with chemicals.
3. Production waste is not dangerous, but may have value as a bio-fertilizer source.
4. Easier registration by the U.S. Environmental Protection Agency (EPA) (Kennedy *et al.,* 2007).

For example, from 1995 to 2000, new synthetic chemicals required over $185 million and 10 years development period before reaching the market, compared with a biopesticide that required about $6 million and only 3 years (Kennedy *et al.,* 2007). The relationship among biopesticides, research, and product development, using the model proposed by Boyetchko (2008) http://www.agriculture.gov.sk.ca/bio-pesticide, is represented in Fig. 1. The figure represents the complexity of research generally required to fully develop a biologically based pest control product.

Included among categories of bioinsecticides are entomopathogenic microorganisms, organism metabolites and transgenic seeds that are biotechnology derived products for insect pest control. This chapter will review known entomopathogenic microorganisms, organism metabolites and transgenic seed commercialized products developed for control of insect pests.

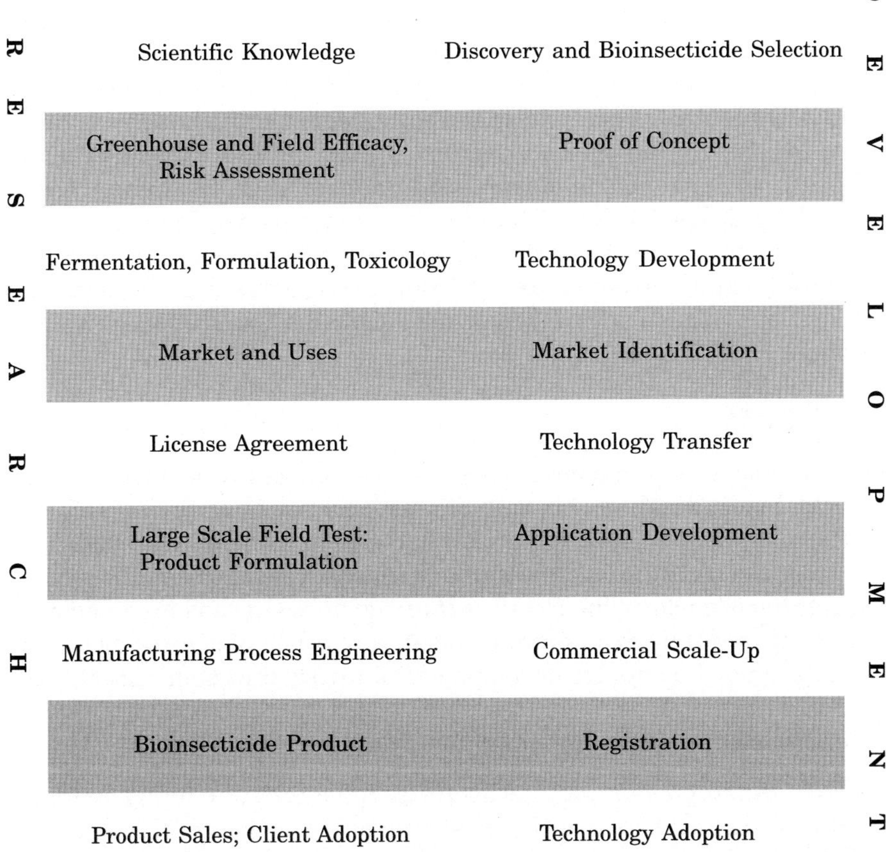

Fig. 7.1. Relationships among biopesticides, research, and product development

ENTOMOPATHOGENIC MICROORGANISMS AS BIOINSECTICIDES

Entomopathogenic microorganisms have been described since 1400's. The research and development of bioinsecticide products continues to increase in response to the growing concern over the use of chemical pesticides and their negative impact on the environment. Scientists recognize that biologically-derived insecticides are a key alternative to broad spectrum chemical insecticides for insect management programs. Researchers realized that specific entomopatogenic microorganisms as well as organism-derived products were more host-specific than chemicals, more quickly broken down in the environment and harmless to non-targeted organisms such as bees, birds, fish and mammals. The two biggest challenges have been to 1) reduce the production cost for biological insecticides to be comparable with the cost of chemicals, and 2) to demonstrate comparable pest control efficacy (Cooping, 2004). More than 1500 species of microorganisms able to infect, intoxicate and/or kill insects have been reported to date. The use of fungi, viruses, bacteria, protozoa and nematodes as control agents of farm and forest insect pests is environmentally more suitable than chemicals. Although currently more expensive, biological insecticides may be less expensive in the long run in terms of deleterious side effects such as human health hazards and destruction of non-targeted organisms. Some organisms are able to persist in the environment extending the time of pest control long after application, making them more attractive to use than conventional insecticides (Kanzok and Jacobs-Lorena, 2006; Grewal and Georgis, 1998; Priest, 2000).

BACTERIA

Over 100 species of bacteria have been recognized that infect and kill insects. Insect death may be a result of either sepsis or toxins produced by the microbe. The most important bacteria for controlling insect pests are from the Bacillaceae family, particularly, from the *Bacillus* genera. Among *Bacillus* bacteria, *Bacillus thuringiensis* (Bt) has provided more biotechnological derived products than any other biological agent and remains a good model for development of future microbial insecticides.

Bacillus thuringiensis-based Products

Bacillus thuringiensis (Bt) has been studied more than many synthetic products. Bt products are the most proven, most successful and most widely used in pest control programs. Bt was discovered in 1901 as the source of disease that was killing large populations of silkworms. It was first used as an insecticide in 1920. The first commercial preparation of Bt for use in agriculture was produced in France in 1938, where a spore-based product was distributed as Sporeine to control flour moths (Iriarte and Caballero, 2001). The mode of action involves a toxic effect. During the sporulation process of the bacterium, an insecticidal protein (crystal-shaped, named Cry protein) is produced. At the end of the sporulation stage, both spores and crystals (protoxin) are released. Once a susceptible immature insect (larva) eats the crystal, specific gut enzymes, which only function in the alkaline conditions of the caterpillar gut, dissolve the crystal producing the active ingredient or toxin. The toxin binds to specific receptors in the larva's gut, which disrupts the pest's digestive tract. Infected larvae stop eating and die 12 hours to 5 days after ingestion. The speed of kill depends on the amount of Bt ingested, the size and variety of the larvae species (Knowles, 1994). Bt was registered by the US EPA in 1961 and its use expanded in the 1980s after the HD-1 strain, var *kurstaki* (Btk) was described. HD-1 showed toxicity 20 times higher than previous Bt strains against several important Lepidoptera pests (Iriarte and Caballero, 2001). Companies across the world started evaluating Btk as an alternative to synthetic chemical insecticides for application to farm and forest environments for Lepidoptera control. Today's HD-1 Bt products have longer persistence, easy-to-mix formulations and consistent efficacy that makes them useful for resistance management. Additional strain discoveries have expanded Bt control programs beyond Lepidoptera to include Diptera and Coleoptera (Cerón, 2001). For Dipteran pest control, Bt var. *israelensis* (Bti) products have been effective in a variety of habitats for control of larvae of several species of mosquito, black fly, midge and fungus gnat. For mosquito control, Bti is applied on the surface of aquatic breeding area. The larvae ingest the insecticide, consisting of Bti spores and crystals; crystal toxin is activated by specific gut enzymes. In this case, the insect's intestine is acidic and the protein crystal of Bti was specifically selected to target these insects because it dissolves in the acid rather than alkaline conditions. As a result, larvae stop feeding and die before they can pupate and reach adulthood (Russell *et al.,* 2009). For Coleoptera control, Bt strains from serovarieties such as var. *tenebrionis* (Btt) has proven their efficacy. As before, this Bt strain was selected because it crystal protein protoxin digests in the gut of the target pest. Bt products have been used as spray formulations for crop protection, especially for organic farming operations (Cerón, 2001).

Advantages of Bt product include the following:

1. Cry proteins are highly selective on only certain species of insects that make them safe for human and other mammals.
2. Bt does not affect beneficial and non-targeted insect populations due to its specificity.
3. Bt does not represent any danger to crops on which it is sprayed and can be used even before or at the harvesting step.
4. Bt killed insects do not affect birds or other animals that consume them as food.
5. Bt does not harm the surrounding environment.

6. Only limited pest resistance to Bt products has been reported to date.
7. Biotechnology production follows high quality standards and established protocols verifying activity and ensuring consistent field performance.
8. Bt products can be used to control more than 130 insects, mostly lepidopteran, dipteran and coleopteran larvae, known as important pests.

It must be noted that Bt products were not an immediate success in terms of economical production and consistent field efficacy. First academic and government research, and then agroindustry research focused development programs to overcome these problems (Srivastava *et al.,* 2009). The number of commercial products available attests to the success of this research and development. Some of the products based on entomopathogenic bacteria available in the international market are listed in Table 7.1.

Table 7.1. Leading products based on entomopatogenic bacteria commercialized by 2009

Bacteria	***Product name***[1]	***Order of Susceptible Insects***
Bt *kurstaki*	Biobit, Dipel, Dipel DF& Dipel ES, Foray, Thuricide **(1)**; Able, Deliver, Javelin WG **(2)**; PHC Beretta **(3)**; Turilav **(4)**; Scutello **(5)**; Biodart **(9)**; Larvorid **(10)**; Thurisave 1 **(11)**; Bio 2001-P, Ecobacilus-P Nuevo **(12)**; Biolcan-Bt **(13)**; Belthirul WP, Labicillus **(14)**; Lipel SP **(17)**; Britz Bt **(19)**; BST 88 **(20)**; Agro Turin **(22)**; B-Tec 32 **(23)**; Bacillus IAB Bt **(24)**; Bt Meristem **(25)**; Baktur **(27)**; Delfin, Novodor 3F **(28)**	Lepidoptera
(recombinant)	Crymax WDG, Lepinox WDG **(2)**	Lepidoptera
Bt *aizawai*	Agree WG, Florbac, Florbac HP WP **(2)** XenTari **(1)** **(14)**	Lepidoptera
Bt *kurstaki*/ *aizawai* (transconjugate)	Condor XL **(2)**	Lepidoptera
Bt *tenebrionis*	Novodor **(1)**	Coleoptera
Paenibacillus popilliae	Doom **(6)** Milky Spore **(7)**	Coleoptera
Bt *israelensis*	Bactimos, Gnatrol & Gnatrol DG, Teknar, Vecto Bac **(1)**; Mosquito Dunks, Mosquito Bits, B.t.i.; Briquets **(8)**; Biodart M **(9)**; BT Horus SC **(15)**; Biagro Bti **(21)**; BectoNeel **(17)**; Coop Oecoplan Biocontrol **(26)**; Skeetal **(27)**; Solbac **(28)**	Diptera

Table 7.1. *(Contd...)*

Table 7.1. (*Contd...*)

Bacteria	*Product name*[1]	*Order of Susceptible Insects*
Bacillus sphaericus	Sphaericide, VectoLex **(1)** Sphaerus SC **(15)**	Diptera
Bacillus thuringiensis	PestoBAC™ **(18)**	Broad spectrum

[1]**(1)** Valent Bioscience Corp., USA (previously Abbott); **(2)** Certis, USA; **(3)** PHC de Mexico, Mexico; **(4)** Laverlam, S.A., Colombia; **(5)** Consorcio de Bélgica (formerly Biobest, N.V.); **(6)** Fairfax Biological Laboratory, Inc., USA; **(7)** St Gabriel Lab., USA, and Gardener's Supply Company, USA; **(8)** Summit Chemical Co, USA Compagnie du Bois Sauvage (40%), Floridienne (by Florinvest 20%) and Domaine d'Argenteuil (40%); **(9)** Ajay BioTech (India) Ltd; **(10)** Exotic Naturals (India); **(11)** Agrotécnica Murciana, Spain; **(12)** Agromed; **(13)** Biológicas Canarias; **(14)** EcoTenda, Spain; **(15)** CEMA Agricultura Biologica, Brazil; BT Horus; **(16)** Agrogreen, Israel; **(17)** Agri Life/ Som Phytopharma Ltd., India; **(18)** Biological Targets Inc., USA; **(19)** Britz Fertilizers, Inc., USA; **(20)** Agrícola El Sol, Guatemala; **(21)** Laboratorio Biagro S.A., Argentina; **(22)** Agrobiológica, S. A. de C. V., Mexico; **(23)** Técnicas de Control Biológico, Spain; **(24)** Tradecorp, S. A., Spain; **(25)** Químicas Meristem S. L., Spain; **(26)** Coop, Switzerland; **(27)** Omya, Switzerland; **(28)** Andermatt Biocontrol & Andermatt Biogarten, Switzerland.

Bacillus sphaericus-based Products

Bacillus sphaericus is a naturally occurring spore-forming bacilli found worldwide in soil and aquatic environments (WDE 2002). Like Bt, *B. sphaericus* produces an insecticidal protein during sporulation phase, which is toxic to many species of mosquito larvae after ingestion. *B. sphaericus* also infects larvae of some closely related beetles, as well as species of mosquitoes in the genera *Culex, Aedes, Psorophora, Coquillettidia, Mansonia* and *Anopheles*. Studies demonstrated longer persistence rates of *B. sphaericus* spores compared with Bti in polluted habitats. This may be the reason why *B. sphaericus* products have been applied mainly on organically enriched habitats against *Culex* species, even though it is pathogenic against a variety of species across several genera and is recommended for application to several habitats (Smitley, 1996). Under certain circumstances, *B. sphaericus* has the advantage that it can recycle from larval cadavers (WDE, 2002; Lacey, 2007) providing additional residual activity. A disadvantage of *B. sphaericus* has been the development of resistance in certain populations of *Culex* (Charles *et al.,* 2000; Darboux *et al.,* 2007). Bti and *B. sphaericus* have been combined for Dipera control in many pest management programs because dual activity has lead to better mosquito pest control (Lacey, 2007). The list of brand name Bt- and *B. sphaericus*-based bioinsecticides registered by 2008 are listed in Table 7.2.

Table 7.2. List of *Bacillus thuringiensis* and *Bacillus sphaericus* commercial products registered by Environmental Protection Agency (EPA)-USA1, 2008

Btk strain ABTS 1857 (Valent Bioscience Corp.)
- ABG6346 BIN
- Bta slurry
- XenTari AS BIN
- XenTari BIN dry flowable
- XenTari BIN technical powder
- XenTari BIN water dispersible granules

Btk strain ABTS 351
- ABG-6345 technical powder (Valent Bioscience Corp.)
- Britz Bt 25 sulfur dust (Britz Fertilizers Inc)
- Britz Bt dust (Britz Fertilizers Inc)
- Bt 320 sulfur 25 dust (Wilbur Ellis Co.)
- Btk-slurry (Valent Bioscience Corp.)
- Bt sulfur 15-50 dust (Loveland Products, Inc.)
- Dipel 10G BIN granules (Valent Bioscience Corp.)
- Dipel®10G Sweet corn granule BIN (Bonide Products Inc.)
- Dipel 2X BIN (Valent Bioscience Corp.)
- Dipel®Bio garden spray (Bonide Products Inc.)
- Dipel DF dry flowable BIN (Valent Bioscience Corp.)
- Dipel ES BIN (Valent Bioscience Corp.)
- Dipel ES-NT BIN emulsifiable (Valent Bioscience Corp.)
- Dipel SG plus BIN sand granule (Valent Bioscience Corp.)
- Dipel technical powder (Valent Bioscience Corp.)
- Dipel (worm killer) WP BIN (Valent Bioscience Corp.)
- Dipel WDG BIN (Valent Bioscience Corp.)
- Foray 48BC (Valent Bioscience Corp.)
- Foray 48F BIN flowable concentrate (Valent Bioscience Corp.)
- Foray 76B (Valent Bioscience Corp.)
- Ringer vegetable insect attack (Woodstream Corp.)
- SA-50 Brand Dipel dust (Southern Agricultural Insecticides, Inc.)
- Wilbur-Ellis Bt 320 Dust (Wilbur Ellis Co.)

Btk strain BMP123 ([1]Agraquest; [2]Becker Microbial Products Inc.)
- Baritone®[1]
- BMP®(2X WP) [2]
- BMP®(10G) [2]
- BMP®(48 LC) [2]
- BMP®(64-ES) [2]
- BMP®(primary product) [2]

Btk strain EG-2348 (Certis USA, LLC)
- Condor®
- Condor®WP
- Condor®XL
- Condor®Technical powder

Btk strain EG2371*(Certis USA, LLC)
- CoStar®
- Cutlass®Technical powder
- Deliver®
- Deliver 2ee®

Btk strain EG-7826 (Certis USA, LLC)
- Lepinox®
- Lepinox®WDG

Btk strain EG7841 (Certis USA, LLC)
- Crymax®
- Crymax®WP

Btk strain SA-11
- Biobit HP BIN WP (Valent Bioscience Corp.)
- Biobit HPWP II BIN (Valent Bioscience Corp.)
- Bonide Dipel 150 dust for vegetables (Bonide Products, Inc.)
- Dipel WP home & garden insecticide (Bonide Products, Inc.)
- Ferti-Lome Dipel biological worm spray (Voluntary Purchasing Groups, Inc
- Green light Bt worm killer (Green Light Co.);
- Javelin®WG (Certis USA, LLC)
- Javelin®WP (Certis USA, LLC)
- Safer B.t. caterpillar killer concentrate (Safer, Inc.)
- San 420 I WG (Certis USA, LLC)
- Thuricide 48LV forestry (Valent Bioscience Corp.)
- Thuricide 76 LV (Valent Bioscience Corp.)

Table 7.2. (*Contd...*)

Table 7.2. (*Contd...*)

Btk strain SA-12
Bonide *Bacillus thuringiensis* (Bt) moth larvae (caterpillar) control (Bonide Products, Inc.)
BTK 32 (Certis USA, LLC)
SA-50 Brand Thuricide HPC (Southern Agricultural Insecticides, Inc.)
SAN420 I technical concentrate (Certis USA, LLC)
Security Bt dust BIN (Wellmark international)
Thuricide 48 LV (Certis USA, LLC)
Thuricide HPC (Certis USA, LLC)
Thuricide HPCX Home & Garden (Certis USA, LLC)
Thuricide HPWP (Certis USA, LLC)
Thuricide technical concentrate (Certis USA, LLC)

Bti strain AM 65-52
Bacillus thuringiensis israelensis slurry (Valent Bioscience Corp.)
Bti granules (Clarke Mosquito Control Products, Inc.)
VBC-60092 (Bactimos briquette) (Valent Bioscience Corp.)
Vectobac®BL (Valent Bioscience Corp.)
Vectobac®BL-12AS (Valent Bioscience Corp.)
Vectobac®BL-12ASII (Valent Bioscience Corp.)
Vectobac®BL-AS (Valent Bioscience Corp.)
Vectobac®BL-CG granules (Valent Bioscience Corp.)
Vectobac®BL-G granules (Valent Bioscience Corp.)
Vectobac®BL –PWDG (Valent Bioscience Corp.)
Vectobac®Primary powder (Valent Bioscience Corp.)
Vectobac®WDG (Valent Bioscience Corp.)

Bti strain BMP 144
Aquabac™ 200G (AFA Environnement Inc., Canada)
Aquabac™ Mosquito Larvae Control (AFA Environnement Inc., Canada)
BMP 144®2X (Becker Microbial Products Inc.)
BMP 144®200G (Becker Microbial Products Inc.)
BMP 144®3X (Becker Microbial Products Inc.)
BMP 144®400G (Becker Microbial Products Inc.)
BMP 144®DF (Becker Microbial Products Inc.)
BMP 144®DF3000 (Becker Microbial Products Inc.)
BMP 144®DFX (Becker Microbial Products Inc.)
BMP 144®WS-Bti (Becker Microbial Products Inc.)
Bti slurry (Becker Microbial Products Inc.)
Meridian-02 (Becker Microbial Products Inc.)
XS-Bti (Watersavr Global Solutions, Inc.)

Bti strain EG 2215
HP Biological larvicide aqueous suspension (Certis USA, LLC)
HP-D (Certis USA, LLC)
HP-D slurry (Certis USA, LLC)
Technical concentrate (Certis USA, LLC)
Bonide®mosquito beater WSP (Bonide Products, Inc.)
Bti granule larvicide (Certis USA, LLC)
Bti technical powder bioinsecticide (Certis USA, LLC)
Bti mosquito briquets (Certis USA, LLC)
Bti mosquito dunks (Certis USA, LLC)
Summit mosquito baits (Summit Chemical Co.)
Summit mosquito briquets (Summit Chemical Co.)
Chemisco®Insect granules ML (Chemisco)
Healthy outdoors brand sustained release mosquito larvicide (Plant Defense Boosters, Inc.)

Bti strain SA3A (Valent Bioscience Corp.)
Teknar®ABG-6346 BIN
Teknar®AS BIN
Teknar®BIN WDG
Teknar®BIN Dry flowable
Teknar®Bti Technical powder

Table 7.2. (*Contd...*)

Table 7.2. (*Contd...*)

Bt subspecies *tenebrionis* (Btt), strain NB-176 (Valent Bioscience Corp.)
Bacillus thuringiensis subsp *tenebrionis* slurry
Btt Technical powder
Btt Technical powder FC
Btt slurry
Novodor flowable concentrate
Novodor technical

Bt subspecie *aizawai* (Bta) strain CG91 (Certis USA, LLC)
Agree
Agree 50 WP
Technical CGA-237218
Bta strain NB200 (Certis USA, LLC)
Florbac slurry BL

***Bacillus sphaericus* strain 2362 (Valent Bioscience, Corp.)**
B. sphaericus slurry
V60035 CG
Vectolex CG BL
Vectolex WDG BL

Bti (strain EG2215) & *B. sphaericus* (Valent Bioscience Corp.)
V60035 CG

Cry1Ab protein segment (Monsanto)
Btk insect control protein
Attribute insect protection sweet corn
Northrup king insect resistant corn

Btk Cry1Ab (plasmid) (Monsanto)
MON 88017
XMON 810

Btk Cry1A(c) protein as produced by the gene (Monsanto)
MON 531
Bollgard LC gold II cotton
Bollgard II cotton

Btk (Cry1Ac)-Bta (Cry1F) (Dow Agroscience)
Mycogen brand Cry1F / Cry1Ac (synpro)
Construct 281/3006

Bt mo Cry1F event TC5276 corn (Dow Agroscience)
Mycogen brand Bt Cry1F

Bt Cry1F (plasmid) (Dow Agroscience)
Pioneer bran seed with Herculex I
Herculex I
Herculex XTRA

Bt Cry 2Ab (Vector G) (Monsanto)
MON 15947

Btt Cry 3A (Monsanto)
Colorado potato beetle protein
New leaf plus potatoes
Plant pesticide Btt

Bt Cry34b + Cry35Ab proteins
Pioneer brand Bt Cry 34/35B/ (Pioneer)
Herculex XTRA (Monsanto)
Mycogen Bt Cry 34/35 const PHP17662 corn (Dow Agroscience)

Bt Cry 3Bb (Vector Z) (Monsanto)
MON 863 corn rootworm protected
Yieldgard plus corn

Bt Cry 3BbI (Vector Z MIR39) (Monsanto)
MON 88017
MON 88017 XMON 810

Paenibacillus popilliae-based products

In 1948, spores of *Paenibacillus popilliae* (formerly known as *Bacillus popilliae,* Pettersson *et al.,* 1999) became the first microbial agent registered as a pesticide active ingredient in USA. As of October 2004, there was one end-use product, Vectolex (Table 7.2). Spores of *P. popilliae* infect larvae (grubs) of Japanese beetles, eventually killing the larvae and preventing their development into adult beetles. As a pesticide active ingredient, the spores of this bacterium are approved for use on lawns and ornamental plants around residential areas.

FUNGI

Entomopathogenic fungi are an effective tool for use as biological insecticides. Recently, 170 mycoinsecticide products were available commercially worldwide (Faria and Wraight, 2007). Most of the products are based on fungi classified as anamorphic Hypocreales and more specifically on *Beauveria bassiana* (34%), *Metarhizium anisopliae* (36%), *Lecanicillium* spp. (9%) and *Isaria* (formerly *Paecilomyces*) *fumosorosea* (6%) (Faria and Wraight, 2007). Fungal agents as bioinsecticides have similar advantages and disadvantages when compared with bacterial and viral (following section) pathogens as pest control agents. In contrast with bacterial and viral pathogens, fungal agents have different mode of action, target pests, and production techniques. These differences allow the fungal pathogens to target arthropod pests that are not susceptible to previously discussed biological insecticides.

Advantages/Disadvantages

Fungal-based insecticides provide many of the benefits generally recognized to be associated with biological pesticides. They are generally target specific, provide good control efficacy, are considered safe for humans and relatively benign to the environment, and have the potential to cycle in the environment to provide additional residual control benefits beyond the initial application. Zimmermann (2007) recently reviewed the safety of *Beauveria bassiana* and *Beauveria brongniartii* relative to environmental fate of the fungus, effects on non-target organisms, and human health, and concluded that these two fungal species are safe. He also identified earlier publications concerning the safety of microbial insecticides (Hall *et al.,* 1982; Laird *et al.,* 1990; Goettel and Jaronski, 1997; Goettel *et al.,* 2001; Vestergaard *et al.,* 2003; Cooping, 2004). The inherent safety of these agents remains one of the greatest advantages of microbial pesticides. Unfortunately, fungal-based insecticides suffer disadvantages similar to other microbial-insecticides when compared with chemical insecticides. Mycoinsecticides suffer from slow speed of kill (5-10 days), rapid environmental degradation in the field due to desiccation and exposure to sunlight, expensive production techniques, and limited storage stability. As with most microbial agents, successful mycoinsecticides fill specific pest control situations, where environmental preservation or consumer health prevail over other considerations (Montesinos, 2003). Examples of success include applications in glasshouses or to natural habitats, where adverse effects of chemical pesticides out-weight their benefits.

Fungal agents differ from bacterial and viral insecticides in several important characteristics including activity by contact rather than ingestion, the ability to produce alternative propagules for application, hydrophobicity of conidia, and toxin production. These characteristics broaden the considerations available when developing mycoinsecticides in terms of target pests, production techniques, and formulation considerations.

The mode of action of mycoinsecticides is characterized as contact activity because viable spores that contact a susceptible insect will germinate and infect the target insect directly through the cuticle. This differs from bacteria and viruses that must be ingested to initiate infection through the intestine. Zimmermann (2007) describes the infection process in six steps: 1) attachment of the spore to the cuticle, 2) spore germination, 3) penetration through the cuticle, 4) overcoming the host's immune defense response to infection, 5) fungus proliferation within the host (formation of hyphal bodies/blastospores), and 6) saprophytic outgrowth from the host cadaver and production of new conidia. Contact of the target pest can either be direct contact by the spray application or by contact with spray residue (Adu-Mensah, 2002; Behle, 2006). When contacting a susceptible insect, the hydrophobicity of conidia helps them to adhere to the insect cuticle. When environmental conditions are favorable, conidia germinate and specific digestive enzymes dissolve the cuticle allowing the fungus to penetrate into the insect. Once inside, the fungus kills the insect as a result of tissue destruction and occasionally by toxins produced by the fungus. When environmental conditions are suitable, the fungus emerges from the insect's body and produces spores that are released to environment to infect other insects.

Host/target Pest

The contact mode of action of mycoinsecticides, allows them to target insect pests not easily infected by other pathogens. Insects with piercing mouthparts fall into this group of insects which generally are not susceptible to bacterial and viral biopesticides, which need to be ingested to initiate infection. Thus, mycoinsecticides fill a control gap left by other microbial agents. As a group and even for several individual species, fungi have relatively wide host ranges and claim control of insects in ten orders: Hemiptera, Coleoptera, Lepidoptera, Thysanoptera, Orthoptera, Diptera, Hymenotpera, Isoptera, Siphonaptera, and Blattodea. In addition, fungi attach several important arachnids including mites and ticks (Faria and Wraight, 2007). The host range of fungi is wide enough to include a large number of target pests, and provides many opportunities for product development.

The host range of fungi unfortunately includes many beneficial insects such as predatory coccinellid beetles, Hemiptera and arachnids. Applications of mycoinsecticides could, therefore, limit the benefit of this important natural control similar to application of a chemical insecticide. Although beneficial arthropods are susceptible to mycoinsecticides, many are less susceptible than the target pest (Duso *et al.,* 2008, Labbe *et al.,* 2009). Also, some predators are able to detect and avoid preying on infected pray insects (Meyling and Pell, 2006). These characteristics minimize the hazard for beneficial insects, especially when compared to the impact of chemical insecticide applications, which are typically more toxic to beneficial insects than to the pests.

The hydrophobic characteristics of conidia not only contribute to adhesion to target insects, but have implications for formulation into mycoinsecticide products. As a result, products that use conidia, often lend to use of non-polar systems such as emulsifiable concentrate formulations. These formulations tend to be more active than wettable powder formulations made with the same pathogen (Kassa *et al.,* 2004). Oil-based formulations are also suited for ultra low volume (ULV) applications. ULV formulations have been developed and successfully tested for control of pests in large environments such as forests or range land (Kassa *et al.,* 2004).

Toxins

Entomopathogenic fungi are known to produce secondary metabolites that are classified as toxins. Li *et al.,* (2003) classified fungal toxins into two groups according to their structure and function: low molecular weight compounds including cyclodepsipeptides, pigments, organic acids, and others, and higher molecular weight proteins including proteases and protein toxins. Examples of low molecular weight cyclic peptide toxins include Beauvericin and Oosperein (from *Beauveria*), and Destruxin (from *Metarhizium*). Although the hazard of applying toxin producing fungi to crops is recognized, the presence of these toxins has not in itself hindered registration of fungal insecticides. These toxins continue to be evaluated for their potential benefits as therapeutic drugs (Lemmens-Gruber *et al.,* 2009) and hazards as contaminants of food and feed (Jestoi, 2008).

Production

Fungal agents for microbial insecticides can be produced using a variety of commercial techniques. Techniques are broadly categorized as solid-substrate or liquid fermentation. Selection of specific production techniques is often related to specific propagule to be used as the active agent in the biopesticide. Propagues include conidia, blastospores, hyphae, and microsclerocia. Conidia are the most commonly reported as the active ingredient in mycoinsecticides and are often produced using solid substrate techniques. This solid substrate production uses sterilized grain or organic matter as a medium for saprophytic mycelia growth, followed by sporulation to form conidial spores (Bhanu-Prakash *et al.,* 2008; Chen *et al.,* 2009; Murugushankar *et al.,* 2009). These spores are collected and formulated into commercial insecticides. Faria and Wraight (2007) point out that most products do not contain a defined propagule composition and the active ingredient concentration (consisting of a mixture of propagules) is determined based on colony forming units. Many factors, including substrates (El Damir, 2006), equipment (Ye *et al.,* 2006), and environmental conditions such as light exposures (Zhang *et al.,* 2009) impact conidial production. Considerable research endeavors to maximize yield of conidia. Even with continual improvements that satisfy laboratory and field studies, yields often are not sufficient for industrial scale production as required for product commercialization (Posada-Florez, 2008). In contrast with solid substrate production, bioreactors have been successfully used to produce submerged conidia (Kassa *et al.,* 2008; Asaff *et al.,* 2009), blastospores (Jackson *et al.,* 2004; Lozano-Contreras *et al.,* 2007) and microsclerocia (Jackson and Jaronski, 2009) using liquid fermentation media. Liquid fermentation has the advantage over solid substrate production by achieving a cost advantage with scale-up of the process. Many production factors determined by laboratory size reactors readily transfer to larger industrial production equipment. As a disadvantage, repeated sub-culturing can result in changes in fungal characteristics, notably a loss of pest virulence (Nahar *et al.,* 2008). Production of microsclerocia as the active agent represents a new propagule for mycoinsecticides. This structure is intended to be applied to the target environment where the structure sporulates and produces new conidia. These new conidia are intended to cause the infection of the target pest. Mycoinseticides based on microsclerocia will likely be used for soil applications (Jackson and Jaronski, 2009), although other indirect applications may be identified with future research. The ability of fungi to produce a variety of propagules improves the potential for developing a fungus as a mycoinsecticides for specific pest control situations.

Unfortunately, mycoinsecticides have several specific problems that reduce the efficacy of applications. Conidia require favorable environmental conditions to initiate germination. Adequate moisture is generally recognized as a requirement for germination. Thus, if a susceptible insect resides in a dry climate, spores may be unable to germinate and initiate infection. Delays in germination allow time for other adverse factors such as sunlight exposure, high temperatures, or molting of the cuticle which may allow the insect to circumvent the potential infection. Insect behaviors also impact efficacy of mycoinsecticide applications. Insects that tend to be exposed to direct contact with applications tend to be more susceptible to infection. Also, mobile insects would have greater opportunity to contact spray residue, improving the potential efficacy of the application. Insects that are cryptic and docile would tend to reduce the potential for contact and subsequent infection. As a direct response to infection, Blandford *et al.* (1998) reported the Senegalese grasshopper, *Oedaleus senegalensis* (Krauss), expresses a thermal regulatory behavior (bask in sunlight) to raise body temperature in an effort constrain the pathogen and limit its impact.

There continues to be a high potential for development of fungal agents as biological insecticides. Fungi have diverse characteristics from other microbes, allowing them to fill some control gaps left by other pathogens. Research continues to identify new characteristics (such as the production of microsclerocia) that further improve opportunities for providing effective pest control. Other research continues to identify methods to improve mycoinsecticide efficacy. Efficacy may be improved by applications that target early stages of the pest to prevent population buildup, target pests under moderate environmental conditions, select crops amenable to highly efficient spray applications, making applications without or asynchronously with fungicide applications.

Most fungi that are currently being commercialized in the world are listed in Table 7.3. *Aschersonia aleyroides* not EPA registered in U.S., Product Aseronija, USSR *All Union Inst.* Not longer in Europe (Eilenberg *et al.,* 2009).

VIRUSES

More than 500 insects are susceptible to infection by viruses. Members of the pathogenic family Baculovirus have been isolated from nearly every moth registered as a pest (von Beek, 2007). Based on this vast diversity, it is possible that viruses could be developed as a major strategy for future pest control. Both DNA and RNA viruses are known to cause infections (Table 7.4). However, double stranded DNA baculovirus are used in all commercial insecticides except the RNA virus *Dendrolimus punctatus* CPV, Familly Reoviridae, registered in China (Sun and Peng,

Table 7.3. Products based on entomopathogenic fungi registered and/or commercialized by 2009

Product name, country	*Company*	*Active against*	*Reference/web link*
Beauveria bassiana			
Agronova WG, Colombia	Life Systems Technology, S.A.	*Anthonomus grandis; Brassolis sophorae* *Clavipalpus* spp.; *Cosmopolites sordidus* *Heliothis virescens;* *Hypothenemus hampei; Manduca sexta* *Opsiphanes cassina; Plectrix* spp. *Spodoptera frugiperda;* Others	http://lstsa.com/html/AgroNova/Lst-ing-AgroNova.htm
balEnce™ HF 5ES, USA	Jabb of the Carolinas, Inc.	*Musca domestica*	http://www.jabb.bz/hf5info.html
balEnce™ biological Fly Bait, USA	Jabb of the Carolinas, Inc.	Several species of flies (Diptera)	http://www.jabb.bz/baitpage.html
balEnce™ biological Fly Bait	Jabb of the Carolinas, Inc.	Several species of flies (Diptera)	http://www.jabb.bz/BAITSPECIMENLABELorganic.htm
Baits motel stay awhile-rest forever, USA	Plant Defense Boosters, Inc.	Several species of Coleoptera (Curculionidae)	http://www.plantdefenseboosters.com.index.html
Basisav-1, Cuba	Instituto de Investigaciones de Sanidad Vegetal (INISAV)	*C. sordidus; Cylas formicarius* *Diatraea saccharalis;* *Lissorhoptrus brevirostris* *Pachnaeus litus* Germar	http://www.inisav.cu/
Basisav-2, Cuba	INISAV	*Atta insularis* Güerin	http://www.inisav.cu/
Basisave 32, Spain	Agrotécnica Murciana.	*A. grandis; Bemisia argentifolii* *B. tabaco; C. sordidus; L. brevirostris;* *P. litus*	http://www.terralia.com/productos_e_insumos_para_agricultura_ecologica/index.php?proceso=registro&numero=643&base=2009
Bassianil^MR, Mexico	BioTropic, S.A. de C.V.	*A. grandis; B. argentifolii; B. tabaco* *Compsus* spp.; *Corytucha* spp. *C. sordidus; H. hampei* *Loxotama elegants; Phyllophaga* spp. *Thrips* spp.; Several species from *Astigmata (Acaridae)*	http://www.biotropic.com.mx/pdf/BASSI ANIL_RES170105.pdf

Table 7.3. (*Contd...*)

Table 7.3. (*Contd...*)

Product name, country	***Company***	***Active against***	***Reference/web link***
Beauveria bassiana			
Bassianil™Collaria WP, Colombia	Agricultura Orgánica	*Acromyrmex* spp.; *Ancognatha* spp *A. grandis; Atta* spp.; *Compsus* spp. *Corytucha* sp.; *C. sordidus; H. hampei* *L. elegant; Phyllophaga* spp. *Pseudococcus* spp. (Hemiptera) Sspecies from *Astigmata (Acaridae)* *Thrips* spp.	http://www.controlbiologico.com/bp_beauverial.html
Baubassil™, Colombia	Fungicol Ltda.Productos BiológicosPerkins Ltda.	*Ancognata* spp.; *B. sophorae* *C. sordidus; D. saccharalis* *H. hampei; Leptotharsa* spp. *Metamasius hemipterus* *Ostrinia nubilalis; P. litus* *Rynchophorus palmarum* *Sagalassa valida*	http://www.fungicol.com/web/products_Baubassil.html http://www.agriculturalimpia.com/PaginaNuevaBaubassil.htm http://www.perkinsltda.com.co/
Bazam, Honduras	Escuela Agrícola Panamericana	*A. grandis; B. argentifolii; B. tabaci* *H. hampei; Thrips* spp.Several species of Hemiptera (Aphididae)	http://plaguicidas.senasa-sag.gob.hn:8080/senasaplus/plaguicidasplus.do?action=ver&id=5154
Bb-*Beauveria bassiana*, India	BacTo Agro Culture Care Private Limited	Effective against many insect pests.	http://www.indiamart.com/biocontrolagents/
Bb Plus, South Africa	Biological Control Products SA (Pty) Ltd.	*Calomycterus setarius* *Cyrtepistomis castaneus* *Odontopus calceatus* *Otiorhynchus ovatus; O. rugosotriatus* *O. sulcatus; Pseudococcus* spp.; Several species of: Aphidoidea Hemiptera (Cicadellidae); Bolitophilidae Diadocidiidae; Ditomyiidae Keroplatidae;	http://www.madumbi-bcp.co.za/bbplus.html

Table 7.3. (*Contd...*)

Table 7.3. (*Contd...*)

Product name, country	***Company***	***Active against***	***Reference/web link***
Beauveria bassiana			
		Mycetophilidae Diptera (Sciaridae); *Tetranychus urticae Thrips* spp.	
Bb Weevil, South Africa	Biological Control Products SA (Pty) Ltd.	*C. sordidus*	http://www.biocontrol.co.za/bb%20weevil.htm
Beauvedieca, Costa Rica	Liga Agrícola Industrial de la Caña de Azúcar S.A., Laica-Dieca	Mainly against *D. saccharalis*, among other insect pests.	http://www.laica.co.cr/d_servicios.asp
Beviguard, India	Bio Organic Industries	All types of hard bodied insect pests	http://www.indiamart.com/bioorganic/bio-organic-fertilizers.html#bioinsecticides
Biagro Bb-mosca, Argentina	Laboratorio Biagro S. A.	*M. domestica*	http://www.biagrosa.com.ar/esp/producto_biagromosca.aspx
Biagro Bb-vinchuca, Argentina	Laboratorio Biagro S. A.	*Triatoma infestans.*	http://www.biagrosa.com.ar/esp/producto_biagrovinchuca.aspx
BioExpert SC, Colombia	Life Systems Technology, S. A.	*Ancognatha* spp.; *B. argentifolii; B. tabaco; Clavipalpus* spp. *Collaria columbiensis; Empoasca* spp. *Frankliniella occidentalis; Premnotripes* spp.; *Thrips palmi T. panamensis; T. tabaco Trialeurodes vaporariorum*	http://lstsa.com/html/BioExpert/Lsting-BioExpert-ca-1.html
BioFung, Mexico	CESAVEG	*A. grandis; B. argentifolii B. tabaci; Diabrotica* spp. *Lygus* spp.; *Oebalus mexicanus Paratrioza cockerelli; T. vaporariorum;* Other pests	http://www.cesaveg.org.mx/html/plaborganismosbeneficos.html
BioGuard Rich, India	Plantrich Chemicals & Fertilizers Ltd.	*A. grandis; B. argentifolii; B. tabaci H. hampei; Leucopholis* spp.; *Thrips* spp. Several species of :Aphidoidea; Cicadellidae Other pests	http://www.plantrich.com/main/BioPesticides/Bioguard_rich.asp

Table 7.3. (*Contd...*)

Table 7.3. (*Contd...*)

Product name, country	***Company***	***Active against***	***Reference/web link***
Beauveria bassiana			
Biohit, India	Jay Enterprises	*Acromyrmex* sp.; *A. grandis*,; *Atta* sp. *B. argentifolii; B. tabaci* *Coptotermes* spp.; *Dermolepida albohirtum* *Diatraea* spp.; *Erinnyis ello* *Maecolaspis* sp. *Melolontha hippocastani* F. *M. melolontha* L.; *Phoenicoprocta sanguine* *Pseudococcus* spp.; Species of: Mites; Grasshoppers (Orthoptera) Aphidoidea; Several species of ; Cicadellidae; Hemiptera (Coccoidea) Polyphaga (Curculionidae); Eriophyidae Tarsonemidae; Acarina (Tetranychidae) Coleoptera; *Thrips* spp.; Others	http://www.indiamart.com/jayenterprises/bio-fertilizers.html#biohit-fertilizers
Biolisa-Madara, Japan	Nitto Denko	*Monochamus alternates;* *Anoplophora glabripennis*	Dubois *et al.*, 2008
Bio-Power India[2]	T. Stanes & Co., Ltd.	*E. ello; Heliothis* spp.; *Maecolaspis* spp. *M. hippocastani; M. melolontha* *P. sanguinea; Pseudococcus* spp. *Spodoptera* spp.; Several species of: Mites; Grasshoppers Several species of: Cicadellidae Coccoidea; Curculionidae	http://www.tstanes.com/bio_power.html
Botani Gard 22 WP, USA	Laverlam International Corp.	*A. grandis*,; *B. argentifolii; B. tabaci* *Diabrotica* spp.; *Lygus* spp.;*Thrips* spp. Several species of: Aphidoidea Eriophyidae; Tarsonemidae,; Acarina (Tetranychidae) Others	http://www.laverlamintl.com/

Table 7.3. (*Contd...*)

Table 7.3. (*Contd...*)

Product name, country	*Company*	*Active against*	*Reference/web link*
Beauveria bassiana			
Bovemac EC, Brazil	Turfal Indústria eComércio de Produtos Biológicos e Agronômicos Ltda	Several species of Coleoptera	http://extranet.agricultura.gov.br/sislegis-consulta/consultarLegislacao.do;jsessionid=15db3a538ed377d54fe41ea2b6ebfd92863d4dfbafc7c338b045b1cf796cbc79.e3uQb3aPbNeQe38Kah0Tc38Pch50?operacao=visualizar&id=19503
Bovenat PM, Brazil	Natural Rural	*B. argentifolii; B. tabaco; C. sordidus H. hampei.*	http://www.naturalrural.com.br/FPDentro/Tutoriais/?Codigo=44
Boveril WP-PL63/ Boveril organic, Brazil	Italforte BioProductos	*B. argentifolii; B. tabaci; Coptotermes* spp. *C. sordidus; Heterotermes* spp.;*H. hampei Incisitermes* spp.;*Kalotermea* spp. *Reticulitermes* spp.*Zootermopsis* spp.	http://www.itafortebioprodutos.com.br/produto.asp?id_produto=1
Boveriol, Brazil	Tecnicontrol Indústria Comércio de Produtos Biológicos Ltda.	*Coptotermes formosanus Coptotermes* spp.; *Heterotermes* spp. *Incisitermes* spp.; *Kalotermea* spp. *Reticulitermes* spp.; *Zootermopsis* spp.	http://www.123achei.com.br/sites/defensivos-agricolas/piracicaba/0/tecnicontrol-industria-e-comercio-de-produtos-biologicos-ltda.html
Bovetrópico™ WP, Colombia	Soluciones microbianas del trópico	*B. argentifolii; H. hampei; Ceratitis capitata; Thrips* spp. Several species of: Aphidoidea Blatellidae; Eriophyidae; Lepidoptera Tarsonemidae; Tetranychidae	http://www.smdeltropico.com/productos_servicios.html
Broadband EC, South Africa	Biological Control Products SA (Pty) Ltd,	*Plutella xylostella; Thrips* spp. Several species of: Lepidoptera Arsonemidae; Tetranychidae	http://www.madumbi-bcp.co.za/broadband.html
Brocaril™, Colombia	Serfi S. A. (Laverman)	*H. hampei*	http://www.gacetajuridica.com.pe/servicios/normaspdf_2009/marzo/09-03-2009/09-03-2009.pdf

Table 7.3. (*Contd...*)

Table 7.3. (*Contd...*)

Product name, country	*Company*	*Active against*	*Reference/web link*
Beauveria bassiana			
Conidia, USA; Russia	Troy Bioscience; Troy Biosciences Inco. Berdsk Plant of Biological Preparations	Many insects, Coleoptera	http://www.scribd.com/doc/8061799/biopestthai; http://74.125.47.132/search?q=cache:u0hIYbc2a8QJ:www.apvma.gov.au/products/constituents/docs/approved_actives_a-z_18-12-2009.doc+Andermatt+Biocontrol+AG,Metarhizum&cd=70&hl=es&ct=clnk
Eco-Bb^{TM}, South Africa	Plant Health Products (PHP) Pty Ltd	*B. argentifolii*, *H. hampei*, (control) and *T. urticae* (suppress)	http://www.plant-health.co.za/eco-bb.html
Enpro Bever, India	Enpro Bio Sciences Private Ltd.	*Ferrisia virgata* (Cockerell), *Hypogeococcus pungens* (Granara de Willink)*Paracoccus marginatus* Willams *Planococcus citri* (Risso) *Phenaccocus solani* Ferris *Ph. solenopsis* Tinsley.	http://www.indiamart.com/enprobio/biofertilizers-pesticides-chemicals.html#enpro-bever-beauveria-basiana-bio-control-agent
Jas Beesi	Shri Ram Solvent Extractions Pvt. Ltd.	*Chrysobothris femorata* (Olivier) *Heliothis* spp.; *H. pungens; F. virgata* *Leucopholis irrorata* (Chevrolat) *Ps. marginatus; P. citri; Ph.s solani, Ph. solenopsis* *Saperda candida* Fabricius *Scolytus rugulosus* (Muller) *Spodoptera* spp. *Synanthedon exitiosa* (Say) *Sy. pictipes* (Grote and Robinson), *Sy. scitula* (Harris); *Thrips* spp. *Phyllophaga* spp. ; Several species of Aphidoidea,	http://www.neemplus.com/bio_pesticide.html

Table 7.3. (*Contd...*)

Table 7.3. *(Contd...)*

Product name, country	***Company***	***Active against***	***Reference/web link***
Beauveria bassiana			
Larvocel^{tm}, India	Excel Industries Ltd.	*Bemisia* spp.; *F. virgata; H.pungens* *P. citri; P. marginatus; Ph. solani* *Ph. solenopsis; Thrips* spp.	http://www.excelind.co.in/Larvocel .htm
Lopst^{TM}, India	Exotic Naturals	*Helicoverpa* spp.; *Lygus lineolaris* *H.pungens; F. virgata;* *Leucopholis irrorata* (Chevrolat), *O. nubilalis; Ostrinia furnacalis* Guenee *P. marginatus; P. citri; Ph. solani;* *Ph. solenopsis; S. candida* *Scolytus rugulosus; Spodoptera* spp. *Sy.exitiosa; Sy. pictipes; Sy.* scitula	http://www.agrinaturals.com/ lopst.html
Micosen, Peru	SENASA	*H. hampei*	http://www.gacetajuridica.com.pe/ servicios/normaspdf_2009/marzo/ 09-03-2009/09-03-2009.pdf
Mirabiol, Nicaragua	Unión de Cooperativas Agrícolas Miraflor	*H. hampei*	http://orton.catie.ac.cr/cgi-bin/wxis. exe/?IsisScript=CENIDA.xis&metho d=post&formato=2&cantidad=1&e xpresion=mfn=025168
Multiplex Baba, India	Multiplex Group	*B. argentifolii; Diabrotica* spp.; *H. hampei; Lygus* spp.	http://www.multiplexgroup.com/ MultiplexBaba.htm
Mycotrol^{TM} ES / Mycotrol^{TM} O[1], USA	Laverlam Inter. Corp.,	*Xyleborus fornicates; Thrips* spp.; Several species of ;Aphidoidea. Soft bodied insects including species of Aphidoidea; Mealybugs; *Thrips* spp. Mites; Scarab beetles; Whiteflies Psyllids; Lepidoptera; Weevils (see Table 7.6)	http://www.laverlamintl.com/
Naturalis-L^{TM} / H&G ^{TM}, USA[1]	Troy Biosciences, Inc.	*A. grandis; B. argentifolii; B. tabaco*	http://www.troybiosciences.com/

Table 7.3. *(Contd...)*

Table 7.3. (*Contd...*)

Product name, country	***Company***	***Active against***	***Reference/web link***
Beauveria bassiana			
Ostrinil™,France	Natural Plant Prot.	*Chamaeropus humilis* *L. decemlineata; O. nubilalis* *Paysandisia archon* (Burmeister) (Lepidoptera: Crambidae)	http://www.abim.ch/documents/presentations2008/session5/3_Besse_ABIM-2008.pdf
PHC™ Bea Tron™, Mexico	Plant Health Care de México	700 insect species, mainly for control of *Bemisia* spp. *Phylophaga* spp.	http://www.phcmexico.com.mx/pdfs/biopesticidas/Bea%20Tron.pdf
Racer-Bb, India	Agri Life & Som Phyto-pharma, India, Ltd.	For control of chewing pests like: *Helicoverpa* spp. *Spodoptera* spp.Leaf folder Stem borerFruit borer etc. (see Table 7.6)	http://www.somphyto.com/biopesti_microracerbb.html
Teraboveria, Guatemala	Agrícola El Sol	*Aeneolamia* spp.; *Agriotes* spp. *Anomala* spp.; *Anthonomus* spp. *Bemisia* spp.; *Cholus* spp.; *Colaspis* spp. *Conedorus* spp.; *C. sordidus;* *Diabrotica* spp. ; *H. hampei* *L. decemlineta; M. hemipterus* *Phyllophaga* spp.; *Prosapia* spp. *Schistocerca* spp.,	http://www.agricolaelsol.com/teraboberia.html
Trichobass-L /-P, Spain	Trichodex, S.A.	*Bemisia tabaci; Trialeurodes vaporiorum* Many other insect pests.	http://www.terralia.com/productos_e_insumos_para_agricultura_ecologica/index.php?proceso=registro&numero=1334&base=2009
Troy Boverin, USA, global	Troy biosciences Inc.	*A. grandis; B. argentifolii; B. tabaco*	http://www.troybiosciences.com/
No name, Australia	Miller Chemical & Fertilizer Pty Ltd	Many insects, mainly Coleoptera	http://74.125.47.132/search?q=cache:u0hIYbc2a8QJ:www.apvma.gov.au/products/constituents/docs/approved_actives_a-z_18-12-2009.doc+Andermatt+Biocontrol+AG,Metarhizum&cd=70&hl=es&ct=clnk

Table 7.3. (*Contd...*)

Table 7.3. (*Contd...*)

Product name, country	***Company***	***Active against***	***Reference/web link***
Beauveria bassiana			
No-name, Brazil	Embrapa Tabuleiros Costeiros	*R. palmarum.*	http://www.catalogosnt.cnptia.embrapa.br/catalogo20/catalogo_de_productos_y_servicios/arvore/CONT000fn87hvpb02wyiv8003d0p3nz4oczl.html
No-name, Brazil	Instituto Agronómico de Pernambuco (IPA)	Many insects, mainly Coleoptera	http://www.ipa.br/pesquisa_laboratorio.php
No-name, Brazil	PESAGRO-RIO Laboratório de Controle Biológico	Many insects, mainly Coleoptera	http://www.pesagro.rj.gov.br/
No-name, Costa Rica	Corporación de Desarrollo Orgánico Brantley S.A.	*Aeneolamia* sp.; *Cosmopolites* sp. *Dysmicoccus brevipes; Heliothis* spp. *H. hampei; Metamasius* spp. *Mochis latipes; Phyllophaga* spp. *Pseudoplusia includens, Rhynchophorus* spp.; *Spodoptera* spp. *Trichoplusia ni; Thrips* spp.	http://www.codeorsa.com/index.php?page=bioplaguicidas
No-name, India	Fertilizers and Pesticides Co.	Many species of: Diptera Coleoptera	http://www.fertilizersandpesticides.com/bio-control-agents.html#tri
No-name, India	Mani Dharma Biotech Pvt Ltd.	Many species of: Diptera; Coleoptera Lepidoptera	http://www.indiamart.com/manidharmabiotech/biofertilizers.html#beauveria-bassiana
No-name, Panama	Aboquete, S.A.	Several insect pests	http://www.aboquete.com/aboquete/inicio-2/
No name, Peru	SENASA	*C. sordidus; H. hampei*	http://www.senasa.gob.pe/0/modulos/JER/JER_Interna.aspx?ARE=0&PFL=2&JER=42
Beauveria brongniartii			
Beaupro™, Switzerland	Andermatt Bio-control AG, Biogarten.	Coleoptera (Scarabaeidae)	https://www.fibl-shop.org/shop/pdf/1032-hilfsstoffliste.pdf

Table 7.3. (*Contd...*)

Table 7.3. (*Contd...*)

Product name, country	***Company***	***Active against***	***Reference/web link***
Beauveria Schweizer, Switzerland	Schweizer	Coleoptera (Scarabaeidae)	https://www.fibl-shop.org/shop/pdf/1032-hilfsstoffliste.pdf
Betel, France	Natural Plant Prot.	*Hoplochelus marginalis*	http://www.prpv.org/index/Phytosanitairement%20V%C3%B4tre%20n%C2%B031.pdf
Biolisa-Kamikiri, Japan	Nitto Denko	Coleoptera (Cerambycidae)	Kunimi, 2007
Melocont™ Pilzgerste, Austria	Kwizda Agro GmbH	*M. melolontha*	http://www.kwizda-agro.at/index.asp?subject=products&topic=crop control
Enthomophthora virulenta			
Vektor™, Colombia, Peru	BioTropic, S.A. de C.V.; Serfi S. A.	*Anthicloris viridis; Bemisia* sp. *Planococcus* spp.; *Pseudococcus* spp. *T. urticae; T. cinnabarinus; Thrips* spp.	http://www.biotropic.com.mx/pdf/VEKTOR_RES170105.pdf
Hirsutella thompsonii			
MiteHit, India	Plantrich Chemicals & Fertilizers Ltd.	*T. urticae; T. cinnabarinus*	http://www.plantrich.com/main/BioPesticides/MITEHIT.asp.asp
No name, Peru	SENASA	*Toxoptera citricida*	http://www.senasa.gob.pe/0/modulos/JER/JER_Interna.aspx?ARE=0&PFL=2&JER=42
Isaria fumosorosea (formerly *Paecilomyces fumosoroseus*)			
Fumo, India	Bio Organic Industries	*Bemisia* sp.; *Liriomyza huidobrensis* *Planococcus* spp.; *P. xylostella* *Pseudococcus* spp.; Other insect pests	http://www.indiamart.com/bioorganic/bio-organic-fertilizers.html#bioinsecticides
Fumosil™, Colombia	Fungicol Ltda.	*Bemisia* sp.; *L. huidobrensis* *Planococcus* spp.; *Pseudococcus* spp. *T. urticae; T. cinnabarinus; Thrips* spp. *Trialeurodes* spp.	http://www.fungicol.com/web/productos_Fumosil.html

Table 7.3. (*Contd...*)

Table 7.3. (*Contd...*)

Product name, country	*Company*	*Active against*	*Reference/web link*
Multiplex Micomite, India	Multiplex Group	*Oligonychus coffeae; T. urticae T. cinnabarinus*	http://www.multiplexgroup.com/MultiplexMyconite.htm
Paecilotrópico™ Pf WP, Colombia	Soluciones microbianas del tropic	*B. argentifolii; B. tabaco; Margarodes vitium; T. urticae T. cinnabarinus*	http://www.smdeltropico.com/productos_servicios.html
PFR-97, USA	Certis	*B. argentifolii; B. tabaci; T. urticae T. cinnabarinus; Thrips* spp.	http://www.certisusa.com/pest_management_products/bioinsecticide/pfr-97_microbial_insecticide.htm
PHC™ Pae Tron™, Mexico	Plant Health Care de México	*B. argentifolii; B. tabaci*	http://www.phcmexico.com.mx/pdfs/biopesticidas/Pae%20Tron.pdf
Priority[2], India	T.Stanes & Co., Ltd.	*Aculus schlectendali; Byrobia rubrioculus Panonychus ulmi ; T. urticae.*	http://www.tstanes.com/priority.html
PreFeRal™, Belgium	Biobest N.V. Biological Systems	*B. argentifolii; B. tabaci*	http://www.biobest.be/images/uploads/public/6476187639_PreFeRal%20WG.pdf
Successor™, Colombia	Life Systems Technology, S.A.	*T. cinnabarinus T. urticae*	http://lstsa.com/html/Successor/Lsting-Successor-ca-1.html
No-name, Panama	Aboquete, S.A.	Several insect pests	http://www.aboquete.com/aboquete/inicio-2/
No name, Peru	SENASA	*A. grandis; B. argentifolii; B. tabaci*	http://www.senasa.gob.pe/0/modulos/JER/JER_Interna.aspx?ARE=0&PFL=2&JER=42
***Isaria* sp.** (formerly *Paecilomyces* sp.)			
PaciHit Rich, India	Plantrich Chemicals & Fertilizers Ltd.	*A. grandis; B. argentifolii; B. tabaci Trhips* spp.	http://www.plantrich.com/main/BioPesticides/pacihit_rich.asp
Lagenidium giganteum			
Laginex™ AS / Technical, USA	AgraQuest Inc.	Mosquitoes	http://www.agraquest.com/agrochemical/products/

Table 7.3. (*Contd...*)

Table 7.3. *(Contd...)*

Product name, country	***Company***	***Active against***	***Reference/web link***
Lecanicillium longisporum (formerly *Verticillium lecanii)*			
Vertalec, Switzerland	Leu + Gygax AG; Hatto und Patrick Welte	Many aphid species (except *Macrosiphoniella sanborni*).	http://www.blw.admin.ch/psm/wirkstoffe/index.html?lang=fr&item=808
Vertalec, The Netherlands	Koppert Biological Systems	Many aphid species (except *M. sanborni*).	http://www.koppert.com/pests/aphid/products-against-aphids/detail/vertalec-2/
Vertirril organic, Brazil	Italforte Bioproducts	Many insects, mainly Coleoptera and Lepidoptera	http://www.itafortebioprodutos.com.br/produto_especial.asp?tipo=2
Lecanicillium muscarium (formerly *V. lecanii*)			
Mycotal, The Netherlands ;Switzerland	Koppert Biological Systems;Omya (Schweiz) AG	*A. grandis; B. argentifolii; B. tabaci Thrips* spp.; *T. urticae; T. cinnabarinus*.	http://www.koppert.com/products/products-pests-diseases/products/detail/mycotal-1/ http://www.blw.admin.ch/psm/wirkstoffe/index.html?lang=fr&item=808"
***Lecanicillium* sp**. (formerly *V. lecanii*)			
Bio-Catch[2], India	T.Stanes & Co., Ltd.	*B. argentifolii; B. tabaci; T. urticae T. cinnabarinus; Thrips* spp.	http://www.tstanes.com/bio_catch.html
BioVert Rich, India	Jay Enterprises	*B. argentifolii; B. tabaci; T. urticae T. cinnabarinus; Thrips* spp.	http://www.indiamart.com/jayenterprises/bio-fertilizers.html#biovert-fertilizers
Jas Verti	Shri Ram Solvent Extractions Pvt. Ltd.	Root-knot nematode eggs	http://www.neemplus.com/bio_pesticide.html
Mealikil Plus; Mealikil™ VL, India	Agri Life & Som Phytopharma, Ltd.	Mealy bugs (see Table 7.6)	http://www.somphyto.com/biopesti_micromealikil_plus.html

Table 7.3. *(Contd...)*

Table 7.3. (*Contd...*)

Product name, country	***Company***	***Active against***	***Reference/web link***
MicroGermin Plus, Switzerland	Omya (Schweiz) AG	*B. argentifolii; B. tabaci; Thrips* spp. *T. urticae; T. cinnabarinus.*	http://www.blw.admin.ch/psm/wirkstoffe/index.html?lang=fr&item=808"
Vertilec^{TM} WP, Colombia	Agricultura Orgánica	*B. tabaco; Corytucha* spp.; *Myzus* spp. *Phylophedra* spp.; *Pseudococcus* spp. *Siba flava; T. vaporariorum*	http://www.controlbiologico.com/bp_vertilec.htm
Vertinat ES / Vertinat PM, Brazil	Natural Rural	*B. argentifolii; B. tabaco Orthezia praelonga*	http://www.naturalrural.com.br/produtos/produtos_especificos.asp?CodProduto=240&Caption=Vertinat
Vertipar^{TM}, India	Exotic Naturals	*Aphis gossypii; Coccus viridis Planococcus citri; Thrips* spp.	http://www.agrinaturals.com/vertipar.html
Vertisav, Cuba	Biasav	*B. argentifolii; B. tabaci; M*any species of : AphididaeIxodidae	http://tips.org.uy/SPA/portal/NEGTexto.asp?Entidad=CUB&Numero=6018
Vertisel^{TM}, India	Excel Industries Ltd.	*A. gossypii; B. argentifolii; B. tabaci C. viridis; P. citri; Thrips* spp. Many species of: AphididaeIxodidae	http://www.excelind.co.in/Verticel.htm#title
Vertisol^{MR}, Colombia, Costa Rica, Honduras, Mexico, Peru	BioTropic, S. A. De C. V.	*B. argentifolii; B. tabaci; Thrips* spp. Many species of Aphididae	http://plaguicidas.senasa-sag.gob.hn:8080/senasaplus/plaguicidasplus.do?action=ver&id=4864 http://www.biotropic.com.mx/pdf/VERTISOL_RES170105.pdf
Verzam, Honduras	Escuela Agrícola Panamericana	*B. argentifolii; B. tabaci C. viridis; P. citri; Thrips* spp.; Many species of Aphididae	http://plaguicidas.senasa-sag.gob.hn:8080/senasaplus/plaguicidasplus.do?action=ver&id=5153
Vetri, India	Bio Organic Industries	*A. gossypii; B. argentifolii; B. tabaci C. viridis; P. citri; Thrips* spp. Many species of Aphididae	http://www.indiamart.com/bioorganic/bio-organic-fertilizers.html#bioinsecticides
VL-*Verticillium lilacinus*, India	BacTo Agro Culture Care Private Limited	Many insect pests.	http://www.indiamart.com/biocontrolagents/

Table 7.3. (*Contd...*)

Table 7.3. (*Contd...*)

Product name, country	*Company*	*Active against*	*Reference/web link*
No name, Brazil	Turfal Ind. Com. Prod. Biol.,	Aphididae	http://www.biologico.sp.gov.br/docs/arq/v76_4/michereff.pdf
No name, Peru	SENASA	*Aleurodicus cocois;* Many species of Nematodes[2].	http://www.senasa.gob.pe/0/modulos/JER/JER_Interna.aspx?ARE=0&PFL=2&JER=52
Metarhizium anisopliae			
Bio-Blast Biological Termiticide, USA	EcoScience/ (was Agro power development, Inc)	*Coptotermes* spp.; *Heterotermes* spp. *Incisitermes* spp.; *Kalotermea* spp. *Reticulitermes* spp.; *Zootermopsis* spp.	http://www.epestsupply.com/images/Products/labels/bioblastlabel.pdf
Bio-Magic[2], India	T.Stanes & Co., Ltd.	Root weevils; Plant hoppers Japanese beetle; Black vine weevil SpittlebugWhite grubs (see Table 7.6)	http://www.tstanes.com/bio_magic.html
DeepGreen™ SC, Colombia	Life Systems Technology, S. A.	*Aeneolamia* sp.; *Ancognatha* sp. *C. columbiensis; H. hampei Perkinsiella saccharicid Phyllophaga* spp.; *S. frugiperda Tagosodes aryzicola*	http://lstsa.com/html/DeepGreen/Lst-ing-DeepGreen-ca-1.htm
Destruxin WP, Mexico; Dextrusin 50 WP, Costa Rica	BioTropic, S.A. de C.V.; Bio-Control, S. A.,	Many species of.ColeopteraLepidoptera (*Phyllophaga* spp.)	http://www.biotropic.com.mx/pdf/DESTRUXIN_RES170105.pdf http://www.protecnet.go.cr/insumosys/ConsultarInsumo.asp?cCodigo=4562&sTipoQry=Plaguicidas
Eco Meta	Toyobo do Brasil Ltda	Many species of:ColeopteraLepidoptera (*Mahanarva posticata, Phyllophaga* spp.)	http://extranet.agricultura.gov.br/sislegis-consulta/consultarLegislacao.do;jsessionid=15db3a538ed377d54f7c338b04e41ea2b6ebfd92863d4dfbafc5b1cf796cbc79.e3uQb3aPbNeQe38Kah0Tc38Pch50?operacao=visualizar&id=19503

Table 7.3. (*Contd...*)

Table 7.3. (*Contd...*)

Product name, country	***Company***	***Active against***	***Reference/web link***
Fitosan M, Mexico	CESAVEG	*Phyllophaga* spp.	http://www.cesaveg.org.mx/html/plaborganismosbeneficos.htm
GRANMET G, Italy.	Agrifutur S.R.L.	*Gryllotalpa grillotalpa; Otiorhynchus* spp. Other species of Coleoptera (Eliteridae)	http://www.agrifutur.com/IT/c/granmet-g-31/
Jas Meta	Shri Ram Solvent Extractions Pvt. Ltd.	*Apassalus parvulus; B. longissima Curculio caryae; Nasutiterense exitous, Oryctes rhinoceros*	http://www.neemplus.com/bio_pesticide.html
MA-*Metarhizum anisopliae*, India	BacTo Agro Culture Care Private Limited	Effective against many insect pests.	http://www.indiamart.com/biocontrolagents
MET-92, Guatemala	Agrícola El Sol	*Agriotes* spp.; *Anomala* spp. *Anthonomus* spp.; *Conedorus* spp. *C. sordidus; Diabrotica* spp.; *H. hampei L. decemlineta; Phyllophaga* spp. *Schistocerca* spp.	http://www.agricolaelsol.com/productos.html
Meta, India	Bio Organic Industries	Many insect pests.	http://www.indiamart.com/bioorganic/bio-organic-fertilizers.html#bioinsecticides
Metabiol™ WP, Colombia	Agricultura Orgánica	*Aeneolamia* spp.; *Ancognatha* spp. *Compsus* spp.; *Cosmopolites* spp. *H. hampei; P. saccharicida, Phyllophaga* spp.;Other pests.	http://www.controlbiologico.com/bp_metabiol.html
Metadieca, Costa Rica	Liga Agrícola Industrial de la Caña de Azúcar S.A., Laica-Dieca	*M. posticata;* Other insect pests.	http://www.laica.co.cr/d_servicios.asp
Meta-Guard™, India	Ajay Bio-Tech Ltd.	Termites	http://ajaybio.in/prodbiostop.htm
Metanat, Brazil	Natural Rural	*Ancognatha sp; Cosmopolites* spp.; *H. hampei; P. saccharicida*	http://www.naturalrural.com.br/FPDentro/Tutoriais/?Codigo=44

Table 7.3. (*Contd...*)

Table 7.3. (*Contd...*)

Product name, country	***Company***	***Active against***	***Reference/web link***
Metapro™, Switzerland	Andermatt Bio-control AG, Biogarten.	Coleoptera (Scarabaeidae)	https://www.fibl-shop.org/shop/pdf/1032-hilfsstoffliste.pdf
Metarhizum Schweizer, Switzerland	Schweizer	Coleoptera (Scarabaeidae)	https://www.fibl-shop.org/shop/pdf/1032-hilfsstoffliste.pdf
Metaril™, Colombia	Productos Biológicos Perkins Ltda.	*Ancognata* spp.; *B. sophorae; C. sordidus D. saccharalis; H. hampei Leptotharsa* spp.; *M. hemipterus O. nubilalis; P. litus; R. palmarum S. valida*	http://www.perkinsltda.com.co/
Metarril M – 102, Brasil	Italforte BioProductos	*Deois* spp.; *M. fimbriolata; M. posticata Zulia* spp.	http://www.itafortebioprodutos.com.br/produto.asp?id_produto=3
Metarriz Biocontrol, Brazil	Biocontrol Sistema de Controle Biológico LTDA-EPP	*Aeneolamia* spp.; *Ancognatha* spp. *C. columbiensis; Euetheola* spp *Euschistus* spp.; *H. hampei; P. saccharicida; Phyllophaga* spp. *Sogata* spp.; *S. frugiperda Tagosodes oryzicola; Tibraca* spp.	http://extranet.agricultura.gov.br/sislegis-consulta/consultarLegislacao.do;jsessionid=15db3a538ed377d54fe41ea2b6ebfd92863d4dfbafc7c338b045b1cf796cbc79.e3uQb3aPbNeQe38Kah0Tc38Pch50?operacao=visualizar&id=19503
Metasav-11, Cuba	Centros de Producción de Entomófagos y Entomopatógenos (CREE)	*C. sordidus; L. brevirostris; Mocis* spp. *Monecphora bicincta fraterna*	http://www.caaez.gob.ve/main/index.php?option=com_content&view=article&id=108&Itemid=69
Metastrópico™ WP, Colombia	Soluciones microbianas del trópico	*Deois* spp.; *M. fimbriolata; M. posticata Zulia* spp.; *Cydamus* sp. *Haematobia irritans; Empoasca kraemeri* Species of Ixodidae	http://www.smdeltropico.com/productos_servicios.html
Methavida, Brasil	Methavida Controle Biológico		http://www.methavida.com.br/

Table 7.3. (*Contd...*)

Table 7.3. (*Contd...*)

Product name, country	***Company***	***Active against***	***Reference/web link***
Multiplex Metarhizium, India	Multiplex Group	*A. parvulus; B. longissima; C. caryae N. exitous; O. rhinoceros*	http://www.multiplexgroup.com/MultiplexMetarshizium.htm
Ostrinil™, France	Arysta LifeScience/ Natural Plant Protection S.A.	*P. archon*	http://www.arysta-eame.com/
PHC™Meta-Trone™, Mexico	Plant Health Care de México	More than 200 insect pest species	http://www.phcmexico.com.mx/pdfs/biopesticidas/Meta%20Tron.pdf
Pacer™ MA, India	Agri Life & Som Phytopharma, Ltd.	Termites.	http://www.somphyto.com/biopesti_micropacerma.htm
Technogreen Metarhizium 35 SL., Costa Rica	Biolaboratorios de Centroamerica S.A . BIOLAB	*Diabrotica* spp.*M. posticata Phyllophaga* spp.	ttp://www.protecnet.go.cr/hinsumosys/ConsultarInsumo.asp?cCodigo=4707&sTipoQry=Plaguicidas
Termox™, India	Exotic Naturals	Ants; Locusts; Fruit flies; Termites (see Table 7.6); *Diabrotica* spp.	http://www.agrinaturals.com/termox.html
Trichomet, Spain	Trichodex, S. A.	*Thrips* spp.	http://www.terralia.com/productos_e_insumos_para_agricultura_ecologica/index.php?proceso=registro&numero=1005&base=2009
Zero QK, Guatemala	Agrícola El Sol	Several species of Blattaria	http://www.agricolaelsol.com/productos.htm
No-name, Brazil	IPA	Many insects, mainly Coleoptera	http://www.ipa.br/pesquisa_laboratorio.php
No-name, Brazil	PESAGRO-RIO Laboratório de Controle Biológico	Many insects, mainly Coleoptera	http://www.pesagro.rj.gov.br/
No-name, Costa Rica	Corporación de Desarrollo Orgánico Brantley S.A.	*Aeneolamia* spp.; *Chavesia* spp. *Collaria* spp.; *D. brevipes Eutheola* spp.; *Hortensia* sp.	http://www.codeorsa.com/index.php?page=bioplaguicidas

Table 7.3. (*Contd...*)

Table 7.3. (*Contd...*)

Product name, country	*Company*	*Active against*	*Reference/web link*
		H. hampei; Prosapia spp. *Phyllophaga* sp.; *Strategus* sp. *Tagosodes* spp. *Thecla basilides Thrips* spp.; *Zulia* spp. Several species of;IsopteraLepidoptera	
No-name, Panama	Aboquete, S.A.	Several insect pests	http://www.aboquete.com/aboquete/inicio-2/
No-name, India	Fertilizers and Pesticides Co.	Wide range of host species.	http://www.fertilizersandpesticides.com/bio-control-agents.html#tri
No-name, India	Mani Dharma Biotech Pvt Ltd.	Grasshoppers; Termites; thrips, etc. (see Table 6)	http://www.indiamart.com/manidharmabiotech/index.html
No name Peru	SENASA	*Schistocerca piceifrons peruviana* *S. interrita*	http://www.senasa.gob.pe/0/modulos/JER/JER_Interna.aspx?ARE=0&PFL=2&JER=42
Metarhizium anisopliae* var. *acridum			
Green Guard SC, Australia	Becker Underwood Pty Ltd,	*Nomadacris septemfasciata* Audinet-Serville	http://74.125.47.132/search?q=cache:0hIYbc2a8QJ:www.apvma.gov.au/products/constituents/docs/approved_actives_a-z_18-12-2009.doc+Andermatt+Biocontrol+AG,Metarhizum&cd=70&hl=es&ct=clnk
Green Muscle™, South Africa	Biological Control Products SA PTY	*N. septemfasciata*	http://www.biocontrol.co.za/greenmuscle.html
No-name, Costa Rica	Corporación de Desarrollo Orgánico Brantley S.A.	*Bemicia* spp.; *Trialeurodes* spp. *Corytucha* sp.; *T. palmi; D. brevipes* Species of:AphididaeIxodidae	http://www.codeorsa.com/index.php?page=bioplaguicidas

Table 7.3. (*Contd...*)

Table 7.3. (*Contd...*)

Product name, country	***Company***	***Active against***	***Reference/web link***
Pochonia chlamydosporia			
No name, Peru	SENASA	*M. incognita*	http://www.senasa.gob.pe/0/modulos/JER/JER_Interna.aspx?ARE=0&PFL=2&JER=52
Fungi mixes			
Amplio™, Spain Mix of: *B. bassiana; M. anisopliae L. lecanii*	Agrícola de Aspe	*A. grandis; B. argentifoli; B. tabaco H. hampei; Thrips* spp.; *T. urticae Pseudococcus* spp.; Several species of: Aphidoidea Cicadellidae; Bolitophilidae; Diadocidiidae; Ditomyiidae Keroplatidae; Mycetophilidae; Sciaridae	http://www.terralia.com/productos_e_insumos_para_agricultura_ecologica/index.php?proceso=registro&numero=400&base=2009
Attacebo™, Colombia Mix of: *B. bassiana*	Fungicol Ltda.	Several species of ants	http://www.fungicol.com/web/productos_Attacebo.html
M. anisopliae; T. viride Biotercel™, India Mix of: Bacteria Fungi	Excel Industries Ltd.	Termites (*Coptotermes* spp.)	http://www.excelind.co.in/pdf/Biotercel%20leaflet.pdf
BMNat, Brazil Mix of: *B. bassiana; M. anisopliae*	Natural Rural	Many species of:Coleoptera; Lepidoptera *Tetranychus* spp.	http://www.naturalrural.com.br/produtos/produtos_especificos.asp?CodProduto=386&Caption=BMNat%20-%20Beauveria%20bassiana%20+%20Metarhizium%20anisopliae
Micosplag, Colombia Mix of: *B. bassiana M. anisopliae; P. lilacinus*	Serfi S.A.	Several species of: ColeopteraLepidoptera	http://www.gacetajuridica.com.pe/servicios/normaspdf_2009/marzo/09-03-2009/09-03-2009.pdf
Million - Multi Strain Bio Pesticide, India Mix of:	Enpro Bio Sciences Private Limited	*A. grandi; B. argentifolii; B. tabaci; H. hampei; Thrips* spp.; *T. urticae*	http://www.indiamart.com/enprobio/biofertilizers-pesticides-chemicals

Table 7.3. (*Contd...*)

Table 7.3. (*Contd...*)

Product name, country	***Company***	***Active against***	***Reference/web link***
B. bassiana; M. anisopliae		*Pseudococcus* spp.,; and several species of: Aphidoidea; Cicadellidae; Noctuidae	.html#million-multi-strain-bio-pesticide
Multiplex Shock, India	Multiplex Group.	Mix of: *A. niger; T. viride; P. lilacinus M. incognita Hoplolaimus indicus; R. reniformis Tylenchorhynchus masoodi*	http://www.multiplexgroup.com/MultiplexShock.html
URPI, Peru	Serfi S.A.	*Mix of: M. anisopliae; P. lilacinus* Several species of Coleoptera	http://www.gacetajuridica.com.pe/servicios/normaspdf_2009/marzo/09-03-2009/09-03-2009.pdf

[1] OMRI listed

[2] Certified for organic farming in India

2007). Additional information about each of these families can be found in publications Miller, 1997; Miller and Ball 1998; Tanada and Kaya 1993, and at websites such as: http://www.inhs.illinois.edu/research/biocontrol/home.html.

Baculoviruses

Baculoviruses can be divided into three sub-groups, nuclear polyhedrosis viruses (NPV), granulosis viruses (GV), and non-occluded viruses (NOV). NPV have been recovered from several insect orders including: Lepidoptera, Hymenoptera, Dipteral Thysanura, and Trichoptera, including many economically important pests (Table 7.4). The DNA containing virions (infective particles) of NPVs can be enveloped singly (SNPV) or multiply (MNPV). Several of these enveloped virions are enclosed in a protein called a polyhedral, which is referred to as an "occlusion body". GVs are morphologically similar to NPVs, except that the virions are singly occluded into small protein bodies called "granules". NOVs are not occluded. For both NPVs and GVs, occlusion bodies are the "active ingredient" for use in virus-based insecticides. When a susceptible insect ingests active virus, the polyhedron protein of the occlusion body dissolves in the insect's intestine, and releases the virions to initiate infection of intestinal cells. Once infected, the insect cells replicate the virus and these replicates infect additional insect cells, repeating the process until the insect is overcome by the infection. A successful infection often results in death of the insect within 5 to 15 days, depending on environmental conditions.

Traditional nomenclature used for baculoviruses provides information about the host and virus. Virus names are a combination of the initials for the scientific name of the host insect combined with the acronym for the virus type. For example, the singly enveloped nuclear polyhedrosis virus from the corn earworm (*Helicoverpa zea*) is referred to as *Hz*SNPV or *Hz*NPV. Similarly, the granulosis virus from codling moth (*Cydia pomonella*) is called *Cp*GV. Because this nomenclature is limited and discovery of new viruses often results in conflicting names, the system has been modified to use two letter initials for virus names, and the above examples would be *Heze*NPV and *Cypo*GV, respectively.

Advantages/Disadvantages

Baculoviruses insecticides provide several distinct advantages over chemical insecticides for pest control. Viruses are highly specific to infect only a few closely related species of insect providing a targeted control strategy. In contrast to a broad-spectrum chemical insecticide, a viral insecticide is able to control the pest and allow the beneficial insect predators and parasites to survive. The beneficial insects are able to reduce the potential of resurgence of the target pest, and to continue control of potential secondary pests. Because infected insects produce large numbers of propagules (occlusion bodies), there is high probability that the disease will cycle to provide additional pest control. Specificity of a virus for the target host also equates to mammalian safety, minimizing potential hazard for humans associated with production, processing, application and crop consumption.

Table 7.4. Select families of viruses known to infect insects with a potential for use as bioinsecticides

Genetic base	***Family***	***Commercial products***
DNA Virus	Baculoviruses, (NPV and GV)	Yes
	Ascoviruses	No
	Iridoviridae	No
	Parvoviridae	No
	Polydnaviridae	No
RNA Virus	Reoviridae	Yes[1]
	Nodaviruses	No
	Picorna-like viruses	No
	Tetraviruses	No

Unfortunately, baculoviruses have distinctive disadvantages that have limited development of many viruses into effective biological insecticides. Because the viruses are obligate parasites, they require living cells for replication. This requires the use of laborious insect colonies (*in vivo*) or expensive cell cultural (*in vitro*) techniques to produce the active agent for formulation into insecticides. Both the *in vivo* and *in vitro* techniques result in high production costs, which prevent economical commercialization of most potential agents. Currently, the most economical production of viruses for insecticides is accomplished using *in vivo* methods. Another disadvantage for development of virus-based products is the limited host range for each virus, which effectively limits potential market for each product. This directly limits potential for economic return and, in turn, limits the incentive for commercialization. Once in the field, efficacy of viral insecticides can be erratic with poor pest control resulting from poor spray coverage such that larvae do not ingest the virus, or from rapid degradation of virus viability caused by exposure to sunlight. Finally, the slow mode of action by virus infections is a disadvantage when compared with chemical insecticides. This delay may allow time for the pest to cause significant crop damage during the time between insecticide application and death of the target insect.

Baculovirus-based Products

Even with these general disadvantages, virus-based insecticides are currently available around the world (Table 7.5). Two viruses were developed and registered as commercial insecticides in the U.S. in the 1970's, Elcar (*Hz*NPV) and Decyde (*Cp*GV). Although these products are no longer available, they established a precedent for current products. Both of these viruses have been re-established as insecticides in the U.S. as Gemstar (Certis USA, Columbia, MD) for *Hz*NPV, and as Cyd-X (Certis USA), Carpovirusine (Arysta Lifescience North America, LLC, Cary, NC), and VirosoftCP4 (BioTEPP, Mont-St-Hilaire, Québec) for *Cp*GV. The *Cp*GV products continue to gain popularity among fruit growers around the U.S. for control of codling moth (*Cydia pomonella*). Outside the U.S., China has emerged as a major producer and consumer of virus-based insecticides, which may have benefited from favorable social, economic and ecological conditions. In China, viruses are produced in factories located in agricultural areas where labor costs are sufficiently low to support the industry (von Beek, 2007).

Perhaps the greatest commercial success of a virus-based insecticide to date has been the control of the velvet soybean caterpillar, *Anticarsia gemmatalis* (Hübner), in Brazil by the application of *Ag*NPV. Beginning in the early 1980's, a government supported program developed *Ag*NPV as an insecticide, and has successfully expanded the program to treat about 1.5 million hectares annually (Moscardi, 1999). The manufacture of *Ag*MNPV insecticide products is the result a unique industry/government partnership. Companies collect infected larvae from the field to provide the occlusion bodies for the commercial insecticides. Samples of subsequent products are sent to EMBRAPA (Brazilian Agricultural Research Corporation) for verification of quality assurance. Once certified, these products are available for application by growers. This program is successful for several reasons. First, government and industry cooperated effectively to develop, promote and certify the quality of the virus insecticides as an effective control. Second the target insect is a primary pest of the crop and is highly susceptible to the viral infection. Third, the soybean crop can tolerate relatively large amounts of indirect (leaf-feeding) damage without significant loss of grain yield. The number of hectares treated annually distinguishes the use of *Ag*MNPV insecticides as a highly successful control program.

Development of virus-based insecticides has provided a platform for debate about the application of genetically modified organisms to the field environment. Currently, all viruses used for commercial insecticides are from natural sources. Yet, the simplicity of the DNA viruses coupled with the current understanding and ability to manipulate genetic material has lead to research efforts to improve the efficacy of natural viruses. Specific efforts were directed at improving the speed a kill by engineering genes encoding insect specific hormones, enzymes or toxins (Hawtin *et al.*, 1992). Also efforts were aimed at expanding the host range of specific viruses by transferring genes to improve infection rates in otherwise marginal target pests. Although technically successful, genetically modified viruses have not been commercialized due to ethical criticism and negative public perceptions.

Table 7.5. Potential and currently available baculoviruses for commercial insecticides for three major world markets

Virus	***Host***	***Europe***	***U. S.***	***China***
Agrotis segetum GV	Turnip Moth	Yes	No	No
Anagrapha falcifera NPV	Celery Looper	No	Yes	No
Cydia pomonella GV	Codling Moth	Yes	Yes	No
Choristoneura fumiferana GV	Spruce Budworm	No	No	No
Plutella xylostella GV	Diamondback Moth	No	No	Yes
Lymantria dispar NPV	Gypsy Moth	No	Yes	No
Harrisina brillians GV	Western Grape Skeletonizer	No	No	No
Heliothis armigera NPV	Cotton Bollworm	Yes	No	Yes
Plodia interpunctella GV	Indian Meal Moth	No	Yes	No
Mamestra brassicae NPV	Cabbage Moth	Yes	No	No
Mamestra configurata NPV	Bertha Armyworm	No	Pending	No
Orgyida pseudotsugata NPV	Douglas Fir Tussock Moth	No	Yes	No
Helicoverpa zea NPV	Corn Earworm	No	Yes	No
Autographa californica NPV	Alfalfa Looper	No	No	Yes
Spodoptera exigua NPV	Beet Armyworm	Yes	Yes	Yes
Spodoptera littoralis NPV	Egyptian Cotton Leafworm	Yes	No	Yes
Bombyx mori NPV	Silk Worm	No	No	No
Adoxophyes orana GV	Summer Fruit Tortrix Moth	Yes	No	No
Gynaephora sp. NPV	Meadow Caterpillar	No	No	Yes
Buxura suppressaria NPV	Tung Tree Geometrid	No	No	Yes
Ectropis oblique NPV	Apple Geometrid	No	No	Yes
Leucania separate NPV	Oriental Armyworm	No	No	Yes
Pieris rapae GV	Small White Butterfly	No	No	Yes
Pseudatelia separate GV	Armyworm	No	No	Yes
Dendrolimus punctatus CPV	Masson Pine Moth	No	No	Yes
Phthorimaea operculella GV	Potato Tuber Moth	No	No	No
Panolis flammea NPV	Pine Beauty Moth	No	No	No
Orcytes rhinoceros NOV	Coconut rhinoceros beetle	No	No	No
Neodiprion sertifer NPV	Eastern Pine Sawfly	No	No	No
Neodiprion lecontei NPV	Redheaded Pine Sawfly	No	No	No
Melanoplus sanguinipes virus	Migratory Grasshopper	No	No	No
Oedaleus senegalensis virus	Senegalese Grasshopper	No	No	No
Tipula paludosa virus	European Crane Fly	No	No	No

List was generally compiled from websites: (http://www.pesticideinfo.org/; http://ec.europa.eu/food/plant/protection/index_en.htm http://www.epa.gov/esticides/biopesticides/ai/viruses.htm); and published resources (Sun and Peng, 2007).

NEMATODES-BASED BIOTECHNOLOGY DERIVED PRODUCT

The use of entomopathogenic nematodes (EPN) as bioinsecticides has been broadly studied since 1930´s. However, significant production, sales, and application as bioinsecticides began 15 years ago. Insecticidal activity of EPN's is caused by the symbiotic bacteria carried by the nematodes. Once a nematode enters a susceptible host, the bacteria are released to infect and kill the insect. Nematodes reproduce rapidly in the cadaver, producing many nematodes after a single infection. Ultimately, the body wall of the dead host insect ruptures and releases thousands of juvenile EPN, which is the insect-infective stage (Kaya, 1985). There are 3 known genera of EPN from the order Rhabdita (previously known as Neoaplectana), *Steinernema, Heterorhabditis,* and *Phasmarhabditis*. They contain the most important species of EPN. Juvenile EPN travel short distances in search of host insects, however, *Heterorhabditis* are able to travel greater distances than *Steinernema*. *Steinernema* juveniles enter the insect through natural openings, such as the mouth, spiracles and anus, and then penetrate into the body cavity, whereas *Heterorhabditis* can also enter by directly crossing the body wall. Part of the anterior area of the intestine of the infective juvenile is modified as a bacterial chamber. In this chamber, the infective juvenile carries cells with symbiotic bacterium. For Steinernematidae, the bacterial symbiont is usually of the bacterial genus *Xenorhabdus,* and for Heterorhabditidae the symbiont genus is *Photorhabdus*. Both Steinernematid and Heterorhabditid nematodes are currently marketed for pest management programs against soil dwelling insect pests. Applications of nematodes are compatible with many chemical pesticides and biopesticides, allowing tank mixes to reduce application costs. EPN can be applied as an inundative application using typical pesticide spray equipment. Many insects that are susceptible to entomopathogenic nematodes are listed in Table 7.6 (Poinar, 1984; Kaya, 1985; Gaugler *et al.,* 1997; Grewal and Georgis, 1998; UC-IPM 2009). Research continues to identify new insects and control environments to broaden the pest control applications.

Advantages/Disadvantages

Among the advantages of using nematodes as bioinsecticides are:

1. The ability of the nematodes to actively search for hosts and cause moratlity within 24-48 h.
2. The potential of dissemination; under appropriate environments, EPN become established, recycle, and their offspring continue to

Table 7.6. Insects susceptible to nematodes infection

Common name	*General description and scientific name*
Armyworms	Lepidoptera: Noctuidae, *Heliothis armigera*, *Pseudaletia unipuncta*, *Spodoptera* spp., Guenée.
Billbugs	Coleoptera: Curculionidae, *Sphenophorous* spp.
Carpenter worms	Known in general as borers, most of them are beetle larvae but are also caterpillars and primitive wasps.
Cat fleas, dog fleas	Siphonaptera: Pulicidae, *Ctenocephalides felis,* Bouché *C. cannis.*
Chafers	Coleoptera: Scarabaeidae (as known as may bug, billy witch, or spang beetle). The most important is cockchafer, *Melolontha melolontha* Fab.
Codling moth	Lepidoptera: Tortricidae, *Cydia pomonella* L.
Crown borers	Larvae of different species of insects including weevils such as *Tyloderma fragariae* Riley (Coleoptera: Curculionidae), and clearwing moth such as *Pennisetia* spp. (Lepidoptera: Sesiidae).
Cutworms	Larvae of many species of moth, mostly from Noctuidae family.
Filth flies	Diptera species includes strong fliers such as house fly, blow flies and flesh flies; and delicate fliers such as drain flies, fruit flies and phorid flies). Coleoptera species includes flea beetles, small, jumping beetles of the leaf beetle from the tribe Alticini (Spinola).
Flea beetles	Coleoptera: Chrysomelidae, *Phyllotreta* spp. Chevrolat.
Fungus gnats, sciarid flies or mushroom flies	Diptera: Sciaridae, *Bradysia* spp. Steffan; *Lycoriella* spp.Frey; and *Sciara* spp. Meigen.
German cockroaches	Blattodea: Blattellidae, *Blattella germanica* L.
Japanese beetle	Coleoptera: Scarabaeidae, *Popillia japonica* Newman.
Leaf miners larvae	Insects which live in and eat the leaf plant tissue. Includes mostly species of Lepidoptera and Diptera and some species of Coleoptera and Hymenoptera.
Leatherjackets	Diptera: Tipulidae, *Tipula* spp. Latreille (crane fly or daddy long legs).
Leopard moth	Lepidoptera: Cossidae, *Zeuzera pyrina* L.
Mole crickets	Orthoptera: Gryllotalpidae (Saussure).

Table 7.6. (*Contd...*)

Table 7.6. (*Contd...*)

Common name	***General description and scientific name***
Phorid flies	Diptera: Phoridae, Curtis. Includes different species of sub-families: Aenigmatiinae, Metopininae, Phorinae, and Sciadocerinae.
Pine weevil	Coleoptera: Curculionidae, *Hylobius abietis* L.
Plum weevil	Coleoptera: Curculionidae, *Conotrachelus nenuphar* Herbst.
Plume moths	Lepidoptera: Pterophoridae (Zeller).
Root weevils	Coleoptera: Curculionidae. The most significant of weevil species are: apopka weevil, *Diaprepes abbreviatus*: obscure root weevil or strawberry root borer, *Nemocestes incomptus* Horn; woods weevil or obscure root weevil, *Sciopithes obscures* Horn; black vine weevil, *Otiorhynchus sulcatus* Fab; strawberry root weevil, *O. ovatus* L.; rough strawberry root weevil, *O. rugosostriatus* Goeze; clay-colored weevil, *O.singularis* Stephens; and *Dyslobus* ssp. LeConte.
Shore flies	Diptera: Ephydridae, *Scatella stagnalis* Fallén.
Slugs	Gastropoda: Stylommatophora. Subfamilies Milacidae, Limacidae and Arionidae. Most important species is *Deroceras reticulatum* Müller.
Stable fly	Diptera: Muscidae, *Stomoxys calcitrans* L.
Stem borers	Includes mostly Coleoptera species, such as *Oberea* spp. Craighead (Cerambycidae) and Lepidoptera, such as *Ostrinia* spp. (Crambidae).
Webworms	Lepidoptera. Most important species includes the bluegrass webworm, *Parapediasia teterrella* Zincken, the cabbage webworm, *Hellula rogatalis* (Hulst), the garden webworm, *Achyra rantalis* Guenée (Crambidae); the eastern tent caterpillar, *Malacosoma americanum* Fab. (Lasiocampidae); the fall webworm, *Hyphantria cunea* Drury (Arctiidae).
Woodborers	Includes mostly species of Coleoptera, such as longhorned beetles (Cerambycidae); metallic wood borers (Buprestidae); bark beetles (Scolytidae); and Weevils (Curculionidae); and Hymenoptera species such as wood wasps (Xiphydriidae) and horntails (Siricidae)
Western flower thrips	Thysanoptera: Thripidae, *Frankliniella occidentalis* Pergande,

Table 7.6. (*Contd...*)

Table 7.6. (*Contd...*)

Common name	***General description and scientific name***
White grubs (beetles)	Coleoptera: Scarabaeidae, mainly *Phyllophaga* spp.beetles) LeConte

[1] In general, *Heterorhabditis bacteriophora* Poinar controls billbugs, black vine weevil, blue grass weevil, carpenter worms, chafers, clover root weevil, crown borers, cutworms, filth flies, fungus gnats, Japanese beetle, root weevils, stable fly, western flower thrips, and white grubs (*Phyllophaga* spp., in addition to *Anomala orientalis* L., *Ataenius spretulus* Haldeman, *Costelytra zealandica* White, *Cotinus nitida* L., *Cyclocephala borealis* Arrow, *Cyclocephala hirta* LeConte, *Cyclocephala lurida* Bland, *Cyclocephala pasadenae* Casey, *Maladera castanea* Arrow, and *Rhizotrogus majalis* Razoumowsky); *H. downesi* controls pine weevil; *H. indica* controls fungus gnats, Japanese beetle, western flower thrips, and the white grub *Cyclocephala borealis* Arrow; *H. marelata* Liu & Berry controls chafers, Japanese beetle, western flower thrips, and the white grub *Cyclocephala borealis* Arrow; *H. megidis* Poinar controls black vine weevil, chafer grub, chafers, Japanese beetle, leatherjackets, root weevils, shore flies, and white grubs (*Anomala orientalis* L., *Cyclocephala borealis* Arrow, *Cyclocephala hirta* LeConte, *Hoplia philanthus* Fuesslin, and *Rhizotrogus majalis* Razoumowsky); *H. zealandica* Poinar controls fungus gnats, Japanese beetle, and white grubs (*Anomala orientalis* L., *Cyclocephala borealis* Arrow, and *Rhizotrogus majalis* Razoumowsky); *Phasmarhabditis hermaphrodita* Schneider controls slugs; *Steinernema abassi* Elawad controls western flower thrips; *S. anomali* controls fungus gnats, Japanese beetle, and shore flies; *S. arenarium* Artyukhovsky controls chafers, and western flower thrips; *S. bicornutum* Tallosi controls western flower thrips; *S. carpocapsae* Weiser controls armyworm, billbugs, blue grass weevil, carpenter worms, cat and dog fleas, citrus root weevil, crown borers, cutworm, filth flies, flea beetles, Japanese beetle, leaf minors, leatherjackets, leopard moth, pine weevil, root weevils, shore flies, stable fly, webworms, woodborers, western flower thrips, white grubs (*Anomala orientalis* L., *Cotinus nitida* L. *Cyclocephala hirta* LeConte, and *Rhizotrogus majalis* Razoumowsky), and caterpillar larvae (as *codling moth); S. feltiae* Filipjev controls black vine weevil, carrot weevil, chafers, clover root weevil, filth flies, fungus gnats, Japanese beetle, leatherjackets, leaf minors, pine weevil, root weevils, shore flies, stable fly, stem borers, western flower thrips, white grub (*Cotinus nitida* L., *Cyclocephala hirta* LeConte, *Hoplia philanthus* Fuesslin, and *Rhizotrogus majalis* Razoumowsky), and several species of ants (repellent effect); *S. glaseri* Steiner controls billbugs, black vine weevil, chafers, Japanese beetle, and white grubs (*Phyllophaga* spp., in addition to *Amphimallon solstitiale* L., *Anomala orientalis* L., *Ataenius spretulus* Haldeman, *Costelytra zealandica* White, *Cotinus nitida* L., (*Cyclocephala borealis* Arrow, *Cyclocephala hirta* LeConte, *Cyclocephala lurida* Bland, *Cyclocephala pasadenae* Casey, *Hoplia philanthus* Fuesslin, *Maladera castanea* Arrow, and *Rhizotrogus majalis* Razoumowsky); *S. kushidai* Mamiya controls Japanese beetle, and white grubs (*Cyclocephala hirta* LeConte, and *Cyclocephala pasadenae* Casey); *S. kraussei* Steiner controls crown borers; *S. longicaudum* controls the white grubs *Anomala orientalis* L.; *S. riobravis* controls chafers, Japanese beetle, mole crickets, plum weevil, root weevils, fungus gnats, and the white grub *Cyclocephala hirta* LeConte; *S. scapterisci* controls Japanese beetle, mole crickets; *S. scarabae* controls Japanese beetle, and white grubs (*Phyllophaga* spp., in addition to *Anomala orientalis* L., *Ataenius spretulus* Haldeman, *Cotinus nitida* L., *Cyclocephala borealis* Arrow, *Cyclocephala hirta* LeConte, *Cyclocephala lurida* Bland, *Cyclocephala pasadenae* Casey, *Maladera castanea* Arrow, and *Rhizotrogus majalis* Razoumowsky) (Nematode information 2009).

control the target insect. Thus, the product cost may be lower in the long run when continued control by the recycling nematode is obtained.

3. The wide host range of a given nematode species or strain that broadens the market potential.
4. Nematodes have not shown significant harmful effect on other beneficial organisms.
5. Nematodes are specific to arthropods; they do no infect plants, birds, mammals or fish.
6. There are minimal effects against non-insect arthropods, such as sowbugs and millipedes. Wounded earthworms are occasionally infected, but not by normal release or application of nematodes.
7. EPN commercial products have been exempted from registration and regulation requirements by U.S. (EPA) and similar agencies in many other countries.
8. They can be mass produced using both *in vivo* and *in vitro* methods.
9. They can be applied using traditional insecticide spraying equipments.

Among the disadvantages of using nematodes as bioinsecticides are:

1. They are more expensive (10-60%) than chemical insecticides.
2. They require free water to move and are susceptible to dehydration.
3. To assist insecticidal activity, pre and post-treatment irrigation is recommended adding steps to the application process.
4. Nematodes do not survive temperatures above 32.2°C, and they become inactive at temperatures below 12.7°C, although each strain and species differs in sensitivity.
5. Packaged for sale, they have a short shelf-life of several weeks, or months if refrigerated. (Gaugler *et al.*, 1997; Lacey *et al.,* 2005; Smart, 1995).

Steinernema (Neoaplectana) Bovien

The genus *Steinernema* is in the family Steinernematidae (Nematoda: Rhabditida). The family also includes one species in the *Neosteinernema* family, *N. longicurvicauda,* which is infective to termites. The family *Steinernema* has 16 described species, but not all are marketed (Table 7.7). *Steinernema* is the most widely researched EPN species commercially produced for insect control. It is readily available for yard and garden use because it is easy to rear and handle. In field applications, *S. carpocapsae* Weiser tends to be most effective against caterpillar larvae. In laboratory and field trials, it has controlled sod webworms, cutworms and certain borers (raspberry crown borer, carpenter worm). It also has been effective against billbug larvae. *S. feltiae* Filipjev is effective against certain fly larvae. It is sometimes applied to control fungus gnats and other soil dwelling insects of greenhouse-grown plants (Table 7.7) (Poinar, 1984; Kaya, 1985; Gaugler *et al.,* 1997; Grewal and Georgis, 1998).

Heterorhabditis Poinar

In 1976, Poinar described a new genus and species of entomopathogenic nematode, *Heterorhabditis bacteriophora* Poinar, which he placed in a new family, Heterorhabditidae (Rhabditida: Nematoda). *Heterorhabditis* has six described species, and is less commercially available compared with *Steinernema,* because it is more difficult to rear and more susceptible to extreme environmental conditions. However, field trials have consistently shown that *Heterorhabditis* is more effective for control of white grubs. In addition, *Heterorhabditis* effectively controls larvae of many nursery pests that feed in the root zone, such as black vine weevil and citrus-infesting root weevils (Table 7.7) (Poinar, 1984; Kaya, 1985; Bedding and Miller, 2008).

Phasmarhabditis Schneider

Phasmarhabditis hermaphrodita Schneider (Nematoda*: Rhabditidae)* is a lethal parasite of *D. reticulatum* and a large range of slug species from the families Milacidae, Limacidae and Arionidae (Wilson *et al.,* 1993) (Table 7.7). *P. hermaphrodita* monoxenically associated with the bacterium *Moraxella* spp. The life cycle of *P. hermaphrodita* is similar to steinernematid EPN: infective stages carry symbiotic bacteria, which kill their host. Unfortunately, the slug cadavers are rapidly removed by scavengers (Foltan and Puza, 2009) and this reduces or eliminates the possibility of dissemination, further reducing its potential as biocontrol agent.

Production

EPN can be reared *in vivo* in insect hosts or they can be mass-produced *in vitro* on solid medium or in liquid medium. For *in vivo* production, a wide range of insects, including several caterpillars, large beetle larvae, and wax moth larvae (also sold commercially as fish bait), are commonly used. For solid medium culture, the media may consist of beef or pork kidney or liver or chicken internal organs. The media usually is made into a paste that is coated onto a porous substrate such as sponge. Paste is sterilized, inoculated with the bacterium, and nematodes are added 24 h later. Infective juveniles are harvested after about 15 days. This method is labor-intensive and typical of cottage industry-like production. Production in liquid medium can be done in small containers or in large fermentation tanks, similar to Bt. Greater numbers of juveniles can be produced per unit area in fermentation tanks, which makes this method especially suited for large-scale commercial production.

Since nematode products are safe to apply and do not pollute the environment, some clients will consider a biological control method, even at a higher application cost. EPN are sold mostly for control of insect pests in high-value horticulture, agriculture, home and garden niche markets, mainly for soil application, and also for termites and roaches. Many trademark names, and suppliers and producer companies are listed in Table 7.7. As a result of the ever changing biological control industry, this list may not reflect all manufactures or suppliers of beneficial nematodes (Gaugler *et al.,* 1997; Grewal and Georgis, 1998; Lacey *et al.,* 2005; Smart, 1995; UC-IPM, 2009).

BIOTECHNOLOGY DERIVED INSECT-RESISTANT PLANTS

Any organism that carries DNA introduced via genetic modification is referred to as transgenic. Some of the first transgenic plants were made to express Bt genes and still are known as transgenic Bt-plants (*i.e*. Bt-maize, Bt-cotton, Bt-tobacco). These genetically modified plants produce their own "insecticide" without changing the crop product (grain, fiber, tobacco, etc.), nutritional nature, or representing any threat to the environment (Evans, 2004; Toenniessen *et al.,* 2003; Sankula and Blumenthal, 2004). Biotechnology derived plants also include herbicide-resistant technologies, such as resistance to glyphosate. The growth of this market has been increasing rapidly over the last decade. For instance, in 2008 the value of the plant biotechnology market increased by 29.5%, to exceed $9 billion. When comparing among crops, maize is at the top of the list of genetically modified seed, representing 48.4% of the total market. Insect-resistant maize seeds contribute to this high percentage (McDougall, 2009; Lemaux, 2009).

The original target insects for transgenic maize were the European corn borer *Ostrinia nubilalis* (Hübner) and corn rootworm *Diabrotica spp.* Each pest causes damage to the plant while feeding in cryptic habitats. Corn borer larvae feed in the whirl of the plant during vegetative growth stages of the plant but tunnel into the stalk and ear during the plant's reproductive growth stages. Rootworm egges are laid

Table 7.7. Biotechnology-developed nematodes-based bioinsecticides in the market (suppliers and producers)

Product name	***Company-country***	***Nematode genera***[1]
Beneficial Nematodes	Buglogical Control Syst, Inc.; Green Spot, Ltd.; Spoerri, Inc., USA	*H. bacteriophora (Hb); S.carpocapsae* (Sc) mix (HbSc)*; S. feltiae* (Sf) mix (HbSf)
Beneficial Nematodes	EcoSolutions, Inc. USA	*S. feltiae*
Beneficial Nematodes	Rincon-Vitova Insectaries Inc., USA	*H. bacteriophora; H. indica H. marelatus; S. carpocapse S. feltiae; Steinernema* spp. *Heterorhabditis* spp. (in bulk)
Beneficial Nematodes	The Beneficial Insect Co,; Extremely Green Gardening Co,; Worm's Way, Inc. USA	*H. bacteriophora* (Hb) *S. carpocapsae* (Sc)
Beneficial Sf Nematodes	Gardens Alive!, USA	*S. feltia*
B-Green	Biobest N.V. Belgium	*H. bacteriophora*
Bio-Garden, BioPack, Bio-Beneficial	Dirt Works, USA	Mix of *Steinernema* spp. *Heterorhabditis* spp.
Bioslug-Schnecken-nematoden	Andermatt Biocontrol / Andermatt Biogarten, Switzerland	*P. hermaphrodita*
Biorend R Ib Melolontha	Idebio, Spain	*H. bacteriophora*
Biorend R Inb Leñosos	Idebio, Spain	Mix of *Steinernema* spp.
Biorend R Nib Invernaderos	Idebio, Spain	Mix of *Steinernema* spp.
BioRoach, Ecomask, Termask	BioLogic Company, USA	*S. carpocapsae*
BioVector	Becker Underwood Inc., USA	*S. glaseri* (WG) *S. riobravis* (MC)
Canadian Beneficial nematodes	Natural Insect Control, Canada	*Steinernema* spp. *Heterorhabditis* spp.
Capsanem	Koppert Biological Systems, The Netherlands	*S. carpocapsae*
Carpocapsae-System	Biobest N.V. Belgium	*S. carpocapsae*
Carponem	Andermatt Biocontrol & Andermatt Biogarten. Switzerland	*S. carpocapsae*
Coop Oecoplan Bio-control Nüzlinge gegen Dickmaulrüsslerlarven	Coop, Switzerland	*Heterorhabditis* spp
Coop Oecoplan Bio-control Nüzlinge gegen kleine Nacktschnecken	Coop, Switzerland	*P. hermaphrodita*
Coop Oecoplan Bio-control Nüzlinge gegen Trauermückenlarven	Coop, Switzerland	*S. feltiae*

Table 7.7. *(Contd...)*

Table 7.7. *(Contd...)*

Product name	***Company-country***	***Nematode genera[1]***
Entonem	CropKing Inc., Koppert, Biological Systems, UK; Greenfire Inc., Welte, Switzerland	*S. feltiae*
Exhibit	Syngenta Bioline, S.A., Spain	*S. carpocapsae* (SC); *S. feltiae* (SF)
Flea Control Nematodes	Gardens Alive!, USA	*S. carpocapsae*
Galanem	Andermatt Biocontrol & Andermatt Biogarten. Switzerland	*H. bacteriophora*
Gnat Guard,	Gardener's Supply Company, USA	Mix of *H. bacteriophora* and *S. feltiae*
Grub Guard	Gardener's Supply Company, USA	Mix of *H. bacteriophora* and *S. carpocapsae*
Grub Patch Lawn Repair Kit	Gardens Alive!, USA	*H. bacteriophora*
Guardian/Lawn Patrol, Gnat Patrol	A-1 Unique Insect Control, USA; Hydro-Gardens, Inc., USA	Mix of *Steinernema* spp. and *Heterorhabditis* spp.
Heteromask	BioLogic Company, Australia	*H. bacteriophora*
Heterorhabditis-System	Biobest N.V. Belgium	*H. megidis*
J-3 Max Hb	The Green Spot, Ltd., USA	*H. bacteriophora (Hb)* *S. carpocapsae* (Sc); mix (HbSc) *S. feltiae* (Sf) mix (HbSf)
Insect parasitic Nematodes	Andermatt Biocontrol AG. Switzerland	*H. bacteriophora; H. megidis; S. carpocapsae*; *S. feltiae*
Kraussei-System	Biobest N.V. Belgium	*S. kraussei*
Larvanem	Koppert; Welte, Switzerland	*H. bacteriophora*, *Heterorhabditis* spp.
Larvanem -M	CropKing Inc., USA Koppert; Greenfire Inc., USA	*H. megidis*
Meginem /Dick-maulrüssler-Nematoden	Andermatt Biocontrol / Andermatt Biogarten, Switzerland	*Heterorhabditis* spp.
Mioplant Natura Nematoden	Migros, Switzerland	*Heterorhabditis* spp.
Nemagreen	Landi reba, Switzerland	*H. bacteriophora*
Nemaplus	Landi reba, Switzerland	*S. feltiae*
Nemastar	Landi reba, Switzerland	*S. carpocapsae*
Nemaslug	BeckerUnderwood Ltd., UK	*P. hermaphrodita*
Nemasys	Becker Underwood Ltd. UK	*S. feltiae* leatherjacket Killer and no Ants *S. kraussei* Vine Weevil Killer; *S. carpocapsae* Caterpillar and Codling Moth Killer *H. megidis* Chafer Grub Killer

Table 7.7. *(Contd...)*

Table 7.7. (*Contd...*)

Product name	***Company-country***	***Nematode genera***[1]
Nematodes to control:	Biocontrole. The Netherlands	*Vine weevil:H. bacteriophora* and *S. kraussei* Grubs: *H. bacteriophora* and *S.megidis* ; Leatherjackets Sciared larvae; Sciared fly *S. feltiae;* Slugs: *P. hermaphrodita;* Woodice (custacea): special formulation carpocapse Mole crickets: *S. carpocapsae*
NemAttack™	Arizona Biological Control, Inc. (ARBICO), USA	*S. carpocapsaeS. feltiae*
NemSeek™	ARBICO-Organics, USA	*H. bacteriophora*
NemaShield™	BioWorks, USA	*S. feltiae*
Nematop	Landi reba, Switzerland	*Heterorhabditis* spp.
Optinem Cydia	Agrifutur S.R.L., Italy.	*S. feltiae*
Optinem Cydia SC	Agrifutur S.R.L., Italy.	*S. feltiae*
Optinem OS	Agrifutur S.R.L., Italy.	*H. bacteriophora*
Optinem Siar	Agrifutur S.R.L., Italy.	*S. feltiae*
Ortho-Biosafe	Monsanto, USA	*S. carpocapsae*
Parasitic nematodes	Peaceful Valley Farm Supply, USA	*H. bacteriophora*
Phasmarhabditis-System	BioBest, Canada;Biobest N.V. Belgium	*P. hermaphrodita*
Predator Nematodes	Greenfire Inc., USA	Mix of *Steinernema* spp. and *Heterorhabditis* spp.
Sanoplant	Switzerland	*S. carpocapsae*
Scanmask	BioLogic Company,USA, IPM Laboratories Inc., USA	*S. feltiae*
SC Nematoden	Boden—Nützlinge, Germany	*S. carpocapsae* (SC) *S. feltiae* (SF)
Steinernema-System	Biobest N.V. Belgium	*S. feltiae*
SuperNemos®	Nemos Horticultural Ltd., Ireland	Mix of *Steinernema* spp. and *Heterorhabditis* spp.
Tecnema-H, TCB	Técnicas de Control Biológico (TCB), Spain	*H. bacteriophora*
Tecnema-S	TCB, Spain	*S. feltiae*
Tecnemac-Palm	TCB, Spain	*S. carpocapsae*
Terranem	Koppert	*H. bacteriophora*
Traunem	Andermatt Biocontrol / Andermatt Biogarten, Switzer-land	*S. feltiae*
Vector	BeckerUnderwood Inc. USA	*S. carpocapsae*
Vegetable R_x Hot Spot, Soil pest R_x	Applied Bio Pest (Growquest, USA)	Mix of *S. carpocapsae* and *S. feltiae*
No name	SENASA, Pru	*H. bacteriophora*

Smart, 1995
http://www.oardc.ohio-state.edu/nematodes/nematode_suppliers.htm
http://www.bugladyconsulting.com/Suppliers%20of%20beneficial%20insects.htm

in the soil in the fall and larvae feed on plant roots the following crop season. Thus, both of these major pests provide unique challenges for control by insecticide applications.

Bt-maize seeds were the first registered and commercially approved (by mid-1990s), and they opened the possibility to use the same biotechnology to benefit cotton and other crops. Genes selected for corn borer-resistant maize were the Cry toxins from Bt. The Bt *kurstaki* (Btk) bacteria produces several toxins active against European corn borer and several other Lepidoptera pests, and the genetic code for these proteins was the first to be transformed into corn including Cry 1Ab for YieldGard Corn Borer, and Knockout (no longer for sale), and Cry 1F for Herculex I, and recently Agrisure CB (Table 7.8). Success of these early transformed plants led the way for the transgenic seed market to grow over the years. Many companies follow a similar strategy for research and development of transgenic crops, as described by Monsanto (Table 7.9), and now offer hybrids that provide protection against European corn borer. With the exception of Herculex, these hybrids provide protection against first and second generation European corn borer and suppression against fall armyworm and stalk borer. Herculex also provides protection against western bean cutworm, fall armyworm, and cutworm. More recently, new transgenic lines with beetle active Cry proteins have been developed to target the corn rootworm pest complex including the Cry 3Bb1 and the Cry34/35Ab1 that protect against the corn rootworm in YieldGard Rootworm and Herculex RW, respectively (Table 7.8). Thus, biotechnology-derived transgenic plants may expresse three Bt Cry proteins throughout the plant, targeting both lepidopteran and coleopteran pests. Compared with bioinsecticides, developing a transgenic seed to market takes longer (6 to 13 yr), and is a good example for any similar biotechnology and derived product for insect control (McDougall, 2009; Lemaux, 2009).

Critical Requirements for Use of Insecticidal-derived Seeds

The presence of a Bt trait in a hybrid does not increase yield, but only aids in preventing yield loss due to insect damage. Although the cost-benefit for Bt-maize for European growers has been positive, there is little correlation from year to year between yields and densities of ECB. However, this is a viable strategy for CRW control where adult densities in this year's field correlate to eggs in the soil for next year's larvae and subsequent damage to the crop. Not all fields experience a yield loss from these rootworms. Thus, it is important to assess the adult population densities through a scouting program in the fall before

Table 7.8. Biotechnology derived insect-resistant plants expressing Bt cry toxins

Insect-Resistant plants	*Company – Bt toxin*	*Target insects larvae common name*
Bt maize		
Agrisure CB/LL	Syngenta - Cry 1Ab - LL	European corn borer (ECB), southwestern corn borer (SWCB), western bean cutworm (WBCW), black cutworm (BCW), southern cornstalk borer (SCSB), lesser cornstalk borer (LCSB), fall armyworm (FAW), sugarcane borer (SCB), and corn earworm (CEW)
Agrisure® GT/CB/LL	Syngenta - Cry 1Ab - LL and GT	ECB, SWCB, WBCW, BCW, SCSB, LCSB, FAW, SCB, CEW suppression
Agrisure CB/LL/RW	Syngenta - Cry 1Ab and LL	Mexican corn rootworm (MCRW), western corn rootworm (WCRW), northern corn rootworm (NCRW), ECB, SWCB, WBCW, BCW, SCSB, LCSB, FAW, SCB, CEW
Herculex I®	Dow AgroSciences + Pioneer Hi-Bred International Inc. - Cry 1Ab and Cry 1F -RRC2	ECB, SWCB, WBCW, BCW, SCSB, LCSB, FAW, SCB, CEW
Herculex ® XTRA	Cry 1Ab, Cry 1F and Cry34/35Ab1 - RRC2 and LL	ECB, SWCB, WBCW, BCW, SCSB, LCSB, FAW, SCB, CEW, MCRW, WCRW, NCRW
Genuity™ Smart Stax™	Monsanto - Cry 1Ab - RRC2 and LL	ECB, SWCB, WBCW, BCW, SCSB, LCSB, FAW, SCB, CEW
Genuity™ VT Triple PRO™2	Monsanto - Cry 1Ab and Cry 3Bb1 - RRC2 and LL	ECB, SWCB, WBCW, BCW, SCSB, LCSB, FAW, SCB, CEW, MCRW, WCRW, NCRW wireworm (WW), white grub (WG), seed corn maggot (SCM), early flea beetle (EFB), and earworm (EW)
YieldGard Corn Borer®	Monsanto - Cry 1Ab	ECB, and SWCB
YieldGard Root worm®	Monsanto - Cry 1Ab and Cry 3Bb1 - RRC2	ECB, SWCB, WBCW, BCW, SCSB, LCSB, FAW, SCB, CEW, MCRW, WCRW, NCRW
YieldGard Plus®	Monsanto - Cry 1Ab and Cry 3Bb1 - RRC2	ECB, SWCB, WBCW, BCW, SCSB, LCSB, FAW, SCB, CEW, MCRW, WCRW, NCRW, WW, WG, SCM, EFB, and EW
Bt cotton		
Genuity™ Bollgard® II Cotton	Monsanto - Cry2Ab	Cotton budworms (CBDW), cotton bollworms (CDLW), FAW, beet armyworm (BAW), cotton loopers (CTL), cabbage loopers (CBL), saltmarsh caterpillars (SMC) and cotton leaf perforators CLP)

Table 7.8. (*Contd...*)

Table 7.8. (*Contd...*)

Insect-Resistant plants	***Company – Bt toxin***	***Target insects larvae***
Genuity™ Bollgard II® with Roundup Ready® Flex Cotton	Monsanto - Cry1Ac, Cry2 Ab - RRC2 and LL	CBDW, CDLW, FAW, BAW, CTL, CBL, SMC, and CLP
WideStrike® Insect Protection with Roundup Ready® Flex[3]	Dow AgroSciences + Phyto Gen - Cry1Ac, Cry1Fa – RRC2	CBDW, CDLW, FAW, BAW, CTL, CBL, SMC, and CLP

[1]HT= herbicide tolerance, phosphinothricin acetyltransferase (PAT) glyphosate-related herbicides like Roundup Ready Corn 2 (RRC2) or glufosinate-related herbicides like Bayer CropScience's Liberty Link® (LL) or Ignit® (GT). [2]Genera from Lepidoptera: Noctuidae: Western Bean Cutworm *Richia albicosta* (Smith); corn earworm, *Helicoverpa zea* Boddie; stalk borer, *Papaipema nebris* (Guenée); fall armyworm, *Spodoptera frugiperda* (J.E. Smith); beet armyworm, *Spodoptera exigua* (Hübner); black cutworm, *Agrotis ipsilon* (Hufnagel); tobacco budworm, *Heliothis virescens* (Fabricius); cotton bollworm *Helicoverpa armigera* (Hübner); and cotton loopers (*Antarchaea chionosticta* Atherton, *Pseudoplusia includens* (Walker), and *Trichoplusia ni*. Genera from Lepidoptera: Pyralidae: Southwestern corn borer, *Diatraea grandiosella* Dyar; Southern corn stalk borer, *Diatraea crambidoides* (Grote); sugarcane borer, *Diatraea saccharalis* (Fabricius); and lesser cornstalk borer, *Elasmopalpus lignosellus* (Zeller). Genera from Lepidoptera: Cambridae: European corn borer, *Ostrinia nubilalis* Hübner. Genera from Lepidoptera: Gelechiidae: pink bollworm, *Pectinophora gossypiella* (Saunders). Genera from Lepidoptera: Bucculatricidae: cotton leaf perforators *Bucculatrix thurberiella* Busck, and Bucculatrix *gossypii* Turner. Genera from Lepidoptera: Arctiidae: saltmarsh caterpillar, *Estigmene acrea* Drury. Genera from Coleoptera: Chrysomelidae: Northern corn rootworm, *Diabrotica barberi* Smith and Lawerence; Mexican corn rootworm, *Diabrotica virgifera zea* Krysan and Smith; Western corn rootworm, *Diabrotica virgifera virgifera* LeConte. Genera from Coleoptera: Elateridae: wireworms, mainly *Agriotes lineatus* (L.), *A. obscurus* (L.) and *A. sputator* (L.). Genera from Coleoptera: Scarabaeidae: white grub, *Phyllophaga* spp. Genera from Coleoptera: Chrysomelidae: early flea beetle, mainly *Chaetocnema ectypa* Horn. Genera from Diptera: Anthomyiidae: seed corn maggot, *Delia platura* (Meigen). [3]Import approval granted only for Canada, Mexico and Japan.

recommening planting Bt-maize seeds the following growing season (Gómez-Barbero *et al.,* 2008).

Bt maize hybrids vary in other agronomic traits such as yield, moisture, drought tolerance, and disease resistance, these factors must be considered as well. Thus, a key issue is to identify fields that will be most responsive to the Bt technology.

Considerations for Bt-biotechnology crop usage must include:

1. Statistics of pest damage in a field in the past.
2. Behavior among insect generations (mid-west USA, first generation ECB prefer early planted corn, whereas second generation prefers

Table 7.9. Phases to move biotechnologically-derived plants into the market (Monsanto)

	Discovery	***Phase I***	***Phase II***	***Phase III***	***Phase IV***
Purposes	Gene/Traits Identification	Proof of concept	Early development	Advanced development	Pre-launch Regulatory
R&D	High through put screening and model crop testing	Gene optimization/ crop transformation	Trait development, pre-regulatory data, large-scale transformation	Trait integration, field testing, regulatory data generation	sub-mission, seed bulk-up, pre-marketing
Average duration	24-48 mo	12-24 mo	12-24 mo	12-24 mo	12-36 mo
Success probability	5%	25%	50%	75%	90%
Candidates	Tens of 1000s	1000s	10s	<5	1 potential product

http://www.genuity.com/Innovation/Phases-of-Innovation.aspx

late maturing corn and control is important to prevent plant lodging and ear drop).

3. Fields located with higher pest incidence (areas along streams and rivers).
4. Fields managed for high yield or at higher plant populations.
5. Maize fields grown for grain will benefit more than those grown for silage.

This last criterion is due that maize for silage purposes tends to be harvested before European corn borer causes ear droppage and stalk breakage, which are a significant component of losses caused by the pest (Lemaux, 2009).

The culture of genetically modified (GM) crops is changing agriculture practices in a number of developed and developing countries. The main Bt crop being grown by small farmers in developing countries is cotton. Bt cotton provides control of tobacco budworm *Heliothis virescens* (Fab.) (Lepidoptera: Noctuidae), pink bollworm *Pectinophora gossypiella* (Saunders) (Lepidoptera: Gelechiidae), and cotton bollworm *Helicoverpa armigera* (Hübner) (Lepidoptera: Noctuidae) (Table 7.9). Problems for controlling budworms by traditional applications of insecticides (resistance to pyrethroid insecticides and the need for 7 insecticide applications) helped to promote adoption of Bt cotton. Prior to Bt introduction in 1996, three noctuid pests caused over a quarter billion dollars worth of damage to U.S. cotton. In 1996, about 75% of all Alabama cotton acres were planted to Bt varieties and by 1998, Bt cotton accounted for over a quarter of U.S. harvested cotton acreage (Frisvold *et al.,* 1999). Environmental impacts appear to be positive during the first ten years of Bt crop use (1996–2005). One study concluded that insecticide use on cotton and maize during this period dropped by 35.6 million kg of insecticides, similar amount of pesticide applied in one year to arable crops in the EU.

Ecological Impact

Using the Environmental Impact Quotient (EIQ) measure of the impact of pesticide use on the environment, the adoption of Bt technology over this ten year period resulted in reduction in the environmental impact associated with insecticide use by 24.3% on the cotton and 4.6% on corn (Lemaux, 2009). Over the years, Bt biotechnology derived crops have been accepted and used by 12 million farmers worldwide. For instance, in Spain, the first planting of Bt maize seeds was in 1998 and reached 20,000 hectares. By 2006, over 53,000 hectares of Bt maize were grown, representing 15% of the country's total maize planting but, where the pest population is high, areas could reach up to 60% (Gómez-Barbero *et al.,* 2008).

In 2008, 24 countries planted biotech crops, in order of decreasing hectarage, USA, Argentina, Brazil, India, Canada, China, Paraguay, Uruguay, Bolivia, Philippines, Australia, Mexico, Spain, Chile, Columbia, Honduras, Burkina Faso, Czech Republic, Romania, Portugal, Germany, Poland, Slovakia and Egypt. Developing countries out-numbered industrial countries by 15 to 10, and this trend is expected to continue in the future with 40 countries, or more, expected to adopt biotech crops by 2015 (Lemaux, 2009; Demain, 2007; McDougall, 2009).

Under an ecological point of view, this practice:

1. Resulted in the reduction of 10 million metric tons of greenhouse emission gases through fuel saving, by the elimination of 475 million gallons of diesel fuel needed to tillage;
2. Has provided seed cereals of an estimated 1,000,000,000,000 consumed meals; and
3. Has avoided pesticide application by 289,000 metric tons (http://www.genuity.com/Innovation/Driving-Innovation.aspx).

Unlike Bt-potato and Bt-rice varieties, which were developed solely by Monsanto and distributed through seed companies wholly or partly controlled by Monsanto, Bt-maize varieties have been developed and are being marketed by a number of different companies.

Bt cotton Bollgard II and WideStrike have been developed to simultaneously express two different Cry toxins, Cry1Ac and Cry2Ab, and Cry1Ac and Cry1Fa, respectively. Toxin Cry1Ac was significantly more toxic than the other two on *H. armigera*, whereas toxin Cry1Fa was the least toxic and caused no significant mortality. When combined, Cry1Ac and Cry1Fa showed an additive interaction in all proportions analyzed for both pest species, whereas Cry1Ac and Cry2Ab interacted synergistically in mixtures comprising 1:1 or 1:4 of each toxin against *H. armigera*. In *E. insulana*, there was no synergism between Cry1Ac and Cry2Ab but both these toxins showed a high insecticidal activity when administered individually and in mixtures. These observations suggest that each particular toxin or toxin combination expressed in transgenic *Bt*-cotton should be carefully selected depending on the most important pest species present in each geographical area (Ibargutxi *et al.,* 2008; Siebert *et al.,* 2008).

Advantages/Disadvantages

There are several advantages in expressing Bt toxins in transgenic Bt crops:

1. The level of toxin expression is sufficient to the pest control.
2. Only those insects that feed the plant system are intoxicated/ treated.
3. The toxin expression can be modulated by using tissue-specific promoters,
4. The protein replaces the use of synthetic pesticides in the environment worldwide (Lemaux, 2009).

Disadvantages of expressing Bt toxins in transgenic Bt crops may be:

1. The ecological pressure to speed insect resistance development.
2. Risk of losing the native seed germoplasm.

Since the approval and registration of Bt-seeds, a concern has been the potential of pest resistance development, and an insect resistance management (IRM) program was implemented. To date, the main component of resistance management programs involves a mandated requirement for farmer to plant at least 20 % of the crop with non-Bt seed, to create a refuge for the pest to maintain a non-resistant gene pool. The goal is to maintain the pest's susceptibility to the Bt toxin. For example, the IRM plan of U.S. Environmental Protection Agency for growing Bt maize mandates a non-Bt corn refuge. Recent field-based monitoring data showed increased frequency of Bt resistance-related genes within *Helicoverpa zea* (Boddie) populations only, No increases were found for five other major pests tested from Australia, China, Spain and the United States. These results suggest that refuges strategy is successfully delaying the development of insect resistance to Bt toxins (Tabashnik *et al.,* 2008).

GE insect resistant Bt-rice has not been approved for cultivation due to concerns about the risk associated with of outcrossing of Bt genes to wild and weedy rice. Eastern India is considered the birthplace of rice and shows the maximum genetic diversity of rice. Genetic contamination of native plants by cross pollination with Bt engineered plant is an environmental concern fearing the loss of native germplasm. Second concern is the high risk of development of pest resistance to Bt rice. With Bt maize and Bt cotton, implementation and monitoring of natural pest refuges benefited from the wide range of the target pests. By contrast, the major target pests for Bt-rice, the stem borers *Scirpophaga incertulas*, Walker (Lepidoptera: Pyralidae) and *Chilo suppressalis*, Walker (Lepidoptera: Pyralidae) do not have alternative hosts. Environmental impact of planting Bt rice showed no negative effects on predators, parasitoids, non-lepidopterous herbivores, or soil invertebrates, except when natural enemies were fed *Bt*-intoxicated prey. Soil microorganism populations have been shown significantly changes, negatively affecting the soil microbial ecology (Cohen *et al.,* 2008).

MICROBIAL METABOLITES: SPINOSAD

Beyond the production of Bt protein toxins, other bacteria have been used to produce insecticidal metabolites. Actinomycets, (*Pseudomonas entomophila*), *Xenorhabdus,* and *Photorhabdus* bacteria produce insecticidal secondary metabolites including peptides, polyketides, and hybrids of both. The diversity richness of these compounds is reflected by an increasing number of compounds that have been identified from entomopathogenic bacteria (Bode 2009). The biotechnology derived microbial metabolite spinosad is one widely commercialized toxin. Spinosad is the result of a mixture of two metabolites isolated from a naturally occurring soil actinomycete *Saccharopolyspora spinosa* (Crouse and Sparks, 1998). The mixture of these two compounds (spinosyn A and spinosyn D which confers insecticidal activity of the fermentation products) was first observed as a mosquito larvicide. These compounds were then characterized and traced to a family of novel macrocyclic lactones called spinosyns (Thompson *et al.,* 2000). Chemical description (CAS) of spinosyn A is (2-((6-Deoxy-2,3,4-tri-O-methyl-α-

Lmannopyranosyl) oxy)-13-((5-(dimethylamino) tetrahydro-6-methyl-2Hpyran-2-yl)oxy)-9-ethyl-2, 3, 3a, 5a, 5b, 6, 9, 10, 11, 12, 13, 14, 16a, 16 btetradecahydro-14-methyl-1H-asindaceno(3,2-d)oxacyclododecin-7, 15-dione (IUPAC); whereas spinosyn D is (2-((6-Deoxy-2, 3, 4-tri-O-methyl-α-Lmannopyranosyl)oxy)-13-((5-(dimethylamino)tetrahydro-6-methyl-2Hpyran-2-yl)oxy)-9-ethyl-2, 3, 3a, 5a, 5b, 6, 9, 10, 11, 12, 13, 14, 16a,16btetradecahydro-4,14-dimethyl-1H-asindaceno(3,2-d)oxacyclododecin-7,15-dione (IUPAC).

The activity of spinosyns A and D appears to be a novel neurotoxic effect causing hyperexcitation followed by disruption of the insect central nervous system. The disruptions of the insect nicotinic and gamma-aminobutyric acid receptor mechanisms have been shown to bind in several union-receptor areas. Mixed toxins that act on different receptors sites reduce the possibility of resistance development by the target insect, and support the use of spinosad for resistance management (Salgado 1998). Spinosyn A has higher insecticidal activity compared with pyrethroids and some of the organophosphate compounds such as carbamate and cyclodiene (Thompson *et al.,* 2000). Spinosad has been shown to be highly active against a variety of insects including species from the orders of Coleoptera, Diptera, Hemiptera, Homoptera, Hymenoptera, Lepidoptera and Thysanoptera (Racke, 2007).

Production

Spinosyns are produced by liquid fermentation. Vegetative inoculum of the bacteria is grown by a submerged aerobic fermentation process. Aqueous growth media contain proteins, carbohydrates, oils, corn solids, cottonseed flour, soybean flour, glucose, methyl oleate and calcium carbonate. Supernatant (metabolites) is separated from the microbil growth, re-crystallized, dissolved in methanol, centrifuged or filtered (to remove solids wastes), concentrated by distillation and converted to salt by mixing with acidified water. Finally, insoluble crystallized spinosad is dissolved the by adding enough alkali to neutralize the solution (Sparks *et al.,* 1998). The first fermentation-derived compound was formulated in 1988. Spinosad is formulated into commercial insecticides that combine the efficacy of a synthetic insecticide with the benefits of a biological pest control organism (Crouse and Sparks, 1998).

Advantages/Disadvantages

The registration and used of spinosad is relatively recent and disadvantages are being identified. A few cases of insect resistance against spinosad have been reported contrary to the prevailing theory described above, and the risk of additional insects developing insecticide resistance to spinosad must be considered (Zhao *et al.,* 2002; Young *et al.,* 2003). Also, the residual activity of spinosad is rather short. Spinosyns degrade by photolysis in water, showing a half-life less than 1 day, and in soil showing a half-life of 9-10 days. Liquid formulations are economical and contain solid spinosad in propylene glycol as a suspension concentrate (SC) (Table 7.10). On the positive side, spinosad formulations have a broad market potential and are registered in over 58 countries for application to 250 crops (USA). Spinosad is applied at low rates ranging from 25 to 200 grams of active ingredient per hectare. Baits applications such as GF-120 require as little as 0.19 to 0.38 grams per hectare (0.00017-0.00034 lbs/acre) of active ingredient (Williams *et al.,* 2005; Mangan *et al.,* 2006) because of the ability of the bait formulation to attract the target insect such as fire ants (Layton, 2009). The commercially formulations uses stabilizers to improve shelf-life, and humectants and adjuvants to improve longevity and attractiveness of the bait once it is applied (Moreno and Mangan, 2002).

Among the advantages of spinosad use as insecticide are:

1. Relatively low toxicity to mammals and birds and has slight to moderate toxicity towards aquatic species.
2. Low insecticidal activity on many beneficial insects.
3. Recommended for Integrated Pest Management programs that rely on natural predators and parasites.
4. Broad spectrum insecticidal activity including many pest species.
5. Certified for organic crop production.

Among the disadvantages of spinosad are:

1. High toxic to oysters and other marine mollusks (Dow 2001).
2. Toxic to Hymenoptera bees and parasitoids such as *Trichogramma* wasps.
3. Repeated applications could lead to some build-up of spinosyns in soil. Regulations in frequency and application rates to avoid resistance.

Spinosyns-derived Products

Use of spinosad products in organic agriculture has been authorized in the U.S., Argentina, Australia, Brazil, Canada, China, European Union, Germany, Japan, Korea, Mexico, New Zealand, Peru, Switzerland, and most Central American countries. The inclusion of some spinosad-based

Table 7.10. Biotechnology-derived metabolites as bioinsecticides products in the market

Product name	***Company-country***	***Pest control***
Audienz®; BioSpin®	d'Omya (Schweiz) AG, Dow AgroSciences, Switzerland	Turf and ornamental insect control of fruit flies, caterpillars, leaf miners, thrips, sawflies, spider mites, and leaf beetle larvae
Bonide Colorado Potato Beetle Beater's; Captain Jacks Dead Bug Brew®	Bonide Products, USA, Home Harvest® Garden Supply/Superior Growers Supply, Inc., USA; Emerald City Garden Supply, UK	Beetles, caterpillars, leafminers, thrips, corn borers, fruit flies, leafrollers, katydids, loopers, borers, shuckworms, and webworms
Boomerang®; Caribstar®; Conserve® SC; Conserve Naturalyte® Insect Control[1]; Entrust[1] 80; MS Superspin; Spintor®; Success® 480 SC; Tracer®;Spinosad home and garden®	Dow AgroSciences LLC; Dow AgroSciences, Japan; Dong Bang Agro Corp, South Korea; Southern Agricultural Insecticides Inc., USA.	Beetles, borers, caterpillars, leaf beetle larvae,leafminers, leafrollers, fire ants, fruit flies, katydids, loopers, sawflies, shuckworms,spider mites, thrips, and webworms
Comfortis®	Elanco Animal Health, Eli Lilly and Co., USA	Fleas (chewable pet medication, 14 weeks and older, not for cats)
Conserve® Fire Ant Bait[1]; Conserve® Professional Fire Ant Bait[1]; Justice Fire Ant Bait[1]	Dow AgroSciences LLC	Fire ants
Conserve Grain Protector	Dow AgroSciences LLC	Grain insect pests
Credence®; Flipper®; Laser®	Dow AgroSciences, Brazil, Argentina, Australia, etc.	Fruit flies, caterpillars, leaf miners, thrips, sawflies, spider mites, fire ants, and leaf beetle larvae.
Garden Safe® Brand Fire Ant Killer[1]	Schultz® Company, USA	Fire ants
Green Light Fire Ant Killer with Spinosad[1]; Green Light® Fire Ant Control with Conserve®[1]	Green Light Co. USA; Dow AgroSciences LLC	Fire ants
Green Light® Lawn & Garden Spray with Spinosad[1]	Green Light Co. USA	Fruit flies, caterpillars, leaf miners, thrips, sawflies, spider mites, fire ants, and leaf beetle larvae.
GF-120[2], GF-120 NF[1]; Naturalure–Naturalyte® fruit fly bait[1]	Dow AgroSciences, LLC; Dow AgroSciences, Japan	Fruit fly
Ferti-Lome® Borer, Bagworm, Leafminer, and Tent Caterpillar Spray	Voluntary Purchasing Groups, Inc., USA	Borer, bagworm, leafminer, and tent caterpillar[3]

Table 7.10. (*Contd...*)

Table 7.10. *(Contd...)*

Product name	***Company-country***	***Pest control***
Ferti-Lome® Come & get it! Fire Ant Killer	Voluntary Purchasing Groups, Inc., USA	Fire ants
Ferti-Lome® Tree & Shrub	Voluntary Purchasing Groups, Inc., USA	Fruit flies, caterpillars, leaf miners, thrips, sawflies, spider mites, fire ants, and leaf beetle larvae.
Monterey Garden Insect Spray[1]	Lawn and Garden Products, Inc. USA	Beetles, caterpillars, leafminers, thrips, corn borers, fruit flies, leafrollers, katydids, loopers, borers, shuckworms, and webworms
Mozkill® 120 SC	Dow AgroSciences, UK	Fleas, acari, mites, ticks
Natular T30[1], Natular XRG[1], Natular XRT[1]	Clarke Mosquito Control Products, Inc., USA	Mosquito and midge larva
Neudorff Bug Bait[1]	W Neudorff GmbH, KG, Germany	Fruit fly, earwigs and cutworms
Ortho Ecosense brand Fire Ant Bait granules[1]	The Ortho Group, USA	Fire ants
Safer® Fire Ant Bait[1]	Woodstream Corp, USA	Fire Ants
Spinoace 12 SC	Bayer CropScience	Beetles, caterpillars, leafminers, thrips, corn borers, fruit flies, leafrollers, katydids, loopers, borers, shuckworms, and webworms
Spinosad	Agrosanto Chemical Company Limited, China	Small caterpillars, leafminers, flies and thrips.
Spinosad	Fujian Sannong Group Co., Ltd., China	Fruit flies, caterpillars, leafminers, thrips, sawflies, spider mites, fire ants, and leaf beetle larvae.
SpY	Dow AgroSciences; Novartis Animal Health Ltd, UK	House flies in farms

This list does not reflect all manufactures or suppliers of metabolites. [1]Approved as organic products by USDA National Organic Standards Board and the Organic Materials Research Institute (OMRI 2009). [2]GF-120 is marketed as Success 0.02 CB in Guatemala and most Central American countries. [3]See Tables 7.2 and 7.55 for insect pest scientific names (Layton, 2009; Brooks, 2008-2009; OMRI 2009).

products in organic agriculture also allows the organic certification for agro-products exportation purposes (Racke, 2007; OMRI 2009). Commercially available spinosad-based products are shown in Table 7.10.

CONCLUSIONS

Maintaining effective pest control is essential for continued crop production to support the needs of a growing world population. Previous commercial efforts to produce microbial-based products over the past 50 years have resulted in a variety of pest control products. The obvious benefits of non-target and environmental safety for utilizing control strategies based on microbial agents provide compelling support to continue research and development in this category of biological control. Competing with chemical pesticides in terms of low costs and consistent efficacy continue to be a problem for many specific microbial systems. It is likely that many pests will be effectively controlled in the future by specifically developed microbial agents. Undoubtedly, product development will continue to rely of new and improved methods in biotechnology. Transgenic crops, broad applications of microbial pesticides (*e.g. Ag*MNPV), and microbial toxins like spinosads now stand as examples that demonstrate effectiveness for developing successful microbial-based biotechnology for large scale insect controls. Continual scientific discoveries in basic biology (discovery of beneficial organisims) combined with improved and novel methods in biotechnology (the ability to manipulate these organisism into useful products) will contribute to future pest control stratagies that are safe and effective. Future research will undoubtedly devise creative pest controls, just as early research on Bt in the 1960's ultimately resulted in development of transgenic crops.

REFERENCES

Adu-Mensah, J. (2002). Primary and secondary uptake of conidia of entomopathogenic fungi and the effect of oxybenzone as a chemical sunlight protector. *Ghana Journal of Science*, **42**: 3-10.

Asaff, A., Escobar, F. and de la Torre, M. (2009). Culture medium improvements for *Isaria fumosorosea* submerged conidia production. *Biochemical Engineering Journal*, **47**: 87-92.

Bedding, R.A. and Miller, L.A. (2008). Use of a nematode, *Heterorhabditis heliothidis,* to control black vine weevil, *Otiorhynchus sulcatus*, in potted plants. *Annals of Applied Biology*, **99(3)**: 211-216.

Behle, R.W. (2006). Importance of direct spray and spray residue contact for infection of *Trichoplusia ni* larvae by filed applications of *Beauveria bassiana*. *Journal of Economic Entomology*, **99**: 1120-1128.

Bhanu-Prakash, G.V.S., Padmaja, V. and Siva-Kiran, R.R. (2008). Statistical optimization of process variables for the large-scale production of *Metarhizium anisopliae* conidiospores in solid-state fermentation. *Biosource Technology*, **99**: 1530-1537.

Blandford, S., Thomas, M.B. and Langewald, J. (1998). Behavioural fever in the Senegalese grasshopper, *Oedaleus senegalensis*, and its implications for biological control using pathogens. *Ecological Entomology*, **23**: 9-14.

Bode, H.B. (2009). Entomopathogenic bacteria as a source of secondary metabolites. *Current Opinion on Chemical Biology*, **13(2)**: 224-230.

Boyetchko, S. (2008). Bio-Pesticides – The future of pest control? Agriculture. Government of Saskatchewan. Canada.

Brooks, W.C. (2008-2009). Spinosad (Comfortis). The Pet Co. Veterinary Information Network, Inc. Available: http://www.veterinarypartner.com/Content.plx?P=A&A=2768&S=1&SourceID=52

Cerón, J.A. (2001). Productos comerciales nativos y recombinantes a base de *Bacillus thuringiensis*. *In*: Ferré, J. and Caballero P. (*Eds*.), *Bioinsecticidas: fundamentos y aplicaciones de* Bacillus thuringiensis *en el control integrado de plagas*. Phytoma, Spain, pp. 153-168.

Charles, J.F., Silva-Filha, M.H. and Nielsen-LeRoux, C. (2000). *Bacillus sphaericus* on mosquitoe larvae: incidence on resistance. *In*: Charles, J.F. Delécluse, A. and Nielsen-LeRoux, C. (*Eds*.), *Entomopathogenic Bacteria: from Laboratory to Field Application*. Kluwer Academic Publishers, The Netherlands, pp. 237-252.

Chen, C., Wand, Z., Ye, S. and Feng, M. (2009). Synchronous production of conidial powder of several fungal biocontrol agents in series fermentation chamber system. *African Journal of Biotechnology*, **8**: 3649-3653.

Cohen, M.B., Chen, M. Bentur, J.S. Heong, K.L. and Ye, G. (2008). *Bt* rice in Asia: potential benefits, impact, and sustainability. *In*: Romeis, J., Shelton, A.M. and Kennedy, G.G. (*Eds*.), *Integration of Insect-Resistant Genetically Modified Crops within IPM Programs*. Springer, Netherlands, pp. 223-248.

Cooping, L.G. (2004). *The manual of biocontrol agents*. Alton: British Crop Protection Council.

Crouse, G.D. and Sparks, T.C. (1998). Naturally derived materials as products and leads for insect control: the spinosyns. *Reviews in Toxicology*, **2**: 133-146.

Darboux, I., Charles, J.F. Pauchet, Y. Warot, S. and Pauron, D. (2007). Transposon-mediated resistance to *Bacillus sphaericus* in a field-evolved population of *Culex pipiens* (Diptera: Culicidae). *Cell Microbiolology*, **9:** 2022-2029.

de Faria, M.R. and Wraight, S.P. (2007). Mycoinsecticides and mycoacaricides: A comprehensive list with worldwide coverage and international classification of formulation types. *Biological Control*, **43**: 237-256.

Demain, A.L. (2007). The business of biotechnology. *Industrial Biotechnology*, **3(3)**: 269-283.

Dow. (2001). Material Safety Data Sheet for Spinosad Technical. Dow AgroSciences, Indianapolis, IN.

Dubois, T., Lund, J., Bauer, L.S., and Hajek, A.E. (2008). Virulence of entomopathogenic hypocrealean fungi infecting *Anoplophora glabripennis*. *BioControl*, **53**: 517-528.

Duso, C., Malagnini, B., Pozzebon, A., Castagnoli, M., Liguori, M. and Simoni, S. (2008). Comparative toxicity of botanical and reduced-risk insecticides to Mediterranean populations of *Tetranychus urticae* and *Phytoseiulus persimilis* (Acari Tetranychidae, Phytoseiidae). *Biological Control*, **47**: 16-21.

Eilenberg, J., Vitt-Meyling, N. and Bruun-Jensen, A. (2009). Insect pathogenic fungi in biological control: status and future challenge. IOBC wprs Bulletin, orgprints.org. Available:http://scholar.google.com/scholar?hl=en&q=Eilenberg++2009+Aschersonia+&btnG=Search&as_sdt=2000&as_ylo=&as_vis=0.

El Damir, M. (2006). Effect of growing media and water volume on conidial production of *Beauveria bassiana* and *Metarhizium anisopliae*. *Journal of Biological Sciences*, **6**: 269-274.

Evans, S.L. (2004). Producing proteins from genetically modified organisms for toxicology and environmental fate assessment of biopesticides. *In*: Parekh,

S.R. (*Ed.*), *The GMO Handbook: Genetically Modified Animals, Microbes, and Plants in Biotechnology*. Humana Press, New Jersey, USA, pp. 53-84.

Foltan, P. and Puza, V. (2009). To complete their life cycle, pathogenic nematode-bacteria complexes deter scavengers from feeding on their host cadaver. *Behavior Processes*, **80(1)**: 76-79.

Frisvold, G., Tronstad, R. and Mortenson, J. (1999). *Economics of Bt cotton. Arizona's cotton field day, Agricultural & Resource Economics*. The University of Arizona.

Gaugler, R., Lewis, E. and Stuart, R.J. (1997). Ecology in the service of biological control: the case of entomopathogenic nematodes. Ecology in the service of biological control: the case of entomopathogenic nematodes. *Oecologia*, **109**: 483-489.

Gips, T. (1986). What is sustainable agriculture? *In*: Allen, P. and van Dusen, D. (*Eds.*), Global Perspectives on Agroecology and Sustainable Agricultural Systems. *Proceedings of the 6th International Scientific Conf. of the Intern. Federation of Organic Agriculture Movements,* UC Santa Cruz, CA, Vol. 1. pp. 63-74.

Goettel, M.S., Hajek, A.E., Siegel, J.P. and Evans, H.S. (2001). Safety of fungal biocontrol agents. *In*: Butt, T.M., Jackson, C. and Megan, N. (*Eds.*), *Fungi as biocontrol agents*: *Progress, problems and potential*. Wallingford: CAB International, pp. 347-376.

Goettel, M.S., and Jaronski, S.T. (1997). Safety and registration of microbial agents for control of grasshoppers and locusts. *Memoirs of the Entomological Society Canada*, **171**: 83-99.

Gómez-Barbero, M., Berbel, J. and Rodríguez-Cerezo, E. (2008). *Nature Biotechnology*, **26(4)**: 384-386.

Grewal, P. and Georgis, R. (1998). Entomopathogenic Nematodes. *In*: Hall, F.R. and Menn, J.J. (*Eds.*), *Biopesticides: Use and Delivery*. Vol. 5. pp. 271-299.

Hall, R.A., Zimmermann, G. and Vey, A. (1982). Guidelines for the registration of entomogenous fungi as insecticides. *Entomophaga*, **27**: 121-127.

Hawtin, R.E., King, A.L. and Possee, R.D. (1992). Prospects for development of a genetically engineered baculovirus insecticide. *Pesticide Science*, **34**: 9-15.

Ibargutxi, M.A., Muñoz, D., Ruíz-de-Escudero, I. and Caballero, P. (2008). Interactions between Cry1Ac, Cry2Ab, and Cry1Fa *Bacillus thuringiensis* toxins in the cotton pests *Helicoverpa armigera* (Hübner) and *Earias insulana* (Boisduval). *Biological Control*, **47**: 89-96.

Iriarte, J. and Caballero, P. (2001). Biología y ecología de *Bacillus thuringiensis*. *In*: Ferré, J. and Caballero, P. (*Eds.*), *Bioinsecticidas*: *fundamentos y aplicaciones de* Bacillus thuringiensis *en el control integrado de plagas*. Phytoma, Spain, pp. 15-44.

Jackson, M.A. and Jaronski, S.T. (2009). Production of microsclerotia of the fungal entomopathogen *Metarhizium anisopliae* and their potential for use as a biocontrol agent for soil-inhabiting insects. *Mycological Research*, **113**: 842-850.

Jackson, M.A., Payne, A.R. and Odelson, D.A. (2004). Liquid-culture production of blastospores of the bioinsecticidal fungus *Paecilomyces fumosoroseus* using portable fermentation equipment. *Journal of Industrial Microbiology and Biotechnology*, **31**: 149-154.

Jestoi, M. (2008). Emerging fusarium-mycotoxins fusaproliferin beauvericin, enniatins, and moniliformin – a review. *Critical Reviews in Food Science and Nutrition*, **48**: 21-49.

Kanzok, S.M. and Jacobs-Lorena, M. (2006). Entomopathogenic fungi as biological insecticides to control malaria. *Trends in Parasitology*, **22**: 49-51.

Kassa, A., Brownbridge, M., Parker, B.L., Skinner, M., Gouli, V., Gouli, S., Guo, M., Lee, F. and Hata, T. (2008). Whey for mass production of *Beauveria bassiana* and *Metarhizium anisopliae*. *Mycological Research*, **112**: 583-591.

Kassa, A., Stephan, D., Vidal, S. and Zimmermann, G. (2004). Laboratory and field evaluation of different formulations of *Metarhizium anisopliae* var. acridum submerged spores and aerial conidia for the control of locusts and grasshoppers. *Biocontrol*, **49**: 63-81.

Kaya, H.K. (1985). Entomogenous nematodes for insect control in IPM systems. *In*: Hoy, M.A. and Herzog, D.C. (*Eds.*). *Biological control in agricultural IPM systems.* Academic Press, Inc., New York, pp. 320-341.

Kennedy, I.R., Solomon, K.R., Gee, S.J., Crossan, A., Wang, S. and Sanchez-Bayo, F. (2007). Achieving rational use of agrochemicals: environmental chemistry in action. *In*: Kennedy, I.R. *(Ed.), Rational Environmental Management of Agrochemicals. ACS Symposium Series,* **966**: 2-12.

Knowles, B.H. (1994). Mechanism of action of *Bacillus thuringiensis* insecticidal delta-endotoxins. *Advances in Insect Physiology*, **24**: 275-308.

Kunimi, Y. (2007). Current status and prospects on microbial control in Japan. *Journal of Invertebrate Pathology*, **95**: 181-183.

Labbe, R.M., Gillespie, D.R., Cloutier, C. and Brodeur, J. (2009). Compatibility of an entomopathogenic fungus with a predator and parasitoid in the biological control of greenhouse whitefly. *Biocontrol Science and Technology*, **19**: 429-446.

Lacey, L.A. (2007). *Bacillus thuringiensis* serovariety *israelensis* and *Bacillus sphaericus* for mosquito control. *Journal of American Mosquito Control Association*, **23(2)**: 133-163.

Lacey, L.A., Neven, L.G., Headrick, H.L. and Fritts, Jr. R. (2005). Factors affecting entomopathogenic nematodes (Steinernematidae) for control of overwintering codling moth (Lepidoptera: Tortricidae) in fruit bins. *Journal of Economic Entomology*, **98(6)**: 1863-1869.

Laird, M., Lacey, L.A. and Davidson, E.W. (1990). *Safety of microbial insecticides.* CRC Press, Boca Raton, FL, pp. 259.

Layton, B. (2009). Organic fire ant control. *In*: Fire Ants in Mississippi. Department of Entomology, Mississippi State Univ. Available: http://msucares.com/insects/fireants/organic.html

Lemaux, P.G. (2009). Genetically engineered plants and foods: a scientist's analysis of the issues (Part II). *Annual Review of Plant Biology*, **60**: 511-559.

Lemmens-Gruber, R., Kamyar, M.R. and Dornetshuber, R. (2009). Cyclodepsipeptides – potential drugs and lead compounds in the drug development process. *Current Medicinal Chemistry*, **16**: 1122-1137.

Li, J.Q., Zhang, Y.A., Zhang, X.Y., Yang, Z.Q. and Yuan, F. (2003). Progress in studies on toxin of entomopathogenic fungi. *Forest Research*, **16**: 233-239.

Lozano-Contreras, M.E., Elias-Santos, M., Rivas-Morales, C., Luna-Olvera, H.A., Galan-Wong, L.J. and Maldonado-Blanco, M.G. (2007). *Paecilomyces fumosoroseus* blastospore production using liquid culture in a bioreactor. *African Journal of Biotechnology*, **6**: 2095-2099.

Mangan, R.L., Moreno, D.S. and Thompson, G.D. (2006). Bait dilution, spinosad concentration, and efficacy of GF-120 based fruit fly sprays. *Crop Protection*, **25(2)**: 125-133.

McDougall, P. (2009). Facts and figures: The status of global agriculture. Crop Life International. http://www.croplife.org/library/attachments/eab5d7c1-5b80-492b-b52b-d3de8819eb77/2/Facts%20And%20Figures%202009.pdf

Menn, J.J. and Hall, F.R. (1999). Biopesticides. *In*: Hall, F.R. and Menn, J.J. (*Eds*.), *Biopesticides, Use and Delivery*. Humana Press, Totowa, NJ, pp. 1-10.

Meyling, N.V. and Pell, J.K. (2006). Detection and avoidance of an entomopathogenic fungus by a generalist insect predator. *Ecological Entomology*, **31**: 162-171.

Miller, L.K. (1997). The Baculoviruses. Plenum Press, New York.

Miller, L.K. and Ball, L.A. (*Eds*.) (1998). The Insect Viruses. Plenum Press, New York.

Montesinos, E. (2003). Development, registration and commercialization of microbial pesticides for plant protection. *International Microbiology*, **6**: 245-252.

Moreno, D.S. and Mangan, R.L. (2002). Bait matrix for novel toxicants for use in control of fruit flies (Diptera: Tephritidae). *In*: Hallman, G. and Schwalbe, C.P. (*Eds*.), *Invasive Arthropods in Agriculture, Problems and Solutions*. Science Publishers Inc., Enfield, NH, pp. 333-362.

Moscardi, F. (1999). Assessment of the application of baculoviruses for control of Lepidoptera. *Annual Review of Entomology*, **44**: 257-289.

Murugushankar, R., Ilakiya, G., Thivya, D., Vidhya, L., Subaashini, R. and Pramila, B. (2009). Mass production of *Beauveria bassiana* (BALS.) Vuil. on wheat bran, rice husk and tapioca. *Pestology*, **33**: 21-22.

Nahar, P.B., Kulkarni, S.A., Kulye, M.S., Chavan, S.B., Kulkarni, G., Rajendran, A., Yadav, P.D., Shouche, Y. and Deshpande, M.V. (2008). Effect of repeated *in vitro* sub-culturing on the virulence of *Metarhizium anisopliae* against *Helicoverpa armigera* (Lepidoptera: Noctuidae). *Biocontrol Science and Technology*, **18**: 337-355.

Nematode information (2009). List of insects susceptible to various species of entomopathogenic nematodes. Available: http://nematodeinformation.com/?p=153.

OMRI. (2009). The Organic Materials Review Institute. Available: www.omri.org. Updated on: 10/26/2009.

Pettersson, B., Rippere, K.E. Yousten, A.A. and Priest, F.G. (1999). Transfer of *Bacillus lentimorbus* and *Bacillus popilliae* to the genus *Paenibacillus* with emended descriptions of *Paenibacillus lentimorbus* comb. nov. and *Paenibacillus popilliae* comb. nov. *International Journal of Systematic Bacteriology*, **49**: 531-540.

Poinar Jr., G.O. (1984). *Nematodes for biological control of insects*. CRC Press, Inc., Boca Raton, FL, pp. 277.

Posada-Florez, F.J. (2008). Production of *Beauveria bassiana* fungal spores on rice to control the coffee berry borer, *Hypothenemus hampei*, in Colombia. *Journal of Insect Science*. 8 art. No. 41.

Priest, F.G. (2000). Biodiversity of the entomopathogenic, endospore-forming bacteria. *In*: Charles, J.F., Delécluse, A. and Nielsen-LeRoux, C. (*Eds*.), *Entomopathogenic Bacteria*: *from Laboratory to Field Application*. Kluwer Academic Publishers, The Netherlands, pp. 1-22.

Racke, K.D. (2007). A reduced risk insecticide for organic agriculture: Spinosad case study. *A.C.S. symposium series. American Chemical Society,* Washington, DC. **947**: 92-108.

Russell, T.L., Gatton, M.L., Ryan, P.A. and Kay, B.H. (2009). Quality assurance of aerial applications of larvicides for mosquito control: effects of granule and catch tray size on field monitoring programs. *Journal of Economic Entomology*, **102(2)**: 507-514.

Salgado, V.L. (1998). Studies on the mode of action of spinosad: insect symptoms and physiological correlates. *Pesticide Biochemestry and Physiology*, **60**: 91-102.

Sankula, S. and Blumenthal, E. (2004). *Impacts on US agriculture of biotechnology-derived crops planted in 2003-An update of eleven case studies.* National Center for Food and Agricultural Policy.

Siebert, M.W., Babcock, J.M., Nolting, S., Santos, A.C., Adamczyk, J.J., Neese, P.A., King, J.E., Jenkins, J.N. and Mc-Carty, J. (2008). Efficacy of Cry1F insecticidal protein in maize and cotton for control of fall armyworm (Lepidoptera: Noctuidae). *Florida Entomologist*, **91(4)**: 555-565.

Smart Jr., G.C. (1995). Entomopathogenic Nematodes for the Biological Control of Insects. *Supplement to the Journal of Nematology*, **27(4S)**: 529-534. Available: http://www.nysaes.cornell.edu/ent/biocontrol/pathogens/nematodes.html;

Smitley, D.R. (1996). Incidence of *Popillia japonica* (Coleoptera: Scarabaeidae) and other scarab larvae in nursery fields. *Journal of Economic Entomology*, **89(5)**: 1262-1266.

Sparks, T.C., Thompson, G.D., Kirst, H.A., Hertlein, M.B., Larson, L.L., Worden, T.V. and Thibault, S.T. (1998). Biological activity of the spinosyns, new fermentation derived insect control agents, on tobacco budworm (Lepidoptera: Noctuidae) larvae. *Journal of Economic Entomology*, **91**: 1277-1283.

Srivastava, C.N., Prejwltta, M., Preeti, S. and Lalit, M. (2009). A review on futuristic domain approach for efficient *Bacillus thuringiensis* (Bt) applications. *Journal of Entomology Research*, **33(1)**.

Sun, Z.L. and Peng, H.Y. (2007). Recent advances in biological control of pest insects by using viruses in China. *Virologica Sinica*, **22**: 158-162.

Tabashnik, B.E., Gassmann, A.J., Crowder, D.W. and Carriére, Y. (2008). Insect resistance to Bt crops: evidence versus theory. *Nature Biotechnology*, **26(2)**: 199-202.

Tanada, Y. and Kaya, H.K. (1993). Insect Pathology. Academic Press, San Diego, CA.

Thompson, G.D., Dutton, R. and Sparks, T.C. (2000). Spinosad—a case study an example from a natural products discovery programme. *Pest Management Science*, **56**: 696-702.

Toenniessen, G.H., O'Toole, J.C. and DeVries, J. (2003). Advances in plant biotechnology and its adoption in developing countries. *Current Opinion on Plant Biology*, **6**: 191-198.

UC-IPM. (2009). *Pest Management Guidelines: Turfgrass. Cutworms and Armyworms.* How to Manage Pests. UC Pest Management Guidelines. University of California. UC-ANR Publication 3365-T. Available: http://www.ipm.ucdavis.edu/PMG/r785300611.html

Van Beek, N. (2007). Can Africa learn from China's natural insecticides? *Fruit and Vegetable Technology*, **7(6)**: 32-33.

Vestergaard, S., Cherry, A., Keller, S. and Goettel, M. (2003). Safety of hyphomycete fungi as microbial control agnents. *In*: Hokkanen H.M.T. and Hajek, A.E. (*Eds*.). *Environmental impacts of microbial insecticides.* Kluwer Academic Publishers, Dordrecht, The Netherlands, pp. 35-62.

WDE (2002). Washington Department of Ecology. *Aquatic Mosquito Control.* National Pollution Discharge Elimination System. Waste Discharge General Permit. Permit no: WAG » 992000. Effective Date: May 10, 2002.

Williams, T., Cisneros, J., Penagos, D.I., Valle J. and Tamez-Guerra, P. (2005). Ultra-low rates of spinosad in phagostimulant granules provide control of *Spodoptera frugiperda* (Lepidoptera: Noctuidae) in maize. *Journal of Economic Entomology*, **97**: 422-428.

Wilson, M.J., Glen, D.M. and George, S.K. (1993). The rhabditid nematode *Phasmarhabditis hermaphrodita* as a potential biological control agent for slugs. *Biocontrol Science and Technology*, **3**: 503-511.

Ye, S.D., Ying, S.H., Chen, C. and Feng, M.G. (2006). New solid-state fermentation chamber for bulk production of aerial conidia of fungal biolcontrol agents on rice. *Biotechnology Letters*, **28**: 799-804.

Young, H.P., Bailey, W.D. and Roe, R.M. (2003). Spinosad selection of a laboratory strain of the tobacco budworm, *Heliothis virescens* (Lepidoptera: Noctuidae), and characterization of resistance. *Crop Protection*, **22**: 265-273.

Zhang, Y.J., Li, Z.H., Luo, Z.B., Zhang, J.Q., Fan, Y.H. and Pie, Y. (2009). Light stimulates conidiation of the entomopathogenic fungus *Beauveria bassiana*. *Biocontrol Science and Technology*, **19**: 91-101.

Zhao, J-Z., Li, Y.X., Collins, H.L., Gusukuma-Minuto, L., Mau, R.F.L., Thompson, G.D. and Shelton, A.M. (2002). Monitoring and characterization of diamondback moth resistance to spinosad. *Journal of Economic Entomology*, **95**: 430-436.

Zimmermann, G. (2007). Review on safety of the entomopathogenic fungi *Beauveria bassiana* and *Beauveria brongniartii*. *Biocontrol Science and Technology*, **17**: 553-596.

Web links

http://www.agriculture.gov.sk.ca/bio-pesticide
http://www.bugladyconsulting.com/Suppliers%20of%20beneficial%20insects.htm
http://www.genuity.com/Innovation/Driving-Innovation.aspx
http://www.oardc.ohio-state.edu/nematodes/nematode_suppliers.htm
http://www.inhs.illinois.edu/research/biocontrol/home.html

{8}

Improvement of Biological Control Agents through Molecular Strategies

NINFA M. ROSAS-GARCÍA

ABSTRACT

Molecular techniques have improve the efficiency of the bacterium Bacillus thuringiensis through increasing toxicity, broadening the range of target pests, and delaying pest resistance. The genes that produce the crystal toxins of B. thuringiensis have suffered a number of modifications that include gene mutation to produce crystal proteins with enhanced activity, gene integration in which toxic genes from one organism are integrated with another one to extend host range spectrum, chimeric proteins development to understand their mode of action and the biochemical and molecular interactions in order to design the suitable changes to achieve high efficacy or avoid unwanted effects on non-target organisms. Insecticidal genes have been inserted into other organisms which do not possess killing abilities but are useful due to their own characteristics. Among them some bacteria and fungi are also included.

INTRODUCTION

Insects, as part of the fauna have strong and serious implications for man from different points of view. Among the most negative implications could be mentioned the damage caused to agriculture, human health, and deterioration of mankind possessions. The arrival of molecular biology techniques have contributed in many scientific disciplines, extending the knowledge of the basic biological mechanisms that take place in organisms. Given the importance that molecular biology represents, biological control of insect pest has used this resource since many years ago, to improve control agents effectiveness against insect pests.

This chapter pretends to give a wide vision of diverse molecular techniques that have been use to improve the toxic activity of some of the most important microbial agents.

One of the most important control agents worldwide is the bacterium *Bacillus thuringiensis*, for this reason much of this chapter is devoted to this organism. However some other bacteria that lack insect killing abilities, have been found to be able to control insect pests after their genes have been modified producing new proteins with enhanced toxic activities. The improved or enhanced virulent capacity of some important fungal species is also revised in this chapter.

Bacillus thuringiensis

Crystal proteins

Bacillus thuringiensis is a spore-forming, gram-positive bacterium that forms a parasporal crystal protein with insecticidal activity. The natural *B. thuringiensis* strains have been genetically improved with three main objectives in mind, according to Kaur (2000):

(1) To increase toxicity to a particular pest, this includes an increase in magnitude and a faster speed of killing.
(2) To broaden the range of target pest by increasing the copy number of *cry* genes and the type of *cry* genes in a strain.
(3) To delay the onset of pest resistances by including toxins that bind to different sites or have different modes of action.

The insecticidal activity comes from the crystal proteins, also called δ-endotoxins, that are formed by one or more proteins and exhibit a toxic effect experimentally verifiable to a target organism, or these proteins have a significant similar sequence to other previously known Cry protein (Crickmore *et al.*, 1998).

Some Cry proteins show high similarity among their structures, now that Cry proteins studied so far possess three domains, despite of a lower degree of sequence similarity (De Maagd *et al.*, 2003; Bravo *et al.*, 2007). Domain I is the N-terminal domain. It is characterized by a bundle of seven or eight antiparallel α-helices in which the hydrophilic helix-α5 is encircled by the remaining amphipathic helices (Bravo *et al.*, 2007; Schnepf *et al.*, 1998; Morse *et al.*, 2001). This domain inserts into the membrane inducing pore formation (Bravo, 1997; Bravo *et al.*, 2007).

The second domain is composed of three antiparallel β-sheets with a topology of "Greek Key", arranged in a β-prism fold (Morse *et al.*, 2001; de Maagd *et al.*, 2003). It has been thought to be a receptor binding domain (Bravo, 1997), through one or several surface-exposed loops, therefore considered to be a major determinant of toxin specificity (De Maagd *et al.*, 1996).

The third domain is formed of two twisted, antiparallel β-sheets forming a β-sandwich (Schnepf *et al.*, 1998), and it is involved in receptor recognition and/or binding to gut membrane proteins (de Maagd *et al.*, 1996), It determines the level of toxicity to a target insect, and plays an important role in membrane permeabilization, (pore formation) (Schwartz *et al.*, 1997). Some tridimensional structures are shown in Fig. 8.1.

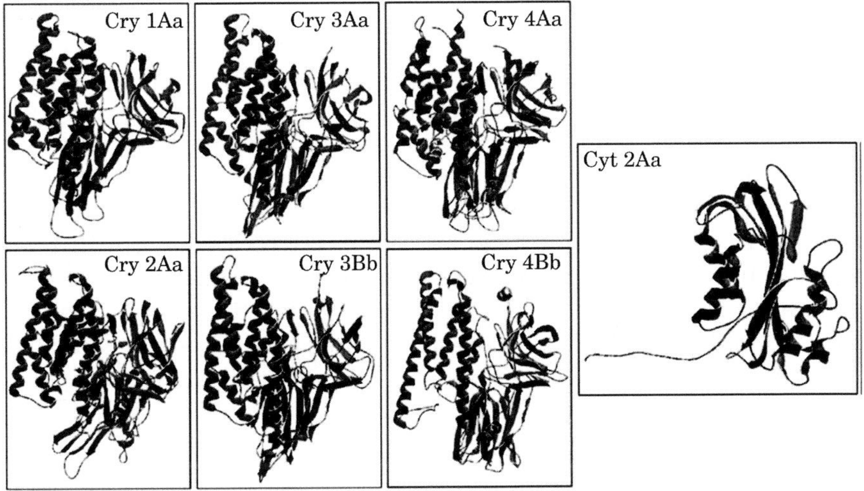

Fig. 8.1. Three-dimensional structures of insecticidal toxins produced by *Bacillus thuringiensis* Cry1Aa, Cry2Aa, Cry3Aa, Cry3Bb, Cry4Aa, Cry4Bb and Cyt2A. "Reprinted from Toxicon 49, Bravo, A., Gill, S.S. and Soberón, M. Mode of action of *Bacillus thuringiensis* Cry and Cyt toxin and their potential for insect control, 423–435, © 2006 Elsevier Ltd. with permission of Elsevier"

These toxins have demonstrated insecticidal capacity toward a wide range of hosts such as insects from orders Lepidoptera, Diptera, Coleoptera, Hymenoptera, and more recently have demonstrated toxic activity towards nematodes, hemipterans, protozoans, etc.

Up to date, improvement of biological control agents, particularly, *B. thuringiensis* had been focused on the search of novel isolates that produce more potent toxins or with a wider spectrum of activity. The search of more efficient strains for pest control has taken the molecular techniques as essential tools for biological control agent improvement.

The study of the δ-endotoxin structure has permitted rational engineering of Cry toxins with desired insecticidal characteristics. Modification of Cry toxins has brought the possibility of obtaining a repertoire of useful toxins for agriculture. The advances in molecular methodologies have allowed gene fusions which have been used as an important resource for production of highly toxic proteins specially when inserted in different organisms with the purpose of extending activity spectrum or increasing toxic activity.

Insecticidal Gene Mutation

One of the key issues in the selection of improved strains is the production of crystal proteins with enhanced activity. Mutations in insecticidal genes are conducted for obtaining strains that can produce crystal proteins with enhanced activity or for acquiring new insecticidal functions. For example the *cry4Ba* gene produced a protoxin whose ~46kDa toxin fragment exhibited no toxicity against genus *Culex*, so its trypsin site was removed by mutating R203 to A by site-directed mutagenesis, the new mutated toxin was called 4BRA and the gene produced a ~66kDa toxic fragment. The putative loop 3 of 4BRA was mutated by site-directed mutagenesis to mimic the corresponding putative loop of Cry4Aa. This exchange increased toxicity of greater than 700-fold against *Culex quinquefasciatus* Say (Diptera: Culicidae) and greater than 285-fold against *C. pipiens* Linnaeus (Diptera: Culicidae) (Abdullah *et al.*, 2003).

Mutation may be related to eliminations of certain gene fragments. The mutation of potential chymotrypsin and trypsin sites in a Cry1Aa toxin resulted in enhanced resistance to degradation by *Choristoneura fumiferana* (Clemens) (Lepidoptera: Tortricidae) proteases showing up to >4-fold increase in toxicity. The elimination of trypsin and chemotrypsin sites near the toxic C-terminus increased crystal toxicity (Bah *et al.*, 2004).

Some mutagenic experiments have not been focused precisely to enhance *cry* gene toxicity, particularly *cry*1 class genes, but mutagenesis in the segments of the aminoacidic sequences have allowed to determine their roles in the toxic activity. These genes have alterations in structure however; they produce toxins very similar to toxins derived from wild-type exhibiting significant functional differences compared to the wild type toxin (Aronson *et al.*, 1995; Schwartz *et al.*, 1997; Girard *et al.*, 2008). This becomes in a key point for the development of targeted and more effective mutants as control agents. The construction of a mutant hybrid protein in *B. thuringiensis* has shown to increase the toxic capacity toward some lepidopterans. The hybrid toxin CryAAC (1Ac/1Ac/1Ca) was mutated by replacing the arginine residue at position 423 (R423) by serine. The new mutant hybrid toxin is slightly more toxic than CryAAC toward *Mamestra brassicae* Linnaeus (Lepidoptera: Noctuidae parenthesis larvae in terms of growth inhibition (Ayra-Pardo *et al.*, 2007).

In a very interesting work, a triple mutation was carried out in residues of domain II of the CryIAb (Cry1Ab) toxin by site-directed mutagenesis. This triple mutant denominated DF-1 (N372A, A282G and L283S) was constructed by substituting residues 282AL283 located in the domain II, between α-helix 8a and α-helix 8 of N372A mutant with 282GS283. Previously N372A was constructed by substituting domain II loop 2 N372 with Gly. The mutant DF-1 exhibited a 36-fold increase in toxicity toward *Lymantria dispar* Linnaeus (Lepidoptera: Lymantriidae) due to a higher binding affinity to a common 210-kDa protein indicating that higher toxicity was not caused by binding to different receptors (Rajamohan *et al.*, 1996). A similar mutation in the α8 loop in domain II of the Cry1Ac toxin indicated that two arginine residues in this region are important in toxicity and binding (Lee *et al.*, 2001).

One of the most important disadvantages in *B. thuringiensis* field application is its instability, which has been mainly attributed to sunlight exposure. Hoti and Balaraman (1993) found that the mutant MB24 derived by chemical mutagenesis of the wild-type strain of *B. thuringiensis* H-14, produced melanin during sporulation. The pigment is produced through a pathway synthesis from L-tyrosine via L-DOPA (L-3,4-dihydroxyphenylalanine) under a suitable growth medium. The intensity of the pigment is higher when potassium mineral salts agar or casein/glycerol agar are used.

Time after, in 2005, Ruan *et al.*, reported the tyrosinase-encoding gene or *mel* gene from *B. thuringiensis* which produces melanin by the action of a tyrosinase. Zhang *et al.,* 2007 took advantage of the melanin produced by the strain of *B. cereus* 58 (Bc58) to address the problem of *B. thuringiensis* sensitivity to UV radiation and suggested the construction of a multifunctional genetically engineered *B. thuringiensis* strain since *B. thuringiensis* and *B. cereus* belong to the same group and are genetically very closely related. This approach could be a novel alternative to extend *B. thuringiensis* field activity.

Gene Integration

The integration of toxin genes derived from one microorganism into another one gives great advantages to control several pests, such as an extended host range spectrum and overcoming possible resistance development, also a wide variety of new gene combinations can be obtained. In addition multiple advantages could be obtained if a selected microorganism has previously shown host specificity, expresses larger amounts of toxins and/or possesses particular characteristics to improve killing actions. One of the most used techniques to construct *B. thuringiensis* strains that produce unique combinations of crystal proteins; some of them with improved insecticidal activities, is the site-specific recombination system. This vector system combines a *B. thuringiensis* plasmid replicon and an indigenous site-specific recombination system that allows selective removal of foreign DNA from the recombinant bacterium after the introduction of the Cry-encoding plasmid (Baum *et al*., 1996). In 1990 Baum *et al*., constructed three cloning vectors based on replication origins derived from resident plasmids of the HD263 and HD73 *B. thuringiensis* strains. The main feature of these shuttle vectors was that they permitted manipulations of cloned genes, and as they were designed with restriction sites, the construction of recombinant plasmids lacking antibiotic resistance gene was possible. The introduction of *Not*I, *Sal*I, and *Sfi*I restriction sites into the plasmid allowed the excision of non-*B. thuringiensis* DNA after the cloning of *cry* genes in *E. coli*, and before the transformation of *B. thuringiensis*.

This system has been used to enhance the production of the crystal protein produced by the strain EG2424, which exhibits toxic activity against coleopterans, and for the strain EG7673 to increase its Cry3 toxin production three or four times more than its parental strain (Baum *et al*., 1996).

Gene integration is also a strategy used for avoiding structural instability of plasmids in the recombinant strain (Kaur, 2000).

Gene integration has been successfully carried out in sporulation deficient mutants (Spo$^-$), as they are very useful since mutants do not produce spores or do not produce mature spores, but produce large amounts of parasporal crystal inclusions that remain encapsulated in the cell, which does not lyse. The sporulation deficient mutants can be obtained by disruption of some of the chromosomal sigma factors genes (Bravo *et al*., 1996) or by blocking some stages of the sporulation process, that could occur before the stage II (Johnson *et al*., 1980; Sastry *et al*., 1983), at the end of stage III or in early stage IV (Sierra-Martínez *et al*., 2004). These kinds of mutants have become of real importance, mainly when insecticide preparations based on *B. thuringiensis* strains must be applied to different environments such as water surfaces or soils to control pests. The introduction of viable spores into these environments is not permissible as they entail a possible risk to spread in an undesirable way. Although there are different techniques to separate spores and crystals, such as centrifugation, UV radiation, two-phase separation, etc., they are impractical and costly for production on large scale, so development of Spo⁻ mutants is a good alternative to produce large amounts of crystal proteins.

The expression of *cryIC* (*cry1Ca*) gene from the non-sporulation-dependent *cryIIIA* (*cry3Aa*) promoter resulted in larger amounts of CryIC (Cry1Ca) accumulated in several *B. thuringiensis* strains (Sanchis *et al.*, 1996).

The introduction of the *cry1C/Ab* gene into a sporulation deficient strain of *B. thuringiensis* resulted in a fusion protein with increased toxicity toward *Heliothis virescens* (Fabricius) (Lepidoptera: Noctuidae) and *Ostrinia nubilalis* (Hübner) (Lepidoptera: Crambidae). This toxicity was presumably due to the presence of the C-terminal half of Cry1Ab lacking 26 amino acid residues including four cysteine residues, which are probably involved in the formation of the protease-resistant crystal structure via disulfide bond formation, increasing in this way solubility of the chimeric protein (Sanchis *et al.*, 1999).

The expression of *cry* genes is frequently associated with the sporulation process, so therefore the spore-deficient mutants are frequently crystal deficient mutants (Cry⁻) (Sierra-Martínez *et al.*, 2004). This important trait has allowed the use of a new strategy to form parasporal bodies in acrystalliferous *B. thuringiensis* strains. The *B. thuringiensis* subsp. *israelensis* 4Q7 strain was used in a process for cloning the whole operon of *cry10A* and using the pSTAB vector. This strategy is produced when this strain is transformed by electroporation. In this way the recombinant strain Bt-pSTAB-*cry10* expressed toxic activity against the fourth instar larvae of *Aedes aegypti* (Linnaeus) (Diptera: Culicidae) (Hernández-Soto *et al.*, 2009). This result may be considered for increasing toxicity of the wild type strains, however widening activity spectrum is also of great importance as in the work in which two novel crystal proteins Cry44Aa1 and Cry30Ba1from *B. thuringiensis* subsp. *entomocidus* were cloned and expressed in an acrystalliferous strain of *B. thuringiensis* subsp. *israelensis* to broad the synthetic combinations of mosquitocidal toxins (Ito *et al.*, 2006).

The development of highly effective *B. thuringiensis* strains becomes of considerable interest, and their potential use to control mosquitoes is also a crucial point in pest management. In an attempt to remedy this issue the construction of a recombinant strain of *B. thuringiensis* subsp. *israelensis* was developed. The recombinant strain that was able to produce Cyt1A, Cry11B and the Bin toxin of *Bacillus sphaericus*, showed the highest toxicity against *C. quinquefasciatus*, while those recombinants that produced combinations of only two toxins exhibited a considerable lower toxicity (Park *et al.*, 2003).

Several research groups have determined that some genes integrated into acrystalliferous strains require the 20-kDa protein for an efficient expression. In an earlier study Adams *et al.*, (1989) reported that the 20-kDa protein acted during the synthesis of CytA (Cyt1Aa) (27-kDa) crystal protein or posttranslationally becoming an important factor for the efficient production of this protein in *E. coli*. However Chang *et al.*, in 1993 concluded that the 20-kDa protein was not essential for the production of CytA (Cyt1Aa) and CryIVD (Cry11Aa) proteins in *B. thuringiensis*, and that their efficient expression was likely attributed to the presence of a second protein, probably a chaperonin, nevertheless, the role of the helper protein P20, has been well documented. In the last years successful attempts to increase the yields of different *B. thuringiensis* toxins at the transcriptional or posttranscriptional level such as CytA1 crystallization, the increased production of Cry4A, Cry11A and even for the truncated Cry1C have been conducted (Visick and Whiteley, 1991; Yoshishe *et al.*, 1992; Wu and Federici, 1993, 1995; Rang *et al.*, 1996).

P20 protein promotes expression in other proteins of *B. thuringiensis* subsp. *israelensis* acting as a molecular chaperon (Shao *et al.*, 2001). This protein also promotes expression of Cry proteins from *B. thuringiensis* subspecies other than *B. thuringiensis* subsp. *israelensis*. For instance Cry2A production is enhanced by P20 up to 15%, (Ge *et al.*, 1998) similarly several Cry1C proteins with mutations that avoid normal crystal formation, have also been improved with P20 (Rang *et al.*, 1996; Park *et al.*, 2000). P20 doubles the yield of Cry1Ac and bigger crystals are formed in recombinant strains (Zheng *et al.*, 1998).

Cloning of a toxin gene in a different host is necessary to conduct studies on the organization and expression of the genes, as well as in gene products identification and their modes of action (Calogero *et al.*, 1989). However, it is also useful to allow the production of one or more toxins into a host to increase its entomopathogenic potential. In the past years the use of integrational vectors for cloning and expressing a toxin gene of *B. thuringiensis* in *B. subtilis* was widely used (Hoch *et al.*, 1967; Ferrari *et al.*, 1983; Gianni and Galizzi, 1986; Shivakumar *et al.*, 1986).

Before 1990, the lack of an efficient transformation method for *B. thuringiensis* made difficult the expression of crystal proteins in different strains. By 1989, Bone and Ellar, reported the transformation of *B. thuringiensis* strains by electroporation. In 1990 Crickmore *et al.* introduced δ-endotoxin genes into native *B. thuringiensis* strains using the electroporation method, although the method is successful it may occur plasmid instability which depends on unique sequences of the vector used, in addition some other plasmids were stable in some *B. thuringiensis* strains but not in others. Although chromosomal integration is a difficult procedure in *B. thuringiensis*, a two-step procedure proposed by Kalman *et al.* (1995), resulted highly efficient.

For example in the first step of this procedure the *cryIC* (*cry1Ca*) gene was cloned into an integration vector containing a *B. thuringiensis* chromosomal fragment coding a phosphatidylinositol-specific phospholipase C which allowed the gene to be targeted to the homologous region of the recipient strain. In the second step a transducing bacteriophage, CP-51 was used to transfer the integrated CryIC (Cry1Ca) into the host strain where it expressed efficiently. In this study the toxic activity of *B. thuringiensis* subsp. *kurstaki* was increased toward *Spodoptera exigua* (Hübner) (Lepidoptera: Noctuidae) with the 50% lethal concentrations six fold lower than those of the parental strains.

The same procedure allowed construction of a new strain denominated BIOT185 constructed from two origin strains possessing Cry8Ea1 and Cry8Ca2 respectively, coexpressed both toxins toxic against the scarabs *Anomala corpulenta* Motschulsky (Coleoptera: Scarabaeidae) and *Holotrichia parallela* Motschulsky (Coleoptera: Scarabaeidae) broadened in this way its insecticidal spectrum (Liu *et al.*, 2010).

Chimeric Proteins

The study of Cry proteins has brought knowledge from their mode of action in the insect midgut to biochemical and molecular interactions in their amino acidic sequence and their structural domains. The knowledge of the three-dimensional structure of the insecticidal Cry toxins is indispensably to understand the determinants of toxin specificity in order to design the suitable changes to achieve high efficacy, broaden selectivity or avoid unwanted effects on non-target organisms (Morse *et al.*, 2001).

The *cry1* gene is the most extensively studied, and the intoxication process caused by Cry1A proteins is well documented elsewhere. The process begins with the specific interaction of the toxin with cell membrane receptors, which have been identified as cadherin-like-proteins, considered as the non-lipid-raft initial Cry1A proteins, aminopeptidase N and alkaline phosphatase which have been described as glycosylphosphatidylinositol GPI-Cry-protein receptors, anchored to the membrane lipid rafts (Avisar *et al.*, 2009). After the toxin-cadherin-like protein interaction occurs, a conformational change in the toxin is induced facilitating the cleavage of the N-terminal α1 helix by the proteases of the insect midgut. The toxin is then oligomerized to a tetrameric pre-pore structure that binds to aminopeptidase N or alkaline phosphatase which directs the tetramer to the lipid rafts occurring membrane insertion (de Maagd *et al.*, 2003; Bravo *et al.*, 2007; Avisar *et al.*, 2009).

The gene fusion, and the introductions or deletions in different regions of the genes, allow the construction of new strains of *B. thuringiensis* capable of producing improved crystal toxins over those of their naturally-occurring counterparts. These techniques lead to a vast and varied production of Cry proteins that can be specifically designed to control pests. These toxins are called chimeric proteins, and a wide variety of combinations may occur when a region in the *cry* gene is combined with the region from another *cry* gene to form a new gene capable of producing a full-length protein. Also some regions from a *cry* toxin gene are introduced in another gene or a region of a *cry* gene is deleted, all of this in order to improve or enhance the crystal insecticidal activity.

The chimeric scanning mutagenesis is a technique that identifies large regions of protein sequence that confer differential specificity to different insect orders. The resulting chimeras may have a different activity profile from the original toxin, increasing or changing toxic activity towards different insect pests (Fig. 8.2).

This valuable tool has allowed identification of determinants of species specificity in the amino acid sequence (Widner and Whiteley, 1990).

The changes in the amino acid sequence of a Cry protein may result in a substantial effect on the toxic activity, as the hybrid protein consisting of predominantly CryIIB (Cry2Ab) amino acids and only 76 amino acid segment of CryIIA (Cry2Aa). The protein CryIIB (Cry2Ab) originally exhibited toxicity only towards *Manduca sexta* Linnaeus (Lepidoptera: Sphingidae), and CryIIA (Cry2Aa) was toxic to *M. sexta* and *A. aegypti*. The toxin CryIIB (Cry2Ab) altered by replacement of a short segment of CryIIA (Cry2Aa) (residues 307 through 382), resulted in a hybrid protein active against mosquitoes (Widner and Whiteley,

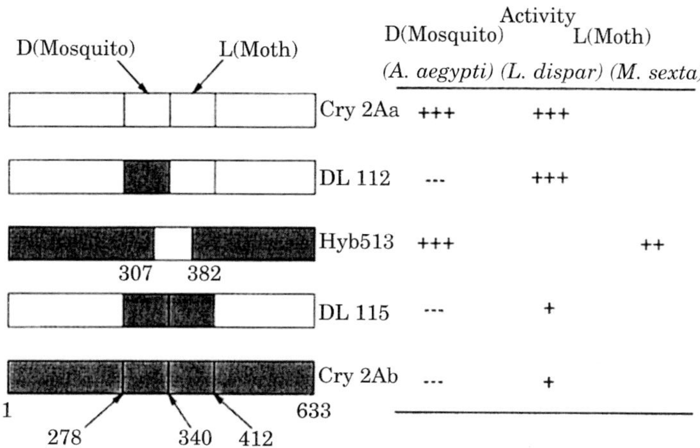

Fig. 8.2. Chimeric protein combinations obtained by chimeric scanning mutagenesis. "Reprinted from Structure, Vol. 9, Morse, R.J., Yamamoto, T and Stroud, R.M. Structure of Cry2Aa Suggests an Unexpected Receptor Binding Epitope, 409–417, ©2001 Elsevier Ltd. with permission from Elsevier"

1990). With such modification, the CryIIB (Cry2Ab) toxin acquired a new toxic activity expanding its insecticidal activity spectrum.

The protein Cry3A was the first coleopteran-active *B. thuringiensis* toxin reported (Rupar *et al.*, 1991), with a high toxicity against *Diabrotica undecimpunctata howardi* Barber (Coleoptera: Chysomelidae) and *Leptinotarsa decemlineata* (Say) (Coleoptera: Chrysomelidae) (Donovan *et al.*, 1992). In spite of this activity, it was not able to control successfully the western corn rootworn, *D. virgifera virgifera* LeConte (Coleoptera: Chrysomelidae). However, a modification conducted by the introduction of a chymotrypsin/cathepsin G protease recognition site enhanced its biological activity. The mutation region was located in the loop between α-helix 3 and α-helix 4 of domain I. This protein was named mCry3A, and its active fragment was rapidly generated and associated with the membrane fraction due to a higher protease sensitivity caused by the addition of the chymotrypsin/ cathepsin G AAPF recognition site. The mCry3A exhibited a greater biological activity towards the corn rootworm larvae than parental Cry3A as it was more rapidly converted into the active fragment due to the enhanced cleavage, and a subsequent binding of the activated toxin (Walters *et al.*, 2008).

An *in vivo* recombination was used to produce *cryIA(b)-cryIC* (*cry1Ab-cry1Ca*) hybrid genes, based on the idea that hybrid toxins derived from toxins with different specificities serve as powerful tools that determine specificity of amino acid sequences, and also allow to relate the structural features of toxins with their specific functions. The hybrid protein CryIA(b)-CryIC(Cry1Ab-Cry1Ca) contained the domains I and II of CryIA(b) and domain III of CryIC and a reciprocal hybrid harbored domains I and II of CryIC and domain III of CryIA(b). After hybrids were tested against *S. exigua*, results indicated that the substitution of domain III of CryIA(b) (Cry1Ab), that exhibits moderate toxicity towards *S. exigua*, greatly increase its toxicity toward this insect. But even the exchanging domain III of CryIA(b) with that of the CryIC protein (which is a more active protein), increased toxicity of CryIA(b) to higher levels than that of CryIC. Domain III of CryIC protein demonstrated to play an important role in the level of toxicity against *S. exigua*.

Another technique called domain swapping consists in the interchange of a domain of one toxin by that of another toxin to form a hybrid. Formation of these hybrids is also possible when toxin arises as a product of recombination between the two encoding genes (De Maagd *et al.*, 1996). This technique could become a powerful tool to modify and increase the variety of active toxins that could be used to control insect pests (De Maagd *et al.*, 1996).

It is thought that the simultaneous production of two or more crystal proteins acting independently on the same insect might avoid or delay insect resistance. This idea motivated the development of a translation fusion product with two individual truncated genes. The gene *cryIAb* toxic against *Heliothis virescens* (Fabricius) (Lepidoptera: Noctuidae) and *Pieris brassicae* Linnaeus (Lepidoptera: Pieridae) and the *cryIC* toxic against *Spodoptera* sp., and *M. brassicae*, were translationally fused. As the comparison of their sequences revealed high homology, the constructed gene in tandem produced an intact and biologically active protein consisting of the N-terminal toxic parts of two different crystal proteins. The toxicity spectrum overlapped those of both crystal proteins. These results can be used as a strategy to enhance insecticidal properties of microorganisms (Honée *et al.*, 1990).

The chimeric Vip3AcAa protein became toxic to the European corn borer *O. nubilalis*, while individual recombinant proteins Vip3Aa1 and Vip3Ac1 did not exhibit toxicity towards this pest. The chimeric gene was created by sequence swapping with the previously known *vip3* and *vip3Aa1*genes. An overlap PCR method was used to generate *vip3Aa1* and *vip3Ac1* chimeric genes. To generate the chimeric gene *vip3AcAa*, two overlapping DNA fragments were amplified by PCR. The two overlapping PCR products were used as a template for the second PCR to generate a full-length hybrid gene. The sequence swapping allows exploring the possibility for toxin improvement by artificial gene recombination (Fang *et al.*, 2007).

ALTERNATIVE CONTROL AGENTS

Bacteria

Other organisms, that originally are non-toxic to insects, can act as alternative control agents if their own characteristics are considered for this purpose. Insecticidal genes can be inserted into these organisms, and thus be able to produce different insecticidal proteins. An interesting case is the use of the plant-colonizing methylotroph *Methylobacterium extorquens*

The cloning of the *cry1Aa* gene in the plant-colonizing methylotroph *M. extorquens* under the control of the methanol dehydrogenase promoter produced a recombinant strain capable of causing rapid feeding inhibition and total mortality to *Bombyx mori* larvae. The importance of using different bacteria expressing *cry* toxins as in *M. extorquens*, rely on their longer persistence, colonizing ability and low-cost production system.

Pseudomonas fluorescens isolated from phylloplane and the endophytic bacterium *Herbaspirillum seropedicae* are bacteria associated to the sugarcane (Downing *et al.*, 2000). On the other hand, *Eldana saccharina* Walker (Lepidoptera: Pyralidae) is the most destructive pest in sugar cane fields in South Africa, limiting sugarcane productivity (Mokhele *et al.*, 2009). As a good strategy to control this pest, a recombinant *P. fluorescens* was constructed. The *cry1Ac7* gene from the *B. thuringiensis* strain 234 cloned under the control of the *tac* promoter was introduced into a broad-host range plasmid and an integration vector. The construct was introduced into the chromosome of *P. fluorescens* becoming a bacterium with a high toxicity toward *E. saccharina* neonate larvae, also higher expression levels were observed indicating a significant difference from the parental strain. *H. seropedicae* produced more toxins when the *cry1Ac7* gene was cloned under the control of the strong promoter Nm^r. Toxicity bioassays conducted with this genetically modified microorganism caused also higher mortality against *E.sacchararina* than the parental strain (Downing *et al.*, 2000).

Alternative control agents have also been considered for dipterans, as these insects are important pests related to human health. Although *B. thuringiensis* subsp. *israelensis* and *B. sphaericus* have proved useful for mosquito control, these bacteria lack some properties to be highly effective in aquatic environments. One of the more important disadvantage is that crystals easily settled down, therefore an excellent candidate is the Gram-negative bacterium *Asticcacaulis excentricus*, because its flotability is a desired characteristic since mosquito larvae fed on the water surface. However, to obtain satisfactory results recombinant strains should be constructed to produce toxic crystals remaining on water surface. An experiment shown in Fig. 8.3 shows how *B. sphaericus* spores begin to sediment after two days of application, being completely sedimented within six days, while *A. excentricus* cells transformed and untransformed remain in suspension up to six days indicating that the recombinant *A. excentricus* cells remained in the upper part of the water column longer than the wild type *Bacillus* strains. (Romero *et al.*, 2001). The transformed bacteria expressed the binary toxin of *B. sphaericus*, with high toxicity toward *Culex* and *Anopheles* mosquito larvae. These bacteria become a promising host for toxic genes to mosquito larvae and a potential control agent as it persists in the larval feeding zone and are UV light resistant (Liu *et al.*, 1996). More recently, *A. excentricus* was transformed with the *cry11Aa* and *p20* genes from *B. thuringiensis* subsp. *israelensis* and the recombinant bacteria exhibited toxicity towards *A. aegypti* (Armengol, *et al.*, 2005).

The nitrogen-fixing filamentous cyanobacterium *Anabaena* that develops in aquatic environments and serves as a food source to mosquito larvae, was transformed to express the Cry4Aa and Cry11Aa *B. thuringiensis* toxins, and the accessory protein P20. The recombinant strain exhibited high toxicity against *A. aegypti* larvae, and crystal proteins probably are protected from UV-B damage due to cyanobacterial pigments that act as UV-B photoprotectors (Manasherob *et al.*, 2002).

The use of these transformed strains may be sufficient to control certain leaf-feeding insects. The use of stronger promoters, genes coding for proteins with enhancing activity, and a higher plasmid copy number could result in phylloplane inhabitant bacteria with increased insecticidal activity (Moar *et al.*, 1994).

Bacillus sphaericus

Bacillus sphaericus produce a spherical spore with a terminal position that synthesize a parasporal inclusion or crystal with toxic activity for a variety of mosquito species (Payme and Davidson, 1984). The strains of *B. sphaericus* exhibit different mosquitocidal activities. Some of them are highly toxic against susceptible mosquito larvae, while other exhibit medium activity or no toxicity against binary toxin-resistant mosquito larvae. Larvae of the genus *Culex* seems to be more susceptible than

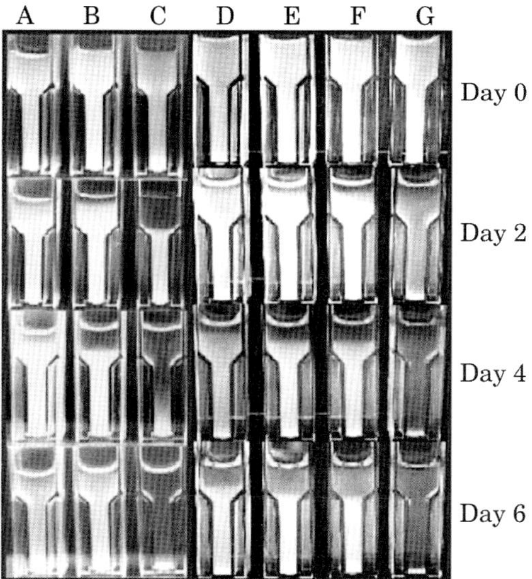

Fig. 8.3. Plastic cuvettes showing flotation characteristics at different time intervals of *Asticaccaulis excentricus, Bacillus sphaericus* and *B. thuringiensis* subsp. *medellin.* A) cells of *A. excentricus* strain C2 untransformed, B) cells of *A. excentricus* strain C2 transformed with plasmid pEA1, C) final whole culture of *B. sphaericus*, D) *A. excentricus* transformed with The *cryIIA* (*cry2Aa*) gene from *B. thuringiensis* was inserted in the leaf-colonizing *B. cereus* by using electroporation. plasmid pSOD2, E) *A. excentricus* strain C2 transformed with plasmid pSOD2, F) *A. excentricus* strain C2 untransformed, G) final whole culture of *B. thuringiensis* subsp. *Medellin*. "Reprinted with permission from Memorias Do Instituto Oswaldo Cruz, Romero, *et al.*, Vol. 96(2) 257-263. © 2001

Anopheles and *Aedes*, which indicates that mosquito species exhibit different susceptibilities (Broadwell and Baumann 1987). The binary toxin composed of a 42-kDa protein and a 51-kDa protein; possesses larvicidal activity (Broadwell *et al.*, 1990). Thanabalu *et al.*, (1991) designed the *mtx* gene as a new mosquitocidal toxin gene, which encodes a protein of 100.5-kDa and 870 amino acids in length and exhibits toxicity against *C. quinquefasciatus*. The *mtx* gene is widely distributed among strains that possess high or low toxicity, with the exception of those of the serotype H26a26b. The highly toxic strains contained the *mtx* gene in addition to the genes that code for the binary toxin (Thanabalu *et al.*, 1991).

The gene integration is also a useful strategy for *B. sphaericus* to combine its good quality with higher toxicity and a wider toxic spectrum. The resultant transgenic bacteria would be expected to have a reduced probability of causing insect resistance, because they possess a mixture of several unrelated toxins that are exposed to the insect. (WHO 1996).

B. sphaericus toxin is mainly active toward insect species of genera *Culex* and *Anopheles*. Bourgouin *et al*., (1990) reported for the first time the transfer of the toxin genes of *B. sphaericus* 1593 into *B. thuringiensis* strains. Two strains of *B. thuringiensis* subsp. *israelensis*, one of them a crystal-minus strain, and the other a crystal-producing strain were transformed with the toxin encoding genes of *B. sphaericus* and their biological activity was tested on three mosquito species. The transformant which expressed only the *B. sphaericus* proteins demonstrated a high toxic activity against *C. pipiens*, even at the same level of activity of that of the reference strain (*B. thuringiensis* subsp. *israelensis*). The transformant which produced crystal proteins of both *B. thuringiensis* subsp. *israelensis*, and *B. sphaericus* exhibited the same highly activity of that of the recipient strain indicating no additive effects between toxins.

On the contrary, Cai *et al.* (2007) expressed the chitinase gene, *chiAC* from *B. thuringiensis* in *B. sphaericus* using the binary toxin promoter. The recombinant strain was 4297-fold more toxic than the wild strain against resistant *C. quinquefasciatus*. This high activity is attributed to the synergy between the chitinase and the binary toxin since chitinase degrades the chitin linkage in the peritrophic matrix, allowing binary toxin comes into contact with cells causing cytopathogenicity.

Another recombinant *B. sphaericus* strain was produced when the plasmid PA7 containing the gen *cyt1Ab1* and its flanking sequences from *B. thuringiensis* subsp. *medellin*, was introduced by electroporation. The production of Cyt1Ab1 protein was more efficient than that of the binary toxin. Although the introduction of this gene did not increase significantly the toxicity of binary toxin towards *A. aegypti* and *C. pipiens* larvae, it partially restored susceptibility of resistant *C. pipiens* and *C. quinquefasciatus* populations to the binary toxin. This fact widens its activity spectrum (Thiéry *et al*., 1998).

A truncated form of the Mtx1 protein (amino acids 30-870) produced by *B. sphaericus* 2297 could be successfully expressed in *E. coli* cells as GST-tMtx1 fusion protein. Previously the *mtx1*gene was expressed as non-fusion protein unsuccessfully, so expression level was highly improved after the putative leader sequence was deleted and the truncated *mtx1* gene was fused to *gst* gene (glutathion-S-transferase) to be more resistant to protease digestion, which could lead to high expression of the gene. The cells expressing GST-tMtx1 exhibit high toxic activity against *C. quinquefasciatus* (Promdonkoy *et al.*, 2004).

The glutathione-S-transferase (GST) fusion system has a wide range of applications since its introduction as a molecular tool for the synthesis of recombinant proteins in bacteria. Among of the most relevant properties is that GST is not sequestered in inclusion bodies and it can be purified without denaturation, for these reason is very useful in studies of protein-protein interaction (Einarson *et al.*, 2007).

Fungi

The characterization of fungal genes which are involved in some way in the insect pathogenic process must be characterized previously to intend to increase their efficacy as biological control agents. The real-time RT-PCR is a sensitive and reliable technique to study gene expression. In the fungus *Metarhizium anisopliae* the expression of four virulence or conidiogenesis-associated genes were analyzed by this method (Fang and Bidochka, 2006):

(1) A subtilisin-like protease encoding gene *pr1*, which is a virulence determinant, it is expressed in fungal mycelia emerged from insect cadavers; also it is expressed during conidiogenesis and the late stages of pathogenesis.
(2) A hydrophobin-encoding gene *ssga* that participates in conidial structure and virulence. This gene is expressed throughout all the developmental stages, with different levels of expression.
(3) A regulator of G-protein signaling gene *cag8* that regulates the GTPase activity of G protein. This gene expresses in conidia before germination and during the last stages of conidiogenesis on insect cadavers.
(4) A nitrogen response regulator gene *nrr1* that regulates nitrogen catabolite derepression. This gene is expressed constitutively throughout germination and conidiogenesis and in the late stages of pathogenesis.

Considering that the extracellular subtilisin-like protease Pr1 is one of the major virulence factor, a genetically improved *M. anisopliae* strain was developed inserting additional copies of the *pr1* gene into the genome of *M. anisopliae* and expressed under the transcriptional control of the *gdp* promoter (*Aspergillus nidulans* promoter) (Fig. 8.4) (St. Leger *et al.*, 1996a).

As expression of Pr1 is involved in penetration process (Hajek and St. Leger, 1994), the constitutive overexpression of this Pr1 could be used as a strategy to enhance strain virulence. In this sense Pr1 was constitutively overproduced once the fungal blastospores are present in the hemolymph of the insect (St. Leger *et al.*, 1988). In *M. sexta* the prophenoloxidase system is activated, and the constitutive expression of Pr1 speeds up the disease process, improving insecticidal activity shortening 25% survival time. In addition, environmental persistence was reduced because the melanized cadavers resulting from infection of the transformants supported very little sporulation (St. Leger *et al.*, 1996b).

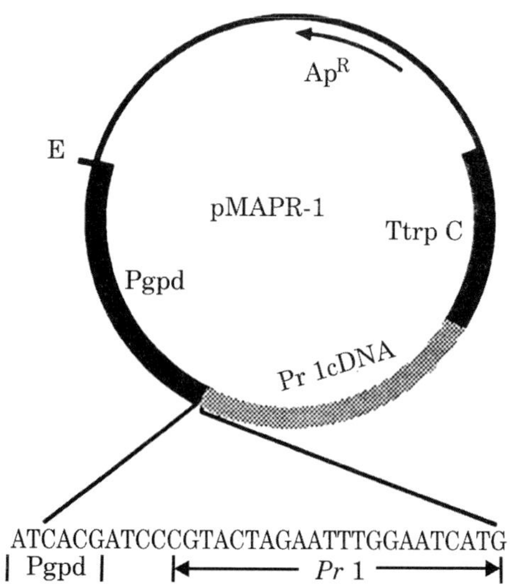

Fig. 8.4. Plasmid for cloning and expression of Pr1 cDNA (pMAPR-1). The cDNA coding for the *M. anisopliae* Pr1 proteinase inserted 3' to the *A. nidulans* promoter sequence and 5' to terminator sequences from *A. nidulans trp C* gene. ApR, ampicillin resistance gene. "Copyright 2001, National Academy of Sciences, U.S.A."

Additionally more subtilisins have been found in *M. anisopliae*, Pr1D, Pr1G, Pr1H, Pr1I, PriJ and Pr1G-K (Freimoser *et al.*, 2003), which represents more possibilities of fungal improvement. But not only proteases have been considered for improvement. Chitinases produced by entomopathogenic fungi play a synergistically role with proteases, and as an important cuticle-degrading enzymes they have been improved by genetic manipulation to over express them enhancing virulence of the transformants when compared to the wild type strain (Fang *et al.*, 2005).

The *Beauveria bassiana* wild type strain Bb0062 was transformed with the binary plasmid pBANF-bar-pAN-Bbchit1, and three transformants exhibited significant higher chitinase activity than the wild type strain due to the overproduction of endochitinase Bbchit1. The overproduction of this enzyme can increase the infection efficiency and consequently accelerate infection (Fang *et al.*, 2005).

CONCLUSIONS

It is a fact that molecular techniques have been extremely useful to improve control agents action. Although many recombinant organisms or toxins have demonstrated to improve their entomopathogenic activity in laboratory tests, it is difficult to incorporate these technologies in the regular application of bioinsecticides so far, especially due to the large number of legal regulations that exist in different countries for use and application of organisms or derivatives genetically modified. However, the future of biological insecticides will largely depend on these trends possibly exceeding expectations for rational and appropriate pest control.

REFERENCES

Abdullah, M.A.F., Alzate, O., Mohammad, M., McNall, R.J., Adang, M.J. and Dean, D.H. (2003). Introduction of *Culex* toxicity into *Bacillus thuringiensis* Cry4Ba by protein engineering. *Applied and Environmental Microbiology*, **69(9)**: 5343-5353.

Adams, L.F., Visick, J.E. and Whiteley H.R. (1989). A 20-kilodalton protein is required for efficient production of the *Bacillus thuringiensis* subsp. *israelensis* 27-kilodalton crystal protein in *Escherichia coli*. *Journal of Bacteriology*, **171(1)**: 521-530.

Armengol, G., Guevara, O.E., Orduz, S. and Crickmore, N. (2005). Expression of the *Bacillus thuringiensis* mosquitocidal toxin Cry11Aa in the aquatic bacterium *Asticcacaulis excentricus*. *Current Microbiology*, **51**: 430-433.

Aronson, A.I., Wu, D. and Zhang, C. (1995). Mutagenesis of specificity and toxicity regions of a *Bacillus thuringiensis* protoxin gene. *Journal of Bacteriology*, **177(14)**: 4059-4065.

Avisar, D., Eilenberg, H., Keller, M., Reznik, N., Segal, M., Sneh, B. and Zilberstein, A. (2009). The *Bacillus thuringiensis* delta-endotoxin Cry1C as a potential bioinsecticide in plants. *Plant Science*, **176**: 315-324.

Ayra-Pardo, C., Davis, P. and Ellar, D. (2007). The mutation $R_{423}S$ in the *Bacillus thuringiensis* hybrid toxin CryAAC slightly increases toxicity for *Mamestra brassicae* L. *Journal of Invertebrate Pathology*, **95**: 41-47.

Bah, A., van Frankenhuysen, K., Brousseau, R. and Masson, L. (2004). *The Bacillus thuringiensis* Cry1Aa toxin: effects of trypsin and chymotrypsin site mutations on toxicity and stability. *Journal of Invertebrate Pathology*, **85**: 120-127.

Baum, J.A., Coylke, D.M., Gilbert, M.P., Jany, C.S. and Gawron-Burke, C. (1990). Novel cloning vectors for *Bacillus thuringiensis*. *Applied and Environmental Microbiology*, **56(11)**: 3420-3428.

Baum, J.A., Kakefuda, M. and Gawron-Burke, C. (1996). Engineering *Bacillus thuringiensis* bioinsecticides with an indigenous site-specific recombination system. *Applied and Environmental Microbiology*, **62(12)**: 4367-4373.

Bone, E.J. and Ellar, D.J. (1989). Transformation of *Bacillus thuringiensis* by electroporation. *Federation of European Microbiological Societies, Microbiology Letters*, **58(2-3)**: 171-177.

Bravo, A. (1997). Phylogenetic relationships of *Bacillus thurin*giensis δ-endotoxin family proteins and their functional domains. *Journal of Bacteriology*, **179(9)**: 2793-2801.

Bravo, A., Gill, S.S. and Soberón, M. (2007). Mode of action of *Bacillus thuringiensis* Cry and Cyt toxins and their potential for insect control. *Toxicon*, **49**: 423-435.

Bourgouin, C., Delécluse, A., de la Torres, F. and Szulmajster, J. (1990). Transfer of the toxin protein genes of *Bacillus sphaericus* into *Bacillus thuringiensis* subsp. *israelensis* and their expression. *Applied and Environmental Microbiology*, **56(2)**: 340-344.

Broadwell, A.H. and Baumann, P. (1987). Proteolysis in the gut of mosquito larvae results in further activation of the *Bacillus sphaericus* toxin. *Applied and Environmental Microbiology*, **53(6)**: 1333-1337.

Broadwell, A.H., Clark, M.A., Baumann, L. and Baumann, P. (1990). Construction by site-directed mutagenesis of a 39-kilodalton mosquitocidal protein similar to the larva-processed toxin of *Bacillus sphaericus* 2362. *Journal of Bacteriology*, **172(7)**: 4032-4036.

Cai, Y., Yan, J., Hu, X., Han, B. and Yuan, Z. (2007). Improving the insecticidal activity against resistant *Culex quinquefasciatus* mosquitoes by expression of chitinase gene *chiAC* in Bacillus *sphaericus*. *Applied and Environmental Microbiology*, **73(23)**: 7744-7746.

Calogero, S., Albertini, A.M., Fogher, C., Marzari, R. and Galizzi, A. (1989). Expression of a cloned *Bacillus thuringiensis* delta-endotoxin gene in *Bacillus subtilis*. *Applied and Environmental Microbiology*, **55(2)**: 446-453.

Chang, C., Yu, Y.-M., Dai, S.-M., Law, S.K. and Gill, S.S. (1993). High-level *cryIVD* and *cytA* gene expression in *Bacillus thuringiensis* does not require the 20-kilodalton protein, and the coexpressed gene products are synergistic in their toxicity to mosquitoes. *Applied and Environmental Microbiology*, **59(3)**: 815-821.

Chen, X.J., Lee, M.K. and Dean, D.H. (1993). Site-directed mutations in a highly conserved region of *Bacillus thuringiensis* δ-endotoxin affect inhibition of short circuit current across *Bombyx mori* midguts. *Proceedings of the National Academy of Sciences, United States of America*, **90**: 9041-9045.

Crickmore, N., Nicholls, C., Earp, D.J., Hodgman, C. and Ellar, D.J. (1990). The construction of *Bacillus thuringiensis* strains expressing novel entomocidal δ-endotoxin combinations. *Biochemical Journal*, **270**: 133-136.

Crickmore, N., Zeigler, D.R., Feitelson, J., Schnepf, E., VanRie, J., Lereclus, D., Baum, J. and Dean, H. (1998). Revision of the nomenclature for the *Bacillus thuringiensis* pesticidal crystal proteins. *Microbiology and Molecular Biology Reviews*, **62(3)**: 807-813.

De Maagd, R.A., Bravo, A., Berry, C., Crickmore, N. and Schnepf, H.E. (2003). Structure, diversity, and evolution of protein toxins from spore-forming entomopathogenic bacteria. *Annual Review of Genetics*, **37**: 409-433.

De Maagd, R.A., Kwa, M.S.G., Van Der Klei, H., Yamamoto, T., Schipper, B., Vlak, J.M., Stiekema, W.J. and Bosch, D. (1996). Domain III substitution in *Bacillus thuringiensis* delta-endotoxin CryIA(b) results in superior toxicity for *Spodoptera exigua* and altered membrane protein recognition. *Applied and Environmental Microbiology*, **62(5)**: 1537-1543.

Donovan, W.P., Rupar, M.J., Slaney, A.C., Malvar, T., Gawron-Burke, M.C. and Johnson, T.B. (1992). Characterization of two genes encoding *Bacillus thuringiensis* insecticidal crystal protein toxic to Coleoptera species. *Applied and Environmental Microbiology*, **58(12)**: 3921-3927.

Downing, K.J., Leslie, G. and Thomson, J.A. (2000). Biocontrol of the sugarcane borer *Eldana saccharina* by expression of the *Bacillus thuringiensis cry1Ac* and *Serratia marcescens chiA* genes in sugar-associated bacteria. *Applied and Environmental Microbiology*, **66(7)**: 2804-2810.

Einarson, M.B., Pugacheva, E.N. and Orlinick, J.R. (2007). Preparation of GST fusion proteins. Cold Spring Harbor Protocols. doi:10.1101/pdb.prot4738.

Fang, W. and Bidochka, M.J. (2006). Expression of genes involved in germination, conidiogenesis and pathogenesis in *Metarhizium anisopliae* using quantitative real-time RT-PCR. *Mycological Research*, **110**: 1165-1171.

Fang W., Leng B., Xiao Y., Jin K., Ma J., Fan Y., Feng J., Yang X., Zhang Y., and Pei, Y. (2005). Cloning of *Beauveria bassiana* chitinase gene *Bbchit1* and its application to improve fungal strain virulence. *Applied and Environmental Microbiology*, **71**: 363-370.

Fang, J., Xu, X., Wang, P., Zhao, J.-Z., Shelton, A.M., Cheng, J., Feng, M.-G. and Shen, Z. (2007). Characterization of chimeric *Bacillus thuringiensis* Vip3 toxins. *Applied and Environmental Microbiology*, **73(3)**: 956-961.

Ferrari, F.A., Nguyen, A., Lang, D. and Hoch, J.A. (1983). Construction and properties of an integrable plasmid for *Bacillus subtilis*. *Journal of Bacteriology*, **154(3)**: 1513-1515.

Freimoser, F.M., Screen, S., Bagga, S., Hu, G. and St. Leger, R.J. (2003). Expressed sequence tag (EST) analysis of two subspecies of *Metarhizium anisopliae* reveals a plethora of secreted proteins with potential activity in insect hosts. *Microbiology*, **149**: 239-247.

Ge, B.X., Bideshi, D., Moar, W.J. and Federici, B.A. (1998). Differential effects of helper proteins encoded by the *cry2A* and *cry11A* operons on the formation of Cry2A inclusions in *Bacillus thuringiensis*. *Federation of European Microbiological Societies, Microbiology Letters*, **165**: 35-41.

Gianni, M. and Galizzi, A. (1986). Isolation of genes preferentially expressed during *Bacillus thuringiensis* spore outgrowth. *Journal of Bacteriology*, **165(1)**: 123-132.

Girard, F., Vachon, V., Préfontaine, G., Marceau, L., Su, Y., Larouche, G., Vincent, C., Schwartz, J.-L., Masson, L. and Laprade, R. (2008). Cysteine scanning mutagenesis of α4, a putative pore-lining helix of the *Bacillus thuringiensis* insecticidal toxin Cry1Aa. *Applied and Environmental Microbiology*, **74(9)**: 2565-2572.

Hajek, A.E. and St. Leger, R.J. (1994). Interaction between fungal pathogens and insect hosts. *Annual Review of Entomology*, **39**: 293-322.

Hernández-Soto, A., Del Rincón-Castro, M.C., Espinoza, A.M. and Ibarra, J.E. (2009). Parasporal body formation via overexpression of the Cry10Aa toxin of *Bacillus thuringiensis* subp. *israelensis* and Cry10Aa-Cyt1Aa synergism. *Applied and Environmental Microbiology*, **75(14)**: 4661-4667.

Hoch, J.A., Barat, M. and Anagnostopoulos, C. (1967). Transformation and transduction in recombination-defective mutants of *Bacillus subtilis*. *Journal of Bacteriology*, **93(6)**: 1925-1937.

Honée, G., Vriezen, W. and Visser, B. (1990). A translation fusion protein product of two different insecticidal crystal proteins genes of *Bacillus thuringiensis* exhibits and enlarged insecticidal spectrum. *Applied and Environmental Microbiology*, **56(3)**: 823-825.

Hoti, S.L. and Balaraman, K. (1993). Formation of melanin pigment by a mutant of *Bacillus thuringiensis* H-14. *Journal of General Microbiology*, **139**: 2365-2369.

Ito, T., Ikeya, T., Sahara, K., Bando, H. and Asano, S. (2006). Cloning and expression of two crystal protein genes, *cry30Ba1* and *cry44Aa1*, obtained from a highly mosquitocidal strain, *Bacillus thuringiensis* subsp. *entomocidus* INA288. *Applied and Environmental Microbiology*, **72(8)**: 5673-5676.

Johnson, D.E., Niezgodski, D.M. and Twaddle, G.M. (1980). Parasporal crystals produced by oligosporogenous mutants of *Bacillus thuringiensis* ($Spo^{-}Cr^{+}$). *Canadian Journal of Microbiology*, **26(4)**: 486-491.

Kalman, S., Kiehne, K.L., Cooper, N., Reynoso, M.S. and Yamamoto, T. (1995). Enhanced production of insecticidal proteins in *Bacillus thuringiensis* strains carrying an additional crystal protein gene in their chromosomes. *Applied and Environmental Microbiology*, **61(8)**: 3063-3068.

Kaur, S. (2000). Molecular approaches towards development of novel *Bacillus thuringiensis* biopesticides. *World Journal of Microbiology & Biotechnolog,* **16**: 781-793.

Lee, M.K., Jenkins, J.L., You, T.H., Curtiss, A., Son, J.J., Adang, M.J. and Dean, D.H. (2001). Mutations at the arginine residues in α8 loop of *Bacillus thuringiensis* δ-endotoxin Cry1Ac affect toxicity and binding to *Manduca sexta* and *Lymantria dispar* aminopeptidase N. *Federation of European Biochemical Societies Letters*, **497**: 108-112.

Liu, J., Yan, G., Shu, C., Zhao, C., Liu, C., Song, F., Zhou, L., Ma, J., Zhang, J. and Huang, D. (2010). Construction of a *Bacillus thuringiensis* engineered strain with high toxicity and broad pesticidal spectrum against coleopteran insects. *Applied Microbiology and Biotechnology*, DOI 10.1007/s00253-010-2479-5.

Liu, J.-W., Yap, W.H., Thanabalu, T. and Porter, A.G. (1996). Efficient synthesis of mosquitocidal toxins in *Asticcacaulis excentricus* demonstrates potential of Gram-negative bacteria in mosquito control. *Nature Biotechnology*, **14**: 343-347.

Manasherob, R., Ben-Dov, E., Xiaoqiang, W., Boussiba, S. and Zaritsky, A. (2002). Protection from UV-B damage of mosquito laqrvicidal toxins from *Bacillus thuringiensis* subsp. *israelensis* expressed in *Anabaena* PCC 7120. *Current Microbiology*, **45**: 217-220.

Moar, W.J., Trumble, J.T., Hice, R.H. and Backman, P.A. (1994). Insecticidal activity of the CryIIA protein from the NRD-12 isolate of *Bacillus thuringiensis* subsp. *kurstaki* expressed in *Escherichia coli* and *Bacillus thuringiensis* and in a leaf-colonizing strain of *Bacillus cereus*. *Applied and Environmental Microbiology*, **60(3)**: 896-902.

Mokhele, T.A., Ahmed, F. and Conlong, D.E. (2009). Detection of sugarcane African stalk borer *Eldana saccharina* Walker (Lepidoptera:Pyralidae) using hyperspectral remote sensing (spectroradiometry). *Sugar Cane International*, **27(6)**: 237-244.

Morse, R.J., Yamamoto, T. and Stroud, R.M. (2001). Structure of Cry2Aa suggests an unexpected receptor binding epitope. *Structure*, **9**: 400-417.

Park, H.-W., Bideshi, D.K. and Federici, B.A. (2000). Molecular genetic manipulation of truncated Cry1C protein synthesis in *Bacillus thuringiensis* to improve stability and yield. *Applied and Environmental Microbiology*, **66(10)**: 4449-4455.

Park, H.-W., Bideshi, D.K. and Federici, B.A. (2003). Recombinant strain of *Bacillus thuringiensis* producing Cyt1A and Cry11B, and the *Bacillus sphaericus* binary toxin. *Applied and Environmental Microbiology*, **69(2)**: 1331-1334.

Payne, J.M. and Davidson, E.W. (1984). Insecticidal activity of crystalline parasporal inclusions and other components of the *Bacillus sphaericus* 1593 spore complex. *Journal of Invertebrate Pathology*, **43**: 383-388.

Promdonkoy, B., Promdonkoy, P., Tanapongpipat, S., Luxananil, P., Chewawiwat, N., Audtho, M. and Panyim, S. (2004). Cloning and characterization of a mosquito larvicidal toxin produced during vegetative stage of *Bacillus sphaericus* 2297. *Current Microbiology*, **49**: 84-88.

Rajamohan, F., Alzate, O., Cotrill, J.A., Curtiss, A. and Dean, D.H. (1996). Protein engineering of *Bacillus thuringiensis* δ-endotoxin: Mutations at domain II of CryIAb enahance receptor affinity and toxicity toward gypsy moth larvae. *Proceedings of the National Academy of Sciences, United States of America*, **93**: 14338-14343.

Rang, C., Bes, M., Lullien-Pellerin, V., Wu, D., Federici, B.A and Frutos, R. (1996). Influence of the 20kDa protein from *Bacillus thuringiensis* ssp. *israelensis* on the rate of production of truncated Cry1C proteins. *Federation of European Microbiological Societies, Microbiology Letters*, **141**: 261-264.

Romero, M., Gil, F.M. and Orduz, S. (2001). Expression of mosquito active toxin genes by a Colombian native strain of the Gram-negative bacterium *Asticcacaulis excentricus*. *Memórias do Instituto Oswaldo Cruz*. **96(2)**: 257-263.

Ruan, L., He, W., Sun, M. and Yu, Z. (2005). Cloning and expression of *mel* gene from *Bacillus thuringiensis* in *Escherichia coli*. *Antonie van Leeuwenhoek*, **87**: 283-288.

Rupar, M.J., Donovan, W.P., Groat, R.G., Slaney, A.C., Mattison, J.W., Johnson, T.B., Charles, J.-F., Dumanoir, V.C. and de Barjac, H. (1991). Two novel strains of *Bacillus thuringiensis* toxic to coleopterans. *Applied and Environmental Microbiology*, **57(11)**: 3337-3344.

Sanchis, V., Agaisse, H., Chaufaux, J. and Lereclus, D. (1996). Construction of a new insecticidal *Bacillus thuringiensis* recombinant strains by using the sporulation non-dependent expression system of cryIIIA and a site specific recombination vector. *Journal of Biotechnology*, **48**: 81-96.

Sanchis, V., Gohar, M., Chaufaux, J., Arantes, O., Meier, A., Agaisse, H., Cayley, J. and Lereclus, D. (1999). Development and field performance of a broad-spectrum nonviable asporogenic recombinant strain of *Bacillus thuringiensis* with greater potency and UV resistance. *Applied and Environmental Microbiology*, **65(9)**: 4032-4039.

Sastry, K.J., Srivastava, O.P., Millet, J., FitzJames, P.C. and Aronson, A.I. (1983). Characterization of *Bacillus subtilis* mutants with temperature-sensitive intracellular protease. *Journal of Bacteriology*, **153(1)**: 511-519.

Schnepf, E., Crickmore, N., Van Rie, J, Lereclus, D., Baum, J., Feitelson, J., Zeigler, D.R. and Dean, D.H. (1998). *Bacillus thuringiensis* and its pesticidal crystal proteins. *Microbiology and Molecular Biology Reviews*, **62(3)**: 775-806.

Schwartz, J.L., Potvin, L., Chen, X.J., Brousseau, R., Laprade, R. and Dean, D.H. (1997). Single-site mutations in the conserved alternating-arginine region affect ionic channels formed by Cry1Aa, a *Bacillus thuringiensis* toxin. *Applied and Environmental Microbiology*, **63(10)**: 3978-3984.

Shao, Z., Liu, Z. and Yu, Z. (2001). Effects of the 20-Kilodalton helper protein on Cry1Ac production and spore formation in *Bacillus thuringiensis*. *Applied and Environmental Microbiology*, **67(12)**: 5362-5369.

Shivakumar, A.G., Gundling, G.J., Benson, T.A., Casuto, D., Miller, M.F. and Spear, B.B. (1986). Vegetative expression of the δ-endotoxin genes of *Bacillus thuringiensis* subsp. *kurstaki* in *Bacillus subtilis*. *Journal of Bacteriology*, **166(1)**: 194-204.

Sierra-Martínez, P., Ibarra, J.E., de la Torre, M. and Olmedo, G. (2004). Endospore degradation in an oligosporogenic, crystalliferous mutant of *Bacillus thuringiensis*. *Current Microbiology*, **48**: 153-158.

St. Leger, R.J., Joshi, L., Bidochka, M.J. and Roberts, D.W. (1996a). Construction of an improved mycoinsecticide overexpressing a toxic protease. *Proceedings of the National Academy of Sciences, United States of America*, **93**: 6349-6354.

St. Leger, R.J., Joshi, L., Bidochka, M.J., Rizzo, N.W. and Roberts, D.W. (1996b). Characterization and ultrastructural localization of chitinases from *Metarhizium anisopliae*, *M. flavoviridae*, and *Beauveria bassiana* during fungal invasion of host (*Manduca sexta*) cuticle. *Applied and Environmental Microbiology*, **62(3)**: 907-912.

St. Leger, R.J., Cooper, R.M. and Charnley, A.K. (1988). The effect of melanization of *Manduca sexta* cuticle on growth and infection by *Metarhizium anisopliae*. *Journal of Invertebrate Pathology*, **52**: 459-470.

Thanabalu, T., Hindley, J., Jackson-Yap, J. and Berry, C. (1991). Cloning, sequencing, and expression of a gene encoding a 100-kilodalton mosquitocidal toxin from *Bacillus sphaericus* SSII-1. *Journal of Bacteriology*. **173(9)**: 2776-2785.

Thiéry, I., Hamon, S., Delécluse, A. and Orduz, S. (1998). The introduction into *Bacillus sphaericus* of the *Bacillus thuringiensis* subsp. *medellin cyt1Ab1* gene results in higher susceptibility of resistant mosquito larva populations to *B. sphaericus*. *Applied and Environmental Microbiology*, **64(10)**: 3910-3916.

Visick, J.E. and Whiteley, H.R. (1991). Effects of a 20-kilodalton protein from *Bacillus thuringiensis* subsp. *israelensis* on production of CytA protein by *Escherichia coli*. *Journal of Bacteriology*, **173(5)**: 1748-1756.

Walters, F.S., Stacy, C.M., Lee, M.K., Palekar, N. and Chen, J.S. (2008). An engineered chymotrypsin/cathepsin G site in domain I renders *Bacillus thuringiensis* Cry3A active against western corn rootworm larvae. *Applied and Environmental Microbiology*, **74(2)**: 367-374.

WHO, (1996). World Health Organization. Report of the WHO informal consultation of the evaluation and testing of insecticides.

Widner, W.R. and Whiteley, H.R. (1990). Location of the dipteran specificity region in a lepidopteran-diperan crystal protein from *Bacillus thuringiensis*. *Journal of Bacteriology*, **172(6)**: 2826-2832.

Wu, D. and Federici, B.A. (1993). A 20-kilodalton protein preserves cell viability and promotes CytA crystal formation during sporulation in *Bacillus thuringiensis*. *Journal of Bacteriology*, **175**: 5276-5280.

Wu, D. and Federici, B.A. (1995). Improved production of the insecticidal CryIVD protein in *Bacillus thuringiensis* using *cryIA*(*c*) prom1oters to express the gene for an associated 20-kDa protein. *Applied Microbiology and Biotechnology*, **42**: 697-702.

Yoshisue, H., Yoshida, K., Sen, K., Sakai, H. and Komano, T. (1992). Effects of *Bacillus thuringiensis* 20-kDa protein on production of the Bti 130-kDa crystal protein in *Escherichia coli*. *Bioscience, Biotechnology and Biochemistry*, **56**: 1429-1433.

Zhang, J., Cai, J., Deng, C. and Ren, G. (2007). Characterization of melanin produced by a wild-type strain of *Bacillus cereus*. *Frontiers of Biology in China*, **2(1)**: 26-29.

Zheng, R. (1998). M.S. thesis. Huazhong Agricultural University, Wuhan, People's Republic of China.

{9}

Enhancing the Virulence of Baculovirus as Biopesticides with Wasp Parasitoid Polydnavirus Genes to Control Lepidopteran Insect Pests

Mario A. Rodríguez-Pérez, Miguel A. Pérez-Rodríguez, Ali Mohammed-ali, Erick De Jesús De Luna-Santillana, Xianwu Guo, and Filiberto Reyes-Villanueva

ABSTRACT

Agricultural ecosystems suffer from chemical contamination with toxic pesticides which have impact on human and animal health. Baculoviruses are being used as biopesticides specifically targeting lepidopteran insects in agriculture and forestry but they have shortages which limit their utility such as a narrow host range and a relatively slow speed-of-kill. Exploitation of hemocyte apoptosis-inducing genes in the polydnaviruses which knock out the anti-viral hemocyte that mediated defensive immune reactions of the insect facilitates their dispersal throughout the hemocoel to target tissues in the fat body, salivary glands, and other tissues. This anti-viral cellular response stops the pathway of viral cell-to-cell transmission in refractory hosts. The tobacco hornworm is normally a semi-permissive host of the AcMNPV baculovirus; the simultaneous presence of the CcPDV polydnavirus transforms the host to a permissive species. The polydnavirus gene products offer the potential to manipulate the baculovirus-driven mortality with a broader spectrum of inter-species activity.

INTRODUCTION

Parasites have evolved sophisticated mechanisms of interaction with insect hosts. First and foremost they must avoid being killed by the host's immune system to establish a successful infection. Evidence from many systems indicates that parasites regulate their host's physiology for their own benefit, evoking changes in their host's cellular and humoral immune systems, hormonal physiology, development, and behavior. One of the most intriguing examples of the sophisticated strategies utilized by parasites to cause immunosuppression of their host was seen in wasp parasitoids that utilize viruses, termed polydnaviruses (PDVs). The name of these viruses is derived from its DNA polydisperse genome, which are injected together with eggs into the host hemolymph when the host is parasitized. These viruses are integrated into the wasp's genomic DNA and are unusual in the fact that the production of viral particles and viral DNA replication are not detected outside the ovaries of female wasps. This occurs only in calyx cells (which are located at the top of the oviducts side) from the ovaries of female parasitic Icneumonidae and Braconidae wasps (Stoltz and Vinson, 1979). For this reason they were called Ichnoviruses which occur in three Ichneumon wasps subfamilies (Campopleginae, Ctenopelmatinae and Banchinae) and the Bracoviruses which are associated with four subfamilies of Braconidae wasps (Cardiochilinae, Cheloninae, Microgastrinae and Miracinae) (Kroemer and Webb, 2004). Because both groups have similar production mechanisms and action mode, they have been classified under the family Polydnaviridae.

Genetic dissimilarity between Ichnovirus and Bracovirus shows they have no evolutionary relationship; however, the life cycles and genomes organization of both polydnavirus are similar, suggesting they have been subjected to common selection pressures which have led to convergent evolution in two genera. Recent analysis of the genes of Polydnaviruses have increased the number of families of known PDV genes. These proteins share no significant sequence homology. The convergent evolution has produced functionally similar immunosuppressive molecules which often cause a phenotype in hemocytes characterized by damage to the cytoskeleton (Glatz *et al.*, 2004; Webb *et al.*, 2006).

One of the primary targets of these viruses appears to be the host insect immune system. When viral particles covering the eggs were inactivated by UV radiation and then artificially injected into caterpillars, they were encapsulated as part of the immune response activation (Shelby and Webb, 1999). The normal response of an insect to an invading metazoan parasite is encapsulation, or the mobilization of blood cells to form a multilayered capsule of hemocytes which encloses the invader. However, after parasitization, host physiology is altered to support the development of parasites, the hemocytes of insects that are attacked by parasitoids carrying PDVs are rendered dysfunctional by the virus (Table 9.1); when a parasite oviposits in another insect, this will result in death, unlike parasitoid wasp, which not kill the host, but the infected insect here will die during or shortly after emergence of parasitoids. Once parasitoid emerges from the cocoon, mates and searches new insects for parasitization to complete their life cycle. In all cases, Polydnaviruses have no adverse effects on parasitoid wasps, but are pathogenic to lepidopteran hosts, as their genomes contain necessary genes to modulate physiological processes in host larvae. The tobacco hornworm *Manduca sexta* (Linnaeus) (Lepidoptera: Sphingidae), and its interactions with the wasp parasitoid *Cotesia congregata* (Say) (Hymenoptera: Braconidae) serve as an important laboratory model for studying how the virus interferes with the normal hemocyte encapsulation response by causing hemocyte death via activation of apototic pathways in those cells which render the host immunocompromised and more susceptible to pathogens such as baculoviruses.

The parasitoids, and particularly, wasps of the genus *Cotesia,* have tremendous economic importance in the biological control of

Table 9.1. Current outline on polydnavirus V1 gene of *Cotesia* spp., the host immune system, and polydnavirus (PDV)-baculovirus (BV) interactions

Insect immune system	
	Cellular Responses: Phagocytosis, Nodulation, Encapsulation, Oxygen radical production. *Humoral Responses:* Phenoloxidase activity, Agglutinins, Antibacterial proteins, Cytokines
***Cotesia* spp**	
	Larval Parasitoids - eggs of adult parasitoid deposited in or on host larvae and progeny develop and emerge from larvae.
PDV	
	Found in hymenopteran families Braconidae and Ichneumonidae. Restricted to endoparasitoids. Genetic symbionts of the wasp as a provirus. Viral replication occurs only in wasp.
CrV1:	
	CrV-1 gene has been originally reported in *Cotesia rubecula* PDV. A protein that suppresses the cellular immune system of the caterpillar host, *i.e.* inhibiting hemocytic encapsulation. *Cotesia congregata* CcPDV (CcV1 gene) involved in causing hemocyte apoptosis and inhibits hemolin functions in parasitized *Manduca sexta* larvae *(Labropoulou et al., 2008).*
Initial changes following parasitization by cotesia	
	Virally encoded early protein induced Homolog of CrV1 expressed Phenoloxidase activity reduced

lepidopteran insect pests in agricultural ecosystems. The genus *Cotesia* is the largest Microgastrine genus in North America. Many of these species also attack pests in Central and South America, as well as in other countries. Animals infected with both *C. congregata* polydnavirus (CcPDV) and baculovirus (*Autographa californica* nucleopolyhedrovirus (AcMNPV)) die faster than larvae injected orally with baculovirus alone; hence, the PDV acts synergistically with baculoviruses to cause rapid insect death (Washburn *et al.*, 2000). Infection of parasitized larvae with the baculovirus results in rapid dissemination throughout the body cavity of the host, due to lack of an anti-viral cellular response in the PDV-infected insects. The immunosuppressive genes identified in the PDV of *C. flavipes* Cameron (Hymenoptera: Braconidae), are being used to develop hybrid baculoviruses carrying PDV genes which inhibit the host immune response and thereby have an expanded host range and enhanced speed-of-kill. One problem associated with the use of baculoviruses for pest control is that the larvae live for several days and continue feeding for a prolonged period before they finally succumb to viral infection. The use of PDV-based viral biopesticides which specifically target lepidopteran insect pests and knock out the host immune system reduces our dependence upon chemical pesticides, which contaminate the environment and threaten human and animal health. Thus, novel bioinsecticides will have a significant positive impact upon a wide variety of agricultural ecosystems (including tomatoes, cotton, corn, and beans) in Mexico, and the USA. The use of toxic pesticides which contaminate the environment can be curtailed with an effective biopesticide. Mechanisms of induction of host immunosuppression are a central theme in toxicology, and the role of viruses in this process is of interest to many toxicologists. The PDVs are unusual in that they exert their immunosuppressive effects without replication in their target cells.

Crop losses would be reduced by the development of viruses with a faster mode of action. Their potential role in insect control in the field is increasing because recombinant baculoviruses expressing toxins and other proteins have been formulated to be more virulent than naturally occurring baculoviruses (Miller, 1997; Szewczyk, *et al.*, 2006). The increased susceptibility of parasitized and PDV-injected nonparasitized larvae to the AcMNPV will give us an avenue for manipulation of the NPVs virulence by incorporating the relevant PDV gene(s) into the baculovirus to enhance toxicity of the viral pathogen. Thus, our basic studies of PDV enhancement of baculovirus efficacy could ultimately have significant applied economical value in the field for control of lepidopteran insect pests. Developing more potent baculoviruses as biopesticides will be extremely valuable to agriculture in countries such as Mexico and the USA. Use of these biologically-based biopesticides will protect the environment from the action and accumulation of toxic chemical pesticides in several countries.

IDENTIFICATION OF GENES CAUSING HOST IMMUNO-SUPPRESSION

Insect's immune system is an effective defense against many pathogens, and triggers a vigorous immune response against endoparasitoids eggs; therefore it is necessary to prevent the immune system action so that parasitism is successful. One potentially important *C. congregata* polydnavirus (PDV) genes is the CcPDV homolog of *C. rubecula* CrV1 (Table 1). In the natural host of the parasitic *wasp C. rubecula*, *Pieris rapae* (Linnaeus) (Lepidoptera: Pieridae), the CrV1 protein disrupts host hemocyte (blood cell) function and inhibits the polymerization of actin molecules inside the cell. The expression of this protein appears to be linked to successful parasitization of the host. Thus, it may be a likely candidate for inducing host hemocyte apoptosis. The PDV CrV1 gene product causes rounding up of hemocytes and other physiological changes in hemocytes of host insect larvae parasitized by *C. rubecula*.

We have strong evidence that a homolog of this gene is present in the genome of *C. congregata* and *C. flavipes* PDVs, in that a CrV1 probe hybridizes to *C. congregata* and *C. flavipes* PDV DNAs in southern blots. Thus, this gene appears to be a candidate for causing host hemocyte apoptosis in parasitized *M. sexta* larvae.

In addition to protein CrV1, other gene products that enable the survival of the parasitoid insect have been identified, for example: protein VHV1.1 is a glycoprotein that is endocytosed by hemocytes, and inhibits the response of encapsulation (Li and Webb, 1994). Proteins TrIV1, TrIV2 induce apoptosis in lepidopteran cells when maintained *in vitro* (Béliveau *et al.*, 2003). BV1 protein of *Toxoneuron* has the function to suppress the immune response of host insect and modulate the endocrine system and development (Lapointe *et al.*, 2005).

EXPRESSION OF PDV GENES IN AcMNPV

Baculoviruses are classified as a group of arthropod-specific viruses, with a rod-shaped nucleocapsid; nowadays, more than 600 types of baculoviruses infecting insects of Lepidoptera, Diptera and

Hymenoptera orders have been described; 90% of them were isolated from species of Lepidoptera (Jehle *et al.,* 2006). There are two virions phenotypes: occlusion derived virus (ODV) and budded virions (BV). ODV are occluded in a crystalline protein matrix: the "occlusion body". ODV and BV contain a single nucleocapsid within an envelope that is derived from the host plasma membrane.

Some species appear to be species-specific, or narrow host-specific, but some NPV such as AcMNPV can infect species of at least 15 Lepidoptera families, however, it does not show the same virulence in all species, thus, the baculovirus host range and susceptibility level is not strictly restricted to the taxonomy status. The transmission of baculoviruses infection primarily occurs through ingestion of viruses that persist in the environment and are transmitted horizontally from host to host. Baculoviruses infect larvae which instars are capable of feeding the occlusion bodies (OB). The OBs are protein structures that contain virions (virus particles) and are the infective unit of the baculoviruses that are critical for spreading the infection among hosts. In the midgut, the virions are rapidly released due to the combined action of alkaline pH and proteases which break the OBs. The virions then pass through the peritrophic membrane of the midgut (Engelhard *et al.*, 1994). In the gut environment, infection begins when the ODV envelope binds and fuses with the brush shaped membrane of insect midgut columnar cells. ODV generates an infection in intestinal cells, a second viral phenotype, the budded virus (BV) which emerges from the lateral plasma membrane. Within the hemocoel, the tracheolar cells are the immediate targets. These respiratory cells penetrate the midgut basal lamina with long cytoplasmic filaments which have intimate association with midgut cells (Haas-Stapleton *et al.*, 2005). The infection of the tracheal system provides the conduit to pass through the basal lamina and spread throughout the insect body.

The tracheal system facilitates the exchange of gases between all tissues and the external environment, through invaginations of the outer wall of the body which fills with air to the tracheae tubes. The tracheae are lined with a chitin-cuticle secreted by the epidermal layers, so it is very similar to the surface of the insect. The tracheae bifurcate through the insect body, in tubes finer until finish in one structure called tracheole. Cells harboring tracheole produce daughter cells in the epidermis tracheal calls traqueoblasts (Engelhard *et al.*, 1994).

Lepidopteran larvae commonly increase resistance to baculoviruses infection as their age. The mechanism responsible for the development of this resistance is still unknown, but this resistance does not occur if the virus-inoculum is hemocoel administered (Engelhard and Volkman, 1995). At each moult, infected cells of the midgut are emptied and discarded in the lumen of the intestine, while regenerative cells are differentiated and form the tissue. Within each larval stage, therefore, the time interval for establishing an AcMNPV-systemic infection becomes progressively smaller and during the molt to the next instar, as AcMNPV can kill susceptible hosts, the BV must infect tracheal epidermal cells within the insect's respiratory system, before midgut cells are sloughed (Washburn *et al.*, 2003). The time it takes to kill the insect baculoviruses host varies from days to weeks depending on temperature, viral dose, and age and species of the insect. During this period between infection and death, insects can continue to feeding and cause crop damage for days or even weeks (Harrison and Bonning, 2001).

AcMNPV hemocoel-infection in non-susceptible species is quickly cleared, little after the secondary infection is established in the larval. This finding provided the first experimental evidence on the immune response to counteract the lepidopteran baculovirus infection (Washburn *et al.*, 1996). Initial events of baculovirus infection are similar to those present in susceptible and non-susceptible hosts, but later it was found that hemocytes surrounding tracheal cells infected and formed capsules. This result is compared with those obtained in larvae immunocompromised due to the presence of parasitism with *C. congregata* or Polydnaviruses injection. The wasp larvae which have been parasitized or inoculated with Polydnaviruses died faster than those non-parasitized and non-inoculated wasp larvae, suggesting that the cellular immune response is a factor that confers resistance to infection by AcMNPV (Washburn *et al.*, 2000).

Polydnaviruses can enhance the effect of pathogenic viruses in causing larval mortality. Washburn *et al.* (2000) demonstrated that naturally parasitized *M. sexta* host larvae, or nonparasitized larvae injected with four wasp equivalents of PDV, are more susceptible to the *A. californica* M nuclear polyhedrosis (AcMNPV) pathogenic virus compared to nonparasitized control larvae treated with the same oral dose of AcMNPV. Normally, nonparasitized tobacco hornworm larvae are refractory to AcMNPV, and require a large dose to succumb to infection, and the spread of infection is blocked due to encapsulation of the midgut lesions by hemocytes that form capsules around the foci in the tracheal epithelium. Parasitization or injection of PDV greatly enhances the mortality of larvae that are subjected to baculovirus infection (Washburn *et al.*, 2000). It would be interesting to find out what PDV gene product(s) induce this change in susceptibility. Recombinant NPVs expressing the appropriate PDV gene product(s) should have increased virulence and a faster speed of kill due to the accelerated speed at which the infection spreads throughout the hemocoel, causing infection of multiple tissues including the fat body.

Studies of Washburn *et al.* (2000) showed that parasitism by the wasp *C. congregata* greatly enhances the virulence of the *A. californica* nuclear polyhedrovirus (AcMNPV) to host tobacco hornworm larvae. A dramatically increased speed-of-kill of the baculovirus was found to be mediated by the inhibitory effects of the parasitoid's polydnavirus on the host insect's immune response, which rendered the host more susceptible to the baculovirus. Polydnaviruses are genetic symbionts of the parasitoid which are produced in the ovary of the female wasp and injected into the host larvae during parasitization of the host. The polydnaviruses cause death (apoptosis) of host blood cells, and this massive cell death suppresses the host immune system, rendering the host more susceptible to the baculovirus. One of our present objectives is to sequence the genome of the polydnavirus of *C. flavipes* (CfPDV), and then use that genetic information to identify genes in the PDV genome which express immunosuppressive gene products which suppress the host insect's immune response to AcMNPV. Our goal is to develop hybrid baculoviruses with a broader host range and faster speed-of-kill. Managing resistance to baculovirus insecticides is critical to the use of viruses as insecticides, and wasp polydnavirus genes appear to be promising candidates for manipulating baculovirus virulence for insect pests.

One likely candidate which suppresses the host immune system is the *C. congregata* polydnavirus homolog of the *C. rubecula* CrV1 gene. This gene product disrupts the actin cytoskeleton of hemocytes so they cannot adhere to a substrate such as a parasitoid egg. Encapsulation of the wasps is prevented by this hemocyte cytological change, which in the tobacco hornworm culminates in apoptosis of host hemocytes of *C. congregata*. This gene is expressed throughout development of the wasps within the host in the *M. sexta- C. congregata* system. In *P. rapae* larvae parasitized by *C. rubecula*, this transcript is expressed during a narrow time frame from 6 to 12 h post-parasitization. In that system, deleterious effects on hemocytes are transient and the cells recover. What then inhibits the encapsulation of wasp larvae has not yet been known. In contrast, in parasitized *M. sexta* larvae the transcript is produced until the wasps emerge from the host, in conjunction with pathological changes in the hemocytes which occur throughout the development of the wasps. Actin staining was carried out and disruption of the actin cytoskeleton was observed. Antibodies to CrV1 bind to hemocytes of all stages of parasitized larvae, in 'hot spots' internally in what appears to be the cytoplasm of the cells.

We administered recombinant AcMNPV which is expressing CrV1 viral proteins (under regulation of late viral promoters) to *Spodoptera exigua* (Hübner) (Lepidoptera: Noctuidae), and *Sesamia nonagrioides* Lefebvre (Lepidoptera: Noctuidae) larvae. We then examined the effects of the AcMNPV virus on larval mortality and pupation. Controls were treated with the same dosage of wild-type AcMNPV virus, and its effects likewise assessed. Recombinant viruses, in supernatant harvested from Sf9 infected cells, produced a significant higher mortality rate in second instar *S. exigua,* and *S. nonagroides* larvae, similar rate to those obtained with wild-type baculovirus. However, we did not investigate if CrV1 homolog genes or other viral proteins might be the PDV gene product which renders nonparasitized larvae more susceptible to virus infection. We also recognize that we did not manipulate the dosage of AcMNPV administered to optimize clear observable effects on hemocyte function and larval mortality. Further research is currently underway such as the assessment of effects on hemocytes. Thus, larvae will be bled at various time points post-treatment and hemocytes stained with FITC-labelled phalloidin to look for evidence of hemocyte blebbing and apoptosis. We also bear in mind that wild-type virus infects hemocytes, so we are comparing the effects of wild-type versus recombinant virus on induction of hemocyte apoptosis. This experiment will clarify whether *in vivo* baculovirus driven expression of CrV1 homologs or other viral proteins triggers host hemocyte apoptosis. Effects of the AcMNPV expressing EP1 (or other viral proteins) on mortality, weight gain, and the rate of successful development to adulthood are also being assessed to determine if larvae infected with the baculovirus expressing PDV proteins are more likely to exhibit mortality compared to those treated with wild type virus.

Our final goal is to find a hybrid PDV-baculovirus construct with enhanced virulence for lepidopteran larvae. Larvae to be tested for recombinant baculovirus virulence will include beet armyworm (a pest on tomatoes), pink bollworm (pest of cotton), and *Heliothis virescens* (Fabricius) (Lepidoptera: Noctuidae) (pest of a variety of crops), and *Helicoverpa zea* (Boddie) (Lepidoptera: Noctuidae) (pest of corn), which is normally a non-permissive host of AcMNPV due to mobilization of a cellular immune response. Enhanced virulence of the virus will likely result in an expansion of the host range of the virus, as well as facilitate rapid death of insects treated with the virus in biological control programs developed to control insect pests using baculoviruses.

In summary, we can conclude that polydnaviruses are potent host immunosuppressors. PDV genes are useful in engineering insect pathogenic viruses to enhance virulence of pathogen. The genome sequencing of PDVs will facilitate characterization of new genes which are detrimental to the host insect. There are still multiple questions to be addressed: Does the PDV act as an endocrine disruptor? Is there a link between behavior, immunity, and neural functioning? Could unravelling bracovirus particle assembly contribute to the design of new vectors for gene therapy? (Bézier *et al.*, 2009).

ACKNOWLEDGEMENTS

We are very grateful to Secretaría de Investigación y Postgrado/ Instituto Politécnico Nacional for the economic support to do this study (Grant 20091300), Mario A. Rodríguez-Pérez holds a scholarship from Comisión de Operación y Fomento de Actividades Académicas/ Instituto Politécnico Nacional.

REFERENCES

Béliveaua, C., Levasseurb, A., Stoltzc, D. and Cusson, M. (2003). Three related TrIV genes: comparative sequence analysis and expression in host larvae and Cf-124T cells. *Journal of Insect Physiology*, **49(5)**: 501-511.

Bézier, A.L., Annaheim, M., Herbinière, J., Wetterwald, C., Gyapay, G., Bernard-Samain, S., Wincker, P., Roditi, I., Heller, M., Belghazi, M., Pfister-Wilhem, R., Periquet, G., Dupuy, C., Huguet, E., Volkoff, A., Lanzrein, B. and Drezen, J.M. (2009). Polydnaviruses of braconid wasps derive from an ancestral nudivirus. *Science*, **323(5916)**: 926-930.

Engelhard, E.K., Kam-Morgan, L.N.W., Washburn, J.O. and Volkman, L.E. (1994). The insect tracheal system: A conduit for the systemic spread of *Autographa californica* M nuclear polyhedrosis virus. *Proceedings of National Academy of Science, United States of America*, **91**: 3224-3227.

Engelhard, E.K. and Volkman, L.E. (1995). Developmental resistance in fourth instar *Trichoplusia ni* orally inoculated with *Autographa californica* M nuclear polyhedrosis virus. *Virology*, **209**: 384-389.

Glatz, R.V., Asgari, S. and Schmidt, O. (2004). Evolution of polydnaviruses as insect immune suppressors. *Trends in Microbiology*, **12**: 545-554.

Haas-Stapleton, E.J., Washburn, J.O. and Volkman, L.E. (2005). *Spodoptera frugiperda* resistance to oral infection by *Autographa californica* multiple nucleopolyhedrovirus linked to aberrant occlusion-derived virus binding in the midgut. *Journal of General Virology*, **86**: 1349-1355.

Harrison, R.L. and Bonning, B.C. (2001). Use of proteases to improve the insecticidal activity of baculoviruses. *Biological Control*, **20**: 199-209.

Jehle, J.A., Blissard, G.W., Bonning, B.C., Cory, J.S., Herniou, E.A., Rohrmann, G.F., Theilmann, D.A., Thiem, S.M. and Vlak, J.M. (2006). On the classification and nomenclature of baculoviruses: A proposal for revision. *Archives of Virology*, **151**: 1257-1266.

Kroemer, J.A. and Webb, -B.A. (2004). Polydnavirus genes and genomes: emerging gene families and new insights into polydnavirus replication. *Annual Review of Entomology*, **49**: 431-456.

Labropoulou, V., Douris, V., Stefanou, D., Magrioti, C., Swevers, L. and Iatrou, K. (2008). Endoparasitoid wasp bracovirus-mediated inhibition of hemolin function and lepidopteran host immunosuppression. *Cellular Microbiology*, **10(10)**: 2118-2128.

Lapointe, R., Wilson, R., Vilaplana, L., O'Reilly, D.R., Falabella, P., Douris, V., Bernier-Cardou, M., Pennacchio, F., Iatrou, K., Malva, C. and Olszewski, J.A. (2005). Expression of a *Toxoneuron nigriceps* polydnavirus-encoded protein causes apoptosis-like programmed cell death in lepidopteran insect cells. *Journal of General Virology*, **86**: 963-971.

Li, X., and Webb, B.A. (1994). Apparent functional role for a cysteine-rich polydnavirus protein in suppression of the insect cellular immune response. *Journal of Virology*, **68(11)**: 7482-7489.

Miller, L.K. (1997). Baculovirus interaction with host apoptotic pathways. *Journal of Cellular Physiology*, **173(2)**: 178-182.

Shelby, K.S. and Webb, B.A. (1999). Polydnavirus-mediated suppression of insect immunity. *Journal of Insect Physiology*, **45**: 507-514.

Stoltz, D.B. and Vinson, S.B. (1979). Viruses and parasitism in insects. *Advances in Virus Research*, **24**: 125-171.

Szewczyk, B., Hoyos-Carvajal, L., Paluszek, M., Skrzecz, I. Lobo de Souza, M. (2006). Baculoviruses-re-emerging biopesticides. *Biotechnology Advances*, **24**: 143-160.

Washburn, J.O., Haas-Stapleton, E.J., Tan, F.F., Beckage, N.E. and Volkman, L.E. (2000). Co-infection of *Manduca sexta* larvae with polydnavirus from *Cotesia congregata* increases susceptibility to fatal infection by *Autographa californica* M Nucleopolyhedrovirus. *Journal of Insect Physiology*, **46**: 179-190.

Washburn, J.O., Kirkpatrick, B.A. and Volkman, L.E. (1996). Insect protection against viruses. *Nature*, **383**: 767.

Washburn, J.O., Trudeau, D., Wong, J.F. and Volkman, L.E. (2003). Early pathogenesis of *Autographa californica* multiple nucleopolyhedrovirus and *Helicoverpa zea* single nucleopolyhedrovirus in *Heliothis virescens*: a comparison of the 'M' and 'S' strategies for establishing fatal infection. *Journal of General Virology*, **84**: 343-351.

Webb, B.A., Strand, M.R., Dickey, S.E., Beck, M.H., Hilgarth, R.S., Barney, W.E., Kadash, K., Kroemer, J.A., Lindstrom, K.G., Rattanadechakul, W., Shelby, K.S., Thoetkiattikul, H., Turnbull, M.W. and Witherell, R.A. (2006). Polydnavirus genomes reflect their dual roles as mutualists and pathogens. *Virology*, **347**: 160-174.

{10}

Environmental Impact and Cost Benefit Analysis of Biological Control Application

Jaime Molina-Ochoa and John E. Foster

ABSTRACT

The implications and interactions of the application of biological control agents in the environment, as well as the risks included by the introduction of exotic natural enemies are briefly reviewed. Description of the role of resistance or quality of the host plant to insect pests, and their interaction with biocontrol agents is provided. A need to increase applications or find new alternatives and environmental impact towards non target insects, and the real cost of its use against the benefits is discussed.

INTRODUCTION

We propose to show the implications, and interactions of the biological control application in the environment, the risks which could include the introduction of exotic enemies, role of the resistance or quality of the host plant, and the necessity of increase applications or finding new alternatives and environmental impact towards non target insects, and the real cost of its use against the benefits it provides.

The United Nations have been adopting the multidimensional development approach and the capability approach since the 90's, both proposed by the economist and Nobel Price Amartya Sen in the framework of his analysis of human development concept. In this framework, health is a constitutive dimension of human development and environmental health can seen as a social condition historically linked to society's industrialization and urbanization (Marsili, 2009).

Agriculture is one of the means to achieve capabilities for the wellbeing and quality of life. This human activity faces limitations caused by the outbreaks of severe infestations of pests and diseases. The use of synthetic pesticides has been the most common method to reduce the pest and disease populations; however, the abuse of use of these chemicals caused impressive damage in the agroecosystem integrants. In response to their negative impacts, social, economic and ecological concerns have demanded alternative methods to reduce the pests and diseases.

The Integrated Pest Management (IPM) is a pest management strategy that focuses on long-term prevention or suppression of pest problems through a combination of practices such as regular pest population monitoring, site or pest inspections, an evaluation of the need for pest control, occupant education, and structural, mechanical, cultural, and biological controls (BC). Under the umbrella of the IPM, BC is the introduction of native enemies of exotic pests into new areas in an attempt to reduce the population sizes of those pests (Hufbauer, 2002).

Usually, the BC is considered an alternative and efficient strategy to reduce pest populations, but the impact of alien species on native organisms is a cause for concern worldwide, with biological invasions commonplace today (Kaufman and Wright, 2009).

Loope and Howarth (2003) emphasized that the proliferation of transportation continues to break down biogeographical boundaries with profound consequences, and an evergrowing volume of transported goods, increasing efficiency and speed, advancing technologies, and trade agreements are key of the phenomenon (Bright, 1999); this is accelerating the rate of biological invasions to a degree without precedent (Stanaway, 2001).

The practice of BC for pest management has been commonly recognized as an effective suppression method for invasive species, and its use was encouraged to reduce dependence on insecticides for the management of invasive insect pests (Kaufman and Wright, 2009); however, BC sometimes also affects the native BC agents.

Based on this background a question comes: is biological control a feasible alternative for a healthy environment at a low cost? Usually, when we talk about a health environment, we expect an environment free of pesticides; however a health environment also implicates the absence of risks for the integrants of the agroecosystems. The native BC agents sometimes are affected by exotic BC agents, and the "healthy environment" is placed in a weak position; one of the most harasser phenomena is the extinction of BC species, and it has a high cost for biodiversity.

The benefits of biological control are those that can provide fairly permanent regulation of devastating agricultural and environmental pests that may be difficult or impossible to manage with more traditional chemical means. However, there are obvious risks. Biological control agents may negatively affect native species directly or indirectly. Historically biological control introductions were not regulated the way they are today, and some horrible mistakes were made in the name of biological control (*e.g.* cane toads in Australia). Even relatively specialized herbivorous insects released for the biological control of invasive weeds can pose risk to related native plants (Hufbauer, 2009).

In order to address a response it is necessary to particularize the cases, and we show some of them because the information available has a broad spectrum.

COMPETITION BETWEEN NATIVE AND INTRODUCED PARASITOIDS OR PREDATORS

There has been much debate about the potential impact of biological control application on nontarget species; many examples seen to show that nontarget species suffer a negative impact from biological control agents, although the quality of evidence varies from anecdotal to relatively quantitative (Stiling and Simberloff, 2000), and the paucity of detailed studies makes it difficult to assess the frequency and severity of nontarget effects (Lynch and Thomas, 2000; Lynch *et al.*, 2002).

Relatively little attention has been given to indirect nontarget effects in which a nontarget species is affected by the biological control agent without suffering direct attack (Huffaker and Kennett, 1966), those effects that do not involve control agents directly attacking nontarget species (Schellhorn *et al.*, 2002).

Schellhorn *et al.* (2002) reported on a study combining experiments, theory and historical information to investigate the decline of a native parasitoid from an agricultural system possibly caused by indirect effects from an exotic parasitoid introduced as a biological control agent. The native parasitoid *Praon pequodorum* Viereck (Braconidae: Hymenoptera), in alfalfa cropping systems suffered a decline due an exotic introduced competitor, *Aphidius ervi* Haliday (Braconidae: Hymenoptera) used against the pea aphids.

At the population level, *P. pequodorum* caused higher parasitism than *A. ervi* when the overall level of parasitism was higher, due a *P. pequodorum* superior with-in host competitive ability. A negative correlation between *A. ervi* and *P. pequodorum* parasitism per plant, suggested strong competition by larvae within host or behavioral avoidance of previously parasitized host by *P. pequodorum.*

Comparing the searching behavior, Schellhorn *et al.*, (2002) considered that *A. ervi* is a superior among-host competitor, searching longer on a plant after an aphid is encountered, moving more rapidly within plants, and attacking and parasitizing more aphids per unit time than *P. pequodorum.* They concluded that indirect nontarget effects of biological control may depend on agricultural practices and the consequent disturbance regime of human-dominant systems.

According to the last statement, With *et al.*, (2002) sustain that the habitat loss and fragmentation are becoming a serious impediment to the biological control of insect pests, particularly within managed systems such as agroecosystems. The habitat fragments support a less diverse community of natural enemies, resulting in lower predation or parasitism rates on pest populations (Roland and Taylor, 1997).

Thus, the potential of predators such as coccinelids to control pest populations in fragmented landscapes may ultimately reflect the extent to which thresholds in landscape structure interfere with aggregative response of predators. With *et al.*, (2002) emphasized that habitat fragmentation may adversely affect the ability of natural enemies to control pest outbreaks in agricultural landscapes by interfering with their search behavior and ability to aggregate in response to prey.

In other hand, Lynch *et al.*, (2002) discussed that even relatively little-preferred nontarget hosts may be at risk of severe population reduction, and perhaps local extinction, from the introduction of a parasitoid biocontrol agent during transient periods just after agent introduction. Extinction of nontargets organisms caused by biological control agents in the past has been believed by several authors, and they sustain that the risk continues (Howarth, 1983; Strong, 1997; Kuris, 2003).

The biocontrol of insect pests may pose a risk to native insects if the biocontrol agent attacks nontarget species (Lynch *et al.*, 2002). However, Van Lenteren and Martin (1999) working with whiteflies are skeptical, critical of the quality of the evidence in general, and they believe extinction or serious population reduction is impossible given the precautions of modern-day biocontrol.

COMPETITION BETWEEN GENERALIST AND SPECIALIST PREDATORS

Biological control theory for predator-prey interactions has been based upon a model communities composed of three discrete trophic levels-plants, herbivores, and predators- in which biological control agents are top consumers and in which different species of predators interact only through competition for shared prey (Rosenheim *et al.*, 1999).

Increased attention has recently been directed to the role of generalist predators as regulators of insect herbivore populations in agroecosystems. Hassell (1978) based in the correspondence between the models and field and laboratory data, discusses the practical implications for biological pest control and suggests how such models may help to formulate a theoretical basis for biological control practices. The dynamics of specialist natural enemies are tightly linked to those of a target pest. This concept has been highlighted by the theory and practice of the biological control.

Hassell (1978) considers in detail several crucial components of predator-prey models: the prey's rate of increase as a function of density, non-random search, mutual interference, and the predator's rate of increase as a function of predator survival and fecundity.

Sheehan (1986) stated the effects of agroecosystem diversification on searching behavior and success of arthropod natural enemies are poorly understood. Crop diversification may increase generalist enemy effectiveness by increasing alternate food or prey availability, as predicted by the enemies hypothesis. But diversification may also reduce enemy searching efficiency and destabilize predator/prey interactions.

Additionally, specialist enemies, often important in biological control programs, may be particularly sensitive to vegetation texture. Pest control by specialist enemies may be more effective in less diverse agroecosystems if concentration of host plants increases attraction or retention of these enemies.

The dynamics of discrete, insect host-parasitoid interactions, having both populations coupled and synchronized with each other, it is implicitly assumed that the parasitoids are effectively specialists on that one host species. However, many natural enemies of insects are polyphagous to some degree and will have rather different dynamical relationships with their prey; this is the case for many parasitoids, staphylinid and carabid beetles, birds and small mammals. A broad diet will tend to buffer the populations of such generalists from fluctuations in abundance of any one of their prey, and give dynamics that are largely uncoupled from that prey (Southwood and Comins, 1976).

The relationships to population density of predations, intraspecific competition and female fertility are the major components in the population dynamics of many species. The way in which these relationships interact, and the resulting effect on the population is conveniently illustrated using a population growth curve, comparing their densities in successive generations (Southwood and Comins, 1976). Most insect populations are attacked by several natural enemies, some polyphagous and other more-or-less monophagous.

Hassell and May (1986) obtained four conclusions related with the generalist and specialist natural enemies in insect predator-prey interactions: i) a specialist can invade and co-exist more easily if acting before the generalists in the life cycle of the prey. ii) A three-species stable system can readily exist where the prey-generalist interaction alone would be unstable or have no equilibrium at all. iii) In some cases the establishment of a specialist leads to higher prey populations than existed previously with only the generalist acting, iv) in some cases, a variety of alternative stable states are possible, either alternating between two-species and three-species states, or between different three-species states.

COMPETITION BETWEEN GENERALIST PREDATORS

Generalist predators have also recently been placed at the center of acrimonius debate over the environmental risks associated with classical biological control, the importation of exotic species of predators to control invasive, usually non-native species of herbivores. Generalist predators may pose substantially enhanced risks of non-target impacts on endemic faunas. The characteristics that make generalists attractive as pest control agents, and in particular their ability to support significant populations by consuming alternate prey, may increase their likehood of producing localized or region al extinctions (Holt and Lawton, 1994; Rosenheim *et al.*, 1999).

Hairston *et al.*, (1960) proposed the context of community-ecology for predator-prey interactions, and it has been adopted as a model for biological control. This context assumes three discrete trophic levels (predators, herbivores, and plants) in which biological control agents are top predators and different species of predators interact only through competition for shared prey. However, Rosenheim *et al.*, (1998) proposed an alternative model in which arthropod communities may comprise more than three tropic levels; trophic levels may be indistinct; predators may consume not only herbivores but also other predators; then biological control agents may therefore be intermediate rather than top predators, and omnivory, cannibalism and intraguild predation are widespread (Rosenheim, 1998; Rosenheim *et al.*, 1995, 1999).

Rosenheim *et al.*, (1999) concluded that a model incorporating higher-order predators and a greater diversity of trophic interactions may prove to be a more fruitful starting point in our search for general rules of pattern and process in the regulation of herbivore populations.

COMPETITION BETWEEN OMNIVORES AND PREDATORS

Omnivores may be more likely to suppress prey populations than strict predators under some circumstances (Eubanks and Denno, 2000). In order to determine the effects of plant quality, and prey abundance on the intensity of interactions involving an omnivorous insect, its two herbivorous prey, and their shared host plant; they found that variation in plant quality, prey abundance, and presence of alternative prey altered the functional response of the omnivorous big-eyed bug, *Geocoris punctipe* Say (Heteroptera: Geocoridae).

Eubanks and Denno (2000) determined that the presence of high-quality parts , such as lima bean pods, reduced the number of prey corn earworm [*Helicoverpa zea* Boddie (Lepidoptera: Noctuidae)] eggs and pea aphids [*Acyrthosiphum pisum* Harris (Hemiptera: Aphididae)] consumed by the big-eyed bug. The pea ahid populations were larger when caged with big-eyed bugs on bean plants with pods than plants without pods. Pods had and indirect but positive effect on the survivorship of herbivorous insects that feed on lima beans. They concluded that plant quality, therefore, mediates the effect of this omnivore on prey suppression.

Other important aspect to consider in the interactions is the supplementation of food. Recently, Shakya *et al.*, (2009) tested the short-term effects of intraguild predation and food supplementation on interactions between two predators, the phytoseid mite *Neoseiulus cucumeris* Oudemans (Acarai: Phytoseiidae), and the anthcorid bug, *Orius laevigatus* Fieber (Hemiptera: Anthocoridae), and their shared prey *Frankliniella occidentalis* Pergande (Thysanoptera: Thripidae), on strawberry plants. All three consumers feed on strawberry pollen, both mites and bugs prey on thrips, and the bug also feeds on the mites (intraguild predation). In structurally simple arenas strong intraguild predation on mites by the bugs was recorded. In whole plant which is a more complex setting, the intensity of intraguild predation differed among the plant structures. Pollen supplementation reduced both intraguild predation on thrips in a structurally simple setting. However, in the whole-plant experiments, the intraguild predation was more intense on pollen-bearing than pollen-free flowers.

Shakya *et al.*, (2009) determined how spatial dynamics, generated when consumers track food sources differently in the habitat and possibly when herbivorous and intraguild prey alter their distribution to escape predation, let to site-specific configuration of interacting populations.

They tested short-term effects of intraguild predation and food supplementation on interactions between two predators, a phytoseiid mite, *N. cucumeris* and the anthocorid bug, *O. laevigatus,* and their shared prey, *F. occidentals*, on strawberry plants. The three specimens feed on strawberry pollen, both mites and bugs prey on thrips, and the bug also feeds on the mites. They concluded that the intensity of resulting trophic interactions was weakened by food supplementation and by increased complexity of the habitat.

INTERACTION OF OMNIVORES-HOST PLANT, AND NATURAL ENEMY

Grosman *et al.*, (2005) remarked that considerably less attention has been devoted to investigating how predators (including omnivores and parasitoids) adapt to new host plants of their phytophagous prey. The adaptation of herbivorous arthropods to novel host plants has been a focus of ecological research for many years, but not in their predators.

The hypothesis of the enemy-free space suggests that plants may be included in the host range of herbivores because of lower predation and parasitism rates on novel host plants. This phenomenon could be important if natural enemies do not follow their prey to the novel host plant, at least not immediately, thus allowing the herbivores to adapt to the novel host plant.

When a phytophagous prey has the opportunity to adapt itself to a new host plant, it may impact on the behavior of the predator or the parasitoid; the presence of certain allelochemicals or metabolites may affect the preference of the natural enemy, playing these new host plant metabolites in the prey a dissuasive role, because the phytophagous contain unpalatable or toxic compounds originating from their new host plant (Grossman *et al.*, 2005). Also, the plant quality affects the survival and diet choice of omnivores (Molina-Ochoa *et al.*, 1999; Coll and Guershon, 2002), and the omnivores may have to adapt to be able to feed on novel host plants as much as herbivores.

Molina-Ochoa *et al*., (1999) studied a tritrophic interaction, host plant, phytophagous pest, and natural enemy. The interactions were studied on an antibiotic variety of maize, Zapalote Chico #2451 P(C3). This variety exhibits a series of allelochemicals that affects the biology of the fall armyworm, *Spodoptera frugiperda* Smith (Lepidoptera: Noctuidae). When the fall armyworm larvae fed on meridic diet supplemented with Zapalote chico silks, their life cycle and their respiratory rates are increased. Increasing in this way the opportunity for parasitization from entomopathogenic nematodes in Petri dish experiments. The entomopathogenic nematodes of the genera, *Steinernema* and *Heterorhabditis* use the carbon dioxide as chemical signal to localize their hosts in soil and cryptic habitats. They found that the use of resistant varieties of corn reduces the concentration required for mean lethal concentrations, using *S. riobrave,* and *S. carpocapsae* all strains.

CONSIDERATIONS

We would like to point out certain considerations instead to give a response to a cost-benefit analysis, avoiding a very simple response or value of the biological control of insect pests in terrestrial environments.

It is important to consider that biological control is not a simple cause-effect lineal phenomenon, it is a net of interactions between the actors of the agroecosystem, including the human being as main modificator.

The application of a biological control agent demands a series of studies in order to establish the possible side effects of its introduction on native fauna or natural enemy of the insect pest. An impressive risk of the introduction of exotic enemies is the extinction of natural enemies.

The host plant, its chemical constitution, and quality may affect the biology of the insect pest, and may also affect directly or indirectly the biology of the parasite, parasitoid, predator or pathogen in consideration.

A resistant plant may also spread metabolites in the soil or in the insect pest affecting the biology of the introduced enemy, sometimes in favor or unfavorably.

The changes in the floristic composition of an ecosystem certainly may affect the size of the insect pest population, diversity, and behavior of natural enemies.

A deep search of native enemies of an insect pest should be conducted in order to determine the possible candidate to be selected to reduce the insect pest population, instead the introduction of an exotic one. The exotic should be selected just in case that the native is unable to reduce the pest population.

The diversity of natural enemies is a resource, and treasure of each country, and a heritage of the world.

REFERENCES

Bright, C. (1999). Invasive species: pathogens of globalization. Foreign Policy, Fall 1999, pp. 50-64.

Coll, M. and Guershon, M. (2002). Omnivory in terrestrial arthopods: mixing plant and prey diets. *Annual Review of Entomology*, **47**: 267-297.

Grosman, A.H., van Breemen, M., Holtz, A., Pallini, A., Molina Rugama, A., Pengel, H., Venzon, M., Zanuncio, J.C., Sabelis, M.W. and Janssen A. (2005). Searching behaviour of an omnivorous predator for novel and native host plants of its herbivores: a study on arthropod colonization of eucalyptus in Brazil. *Entomologia Experimentalis et Applicata*, **116**: 135-142.

Hairston, N.G., Smith, F.E. and Slobodkin, L.B. (1960). Community structure, population control, and competition. *American Naturalist*, **149**: 421-425.

Hassell, M.P. (1978). The dynamics of arthropod predator-prey systems. Princeton University Press, Princeton, New Jersey, USA.

Hassell, M.P. and May, R.M. (1986). Generalist and specialist natural enemies in insect predator-prey interactions. *Journal of Animal Ecology*, **55**: 923-940.

Holt, R.D. and Lawton, J.H. (1994). The ecological consequences of shared natural enemies. *Annual Review of Ecology and Systematics*, **25**: 495-520.

Howarth, F. (1983). Classical biocontrol: Panacea or Pandora's box? *Proceedings Hawaiian Entomological Society*, **24**: 239-244.

Hufbauer, R.A. (2002). Evidence for nonadaptative evolution in parasitoid virulence following a biological control introduction. *Ecological Applications*, **12**: 66-78.

Hufbauer, R.A. (2009). What is biological control? http://lamar.colostate.edu/~hufbauer/Pages/biologicalcontrol.html

Huffaker, C.B. and Kennett, C.E. (1966). Studies of two parasites of the olive scale, *Parlatoria oleae* (Colvee). IV. Biological control of *Parlatoria oleae* (Colvee) through compensatory action of two introduced parasites. *Hilgardia*, **37**: 283-335.

Kuris, A.M. (2003). Did biological control cause extinction of the coconut moth, *Levuana iridescens,* in Fiji? *Biological Invasions*, **5**: 133-141.

Loope, L.L. and Howarth, F.G. (2003). Globalization and pest invasion: where will we be in five years?., pp. 34-39. *In*: Proceedings of the 1st international Symposium on Biological Control of Arthropods. Honolulu, Hawaii, January 2002. United States Department of Agriculture, Forest Service, Morgantown, West Virginia, FHTET-2003-05.

Lynch, L.D., Ives, A.R. Waagem, J.K., Hochberg, M.E. and Thomas, M.B. (2002). The risks of biocontrol: transient impacts and minimum nontarget densities. *Ecological Applications*, **12(6)**: 1872-1882.

Lynch, L.D., and Thomas, M.B. (2000). Nontarget effects in the biocontrol of insects using insects, nematodes and microbial agents: the evidence. *Biocontrol News and Information*, **21**: 117N-130N.

Marsili, D. (2009). Environmental health and the multidimensional concept of development: the role of environmental epidemiology within international cooperation initiatives. *Annali dell' Istituto Superiore di Sanitá*, **45**: 76-82.

Molina-Ochoa, J., Lezama-Gutiérrez, R., Hamm, J.J., Wiseman, B.R. and López-Edwards, M. (1999). Integrated control of fall armyworm (Lepidoptera: Noctuidae) using resistant plants and entomopathogenic nematodes (Rhabditida: Steinernematidae). *Florida Entomologist*, **82**: 263-271.

Roland, J. and Taylor, P.D. (1997). Insect parasitoid species respond to forest structure at different spatial scales. *Nature*, **386**: 710-713.

Rosenheim, J.A. (1998). Higher-order predators and the regulation of insect herbivore populations. *Annual Review of Entomology*, **43**: 421-447.

Rosenheim, J.A., Kaya, H.K., Ehler, L.E., Marois, J.J. and Jaffee, B.A. (1995). Intraguild predation among biological control agents: theory and evidence. *Biological Control*, **5**: 303-335.

Rosenheim, J.A., Limburg D.L. and Colfer R.G. (1999). Impact of generalist predators on biological control agent, *Chrysoperla carnea:* direct observations. *Ecological Applications*, **9**: 409-417.

Schellhorn, N.A., Kuhman, T.R., Olson, A.C. and Ives A.R. (2002). Competition between native and introduced parasitoids of aphids: Nontarget effects and biological control. *Ecology*, **83**: 2745-2757.

Shakya, S., Weintraub, P.G. and Coll, M. (2009). Effect of pollen supplementation on itraguild predatory interactions between two omnivores: the importance of spatial dynamics. *Biological Control*, **50**: 281-287.

Sheehan, W. (1986). Rsponse by specialist and generalist natural enemies to agroecosystem diversification: a selective review. *Environmental Entomology*, **15**: 456-461.

Southwood, T.R.E. and Comins, H.N. (1976). A synoptic population model. *Journal of Animal Ecology*, **45**: 949-965.

Stanaway, M.A., Zalucki, M.P., Gillespie, P.S., Rodriguez, C.M. and Maynard, G.V. (2001). Pest risk assessment of insects in sea cargo containers. *Australian Journal of Entomology*, **40**: 180-192.

Stiling, P. and Simberloff, D. (2000). The frequency and strength of non-target effects of invertebrate biological control agents of plant pests and weeds. *In*: Follet, P.A. and Duan, J.J. (*Eds.*), *Nontarget effects of biological control*. Kluwer Academic Publishers, Dordrecht, The Netherlands, pp. 31-44.

Strong, D.R. (1997). Fear no weevil? *Science*, **277**: 1058-1059.

Van Lenteren, J.C. and Martin, N.A. (1999). Biological control of whitefly. *In*: Albajes, R., Gullino, M.L., Van Lenteren, J.C. and Elad, Y. (*Eds.*), *Integrated pest and disease management in greenhouse crops*. Kluwer, Dordecht, The Netherlands, pp. 202-216.

With, K.A., Pavuk D.M., Worchuck, J.L., Oates, R.K. and Fisher, J.L. (2002). Threshold effects of landscape structure on biological control in agroecosystems. *Ecological Applications*, **12**: 52-65.

Subject Index

α-exotoxin 7
β-exotoxin 7, 154
δ-endotoxin 6, 154
δ-Aga-IV venom 51
20-kDa protein 240

A

Acetamiprid 131
Acremonium 75
Actinomycets 220
Active ingredient 200
Activity spectrum 239, 243
Af NPV 137
Ag NPV 137
Agelaius phoeniceus 160
Agriculture 272
Agroecosystems 274
Alkaline phosphatase 242
Allelochemicals 278
Alphabaculovirus 135
Amber disease 17
Amblyseius swirskii 106
American foulbrood 14
Aminopeptidase N 242
Anabaena 246
Anamorph 66
Antibodies 267
Ants 160
Ao GV 137
Apoptosis 263, 266
Appressorium 80
Arachnids 176
As II and Sh1 toxins 51
Aschersonia 75
Aspergillus 75
Asticcacaulis excentricus 246
Augmentation 44
Azadirachtin 131

B

Bacillus popilliae 148, 155
B. sphaericus 171, 246
B. thuringiensis 148, 153, 168
    subsp. *kurstaki* 131, 134, 153, 169
        aizawai 133, 134
        alesti 133
        berliner 133
        chinesis 151, 153, 154
        galleriae 152, 153
        israelensis 152, 169
        japonensis 151, 152
        kumamotoensis 150, 151
        morrisoni 155
        san diego 149, 153
        tenebrionis 149, 150, 151, 153, 169
        tolworthi 133, 150, 151
Bacteria 2
Baculoviruses 31, 33
Banker plants 103
Bassianolide 84
Beauveria 138, 75

B. bassiana 84, 140, 148, 156, 157, 175, 180, 250
B. brongniartii 149, 175, 188
B. tenella 156
Beauvericin 84, 177
Behavioral responses 100
Beneficial insects 201
Betabaculovirus 135
Binary plasmid 250
Binary toxin 247
Bioinsecticide products 134
Biopesticides 130, 131
Bioreactors 178
Birds 160
Bt cotton 135, 215, 217
Bt genes 210
Bt maize 135, 214, 215
Bt rice 220
Bt toxins 133
Bt transgenic crops 135
Butterflies 124

C

Cactolaccus grandis 104
Cadherin-like-proteins 242
Cag genes 81
Calyx cells 260
Cardiochiles diaphaniae 109
Cellular defense reaction 82
Cellular immune response 265
Chaperonin 240
Chemical pesticides 129
Chimeric proteins 242
Chimeric scanning mutagenesis 242
Chitinase 81
Chitinase gene 248
Coccinella septempunctata 101, 102
Coincidental IGP 98
Coleomegilla maculate 102
Conidia 79, 139
Conidiobolus 138
Cordyceps 75, 138
Cp GV 137
CrV1 gene 266
CrV1 protein 263
Cry 156
Cry gene 151
Cry protein 133, 168, 234
Cry1Ab 135
Cry1Ac 135
Cry3A 150
Cry3Aa gene 154
Cry3B2 150
Cry4 gene 154
Crystal protein 151
Cypoviruses 34

D

Dastarcus helophoroides 159
Delayed-early genes (DE) 45
Delphastus catalinae 101
Deltabaculovirus 135
Delta-endotoxins 133
d-endotoxins 132, 234
Destruxin 84, 177
Diadegma semiclausum 105
Diapauses protein 1-like protein 83
Diuretic hormone 51
DNA integrase-recombinase 48
Dolichogenidea tasmanica 105
Domain I 234
Domain swapping 244

E

Early genes 47
Ecotron 113
Electropherotypes 46
ELISA 105
Encapsulation 260
Encarsia spp 104
Endotoxins 7
Enemy-free space 278
Enhancin 52
Enthomophthora virulenta 189
Entomopathogenic nematodes 159
Entomophaga 138
Entomophthora 138
E. maimaiga 139
Entomopoxvirus 34
Enveloped viruses 36
Epizootics 3

Eretmocerus eremicus 104
Eretmocerus sp. nr. emiratus 101
Erynia 138
Exotic species 276

F

Facultative anaerobes 4
Fastidious organism 13
Flotability 246
Food 277
Food web 100, 101
Formic acid 147
Frankliniella occidentalis 99
Fusarium 75, 88

G

Galandromus occidentalis 99
Galleriae 152, 153
Gammabaculovirus 135
Gdp promoter 249
Gene 150
Gene *cag8* 249
Gene fusion 242
Gene integration 238, 247
Gene *nrr1* 249
Gene *pr1* 249
Gene *ssga* 249
Generalist predators 275, 276
Genes *cry3B* 150
Genetic engineering 106
Geocoris punctipes 101
Geocoris spp. 99
Global warming 110
Glyphosate 210
Granules 200
Granulosis viruses 200
Gst gene 248
Guild 96

H

Health environment 272
Hemocyte 263
Herbaspirillum seropedicae 245
Herbivorous insects 273
Heterorhabditis 204
H. bacteriophora 209, 211
H. megidis 212
Hippodamia convergens 101, 106
Hirsutella 75, 138
H. thompsonii 189
Holometabola 124
Hot spots 267
Hybrid 244
Hybrid genes 243
Hybrid protein 236, 244
Hybrid toxin 237, 243
Hydrophobic interaction 79
Hymenostilbe 75
Hypocreales 175
Hypocrella 75, 138
Hz NPV 137

I

ICP 8
Immediate early genes (IE) 45
Immune system 260
Immunosuppressive genes 262
Industrial process 11
Inert materials 42
Infection 264
Insect cell cultures 36
Insecticidal genes 236
Integrated Pest Management 272
Intermediate genes 47
Intestinal cells 264
Intraguild predation 96
Inundative strategy 44
Iridoviruses 35
Isaria fumosorosea 175, 189

J

Juvenile 204

L

Lagenidium 138
L. giganteum 190
Lariophagus distinguendus 159
Late genes (L) 45, 47
Ld NPV 137
Lecanicillium lecanii 87

L. longisporum 191
L. muscarium 191
Lecanicillium spp. 175
Liquid fermentation 178, 222
Low technology system 11

M

Market 130
Maternal microinjection 109
Mb NPV 137
Melanin 82
Melanotic capsule 15
Metarhizium 75, 138
M. anisopliae 79, 86, 140, 148, 157, 175, 193, 249
M. anisopliae var. *acridum* 197
M. anisopliae var. *anisopliae* 158
Metaseiulus occidentalis 109
Methylobacterium extorquens 245
Micro-encapsulation process 42
Microsclerocia 178
Milky disease 13, 155
Mites 176
Molecular biology techniques 233
Moths 124
Mouse factor 7
Mtx gene 247
Mtx1 protein 248
*Mtx1*gene 248
Multiple nucleopolyhedroviruses 33
Mutagenesis 236
Mycoinsecticides 175
Myriangium 75

N

Nabis spp 99
Natural enemies 102, 109
Nectria 75
Nematodes 204
Neoseiulus cucumeris 100
Neosteinernema longicurvicauda 208
Neozygites 138
Neurotoxin AaHIT 50
Neurotoxin TxP-1 50
Niche shifts 102
Nomuraea 75
Nomurea rileyi 79, 88, 139
Non-occluded viruses 200
Ns NPV 137
Nuclear polyhedrosis viruses 200

O

Obligate entomopathogenic bacteria 4
Obligate parasites 32, 201
Occlusion body 31, 200
Oechalia schellenbergii 112
Oil-based formulations 177
Omnivores 277
Omnivorous IGP 98
Oosperein 177
Op NPV 137
Operon 239
Opportunistic pathogens 4
Orius insidiosus 101
O. laevigatus 100, 104
O. tristicolor 99

P

P 10 gene 49
Paecilomyces 75, 138, 140
P. fumosoroseus 87, 140
Paenibacillus popilliae 174
Pandora 138
Parasitism 96
Parasitoids 97, 148, 261
Paratransgenisis 107
Pathogens 148
p-benzoquinone 147
Peptides 220
Phagostimulants 42
Phase I & II 18
Phase variation 18
Phasmarhabditis 204
P. hermaphrodita 209, 212
Pheidole sp. 160
Phenoloxidase (PO) cascade system 82
Photorhabdus 220
Phytophagous insects 146
Phytoseiulus persimilis 106
Plasmid replicon 238

Plasmids 32
Pochonia chlamydosporia 198
Podonectria 75
Pollen 278
Polydnaviruses 260
Polyhedrin 49
Polyhedrin gene 46
Polyketides 220
Pore formation 235
Potency 43
Predation 96
Predators 97, 148
Promoter 239
Protease 81
Proteins TrIV1 263
Pseudomonas 134
P. entomophila 220
P. fluorescens 245
pSTAB vector 239

R

Recombinant strain 240
Refuge crops 103, 104
Refuges 220
Resistance 264
Resistance management 219
Respiratory cells 264
Rhizophagus grandis 160
RT-PCR 249

S

Saccharopolyspora spinosa 220
San diego 149
Scleroderma guani 159
Scymnus loweii 106
Se NPV 137
Second domain 234
Secondary infection 265
Serotype 7
Shuttle vectors 238
Single nucleopolyhedroviruses 33
Site-specific recombination system 238
Sl NPV 137
Solenopsis geminate 160
Solid substrate 178
Sorosporella 75
Southern blots 263
Spatial scales 100
Specialist 276
Specialist enemies 275
Species 66
Spectrum 237
Spinosad 131, 220
Spinosyns 220
Spore-Forming Bacteria 5
Sporothrix 75
Stable system 276
Steinernema 204, 208
S. carpocapsae 159, 209, 211
S. feltiae 211
S. kraussei 212
S. riobravis 211
S. glaseri 211
Subtilisin-related proteases 81
Subtilisins 250
Surfactants 42
Symbiotic bacteria 204
Synema 86

T

Teleomorph 66
Tetracrium 75
Third domain 235
Ticks 176
Time scales 99
Tolypocladium sp. 75, 89
Torrubiella 75, 138
Tracheae tubes 264
Tracheal system 22, 64
Tracheolar cells 264
Tracheole 264
Transgenic organisms 106
Transgenic plants 214
Transovarial transmission 38
Traqueoblasts 264
TrIV2 263
Trophic cascades 101
Trophic levels 110, 275
Trypsin-related enzymes 81
Type B milky disease 15

U

ULV formulations 177

V

Vertical transmission 37
Verticillium 75, 138
V. lecanii 140
Very late genes (VL) 45
VHV1.1 263
Vip proteins 7
Vip3AcAa 244
Virion 31
Virogenic stroma 37
Viroids 32
Viropexis 37
Viruses 31

W

Wettable formulation 157
Wettable powder formulations 177

X

Xenorhabdus 220

Z

Zelus spp 99
Zoophthora 138
Zygospore 74